Methods in Molecular Biology

For further volumes:
http://www.springer.com/series/7651

Protein Scaffolds

Design, Synthesis, and Applications

Edited by

Andrew K. Udit

Department of Chemistry, Occidental College, Los Angeles, CA, USA

Humana Press

Editor
Andrew K. Udit
Department of Chemistry
Occidental College
Los Angeles, CA, USA

ISSN 1064-3745 ISSN 1940-6029 (electronic)
Methods in Molecular Biology
ISBN 978-1-4939-7892-2 ISBN 978-1-4939-7893-9 (eBook)
https://doi.org/10.1007/978-1-4939-7893-9

Library of Congress Control Number: 2018941577

Printed on acid-free paper

This Humana Press imprint is published by the registered company Springer Science+Business Media, LLC part of Springer Nature.
The registered company address is: 233 Spring Street, New York, NY 10013, U.S.A.

Preface

Proteins may be the ultimate scaffold for building molecular architectures: the varied chemical functionality of amino acid side chains coupled with the wide range possible in protein size makes it viable to generate a vast array of structures that differ either slightly or markedly in chemical and structural properties. These architectures help extremophiles tolerate harsh living conditions, allow marine organisms to stick stubbornly to ships, and impart a warm glow to bioluminescent organisms. Add to this the novel functionality imparted by incorporating unnatural amino acids, or targeted modifications utilizing site-specific bioconjugation methods, and it becomes clear that proteins are malleable platforms for myriad applications.

Recognizing the amazing potential of proteins, our role as scientists to understand, harness, and exploit these wonderful molecules for the betterment of society becomes increasingly indispensable. This volume of *Methods in Molecular Biology*, titled *Protein Scaffolds: Design, Synthesis, and Application*, has collected a wide range of some of the most exciting, contemporary work using various protein structures as platforms for building macromolecules with targeted application. While this volume has not been formally divided into subsections, the chapters are grouped. The first few describe a series of protocols exploiting virus nanoparticles; in particular, their ability to encapsulate makes them especially appealing, either for payload delivery or compartmentalization. The next group of chapters describes using virus particles and other proteins as scaffolds for either building materials or presentation of ligands. The multivalent nature of these structures makes them particularly well suited for this application. The third group discusses site-specific bioconjugation methods; the protocols describing unnatural amino acids and bioorthogonal chemistry add a distinctive dimension, allowing truly targeted modifications. The fourth group of chapters explores elastin and collagen peptides and proteins, with focus on building and characterizing higher-order structures. The last group describes some unique scaffolds and techniques that include peptide amphiphile micelles, assembly of chaperones, and a review of designed ankyrin repeat proteins.

Each chapter provides a self-contained, detailed protocol; readers will be able to readily and fully reproduce the methods described. The *Notes* sections in particular will provide the nuances not typically found in journal articles. It is my sincere hope that there is something in here for everyone, and that your copy of this book eventually becomes discolored and worn from being soaked in buffer while it lays open for you to read on the benchtop.

Los Angeles, CA, USA *Andrew K. Udit*

Contents

Contributors

DANNY ABAD · *Department of Biological Chemistry, University of California, Los Angeles, CA, USA*

YUSUKE AZUMA · *Laboratory of Organic Chemistry, ETH Zurich, Zurich, Switzerland*

JOHN C. BARRETT · *Institute for Molecular Engineering, University of Chicago, Chicago, IL, USA*

SERENE BAYRAM · *Department of Chemistry and Center for Self-Assembled Chemical Structures, McGill University, Montreal, QC, Canada*

AMY SZUCHMACHER BLUM · *Department of Chemistry and Center for Self-Assembled Chemical Structures, McGill University, Montreal, QC, Canada*

STEVEN D. BROWN · *Department of Gastroenterology, University of California—San Diego, La Jolla, CA, USA*

DOUGLAS S. CLARK · *Department of Chemical and Biomolecular Engineering, University of California, Berkeley, CA, USA*

JEROEN J. L. M. CORNELISSEN · *Laboratory of Biomolecular Nanotechnology, MESA + Institute for Nanotechnology, University of Twente, Enschede, The Netherlands*

JAMES K. DEYLING · *Cleveland State University, Cleveland, OH, USA*

MARK B. VAN ELDIJK · *Institute for Molecules and Materials, Radboud University, Nijmegen, The Netherlands*

ADEL M. ELSOHLY · *Department of Chemistry, University of California, Berkeley, CA, USA*

KATHARINE Y. FANG · *Division of Chemistry and Chemical Engineering, California Institute of Technology, Pasadena, CA, USA*

MATTHEW B. FRANCIS · *Department of Chemistry, University of California, Berkeley, CA, USA; Materials Sciences Division, Lawrence Berkeley National Laboratories, Berkeley, CA, USA*

JEREMIAH J. GASSENSMITH · *Department of Chemistry and Biochemistry, University of Texas at Dallas, Richardson, TX, USA*

XI GENG · *Department of Chemistry, Virginia Tech, Hahn Hall South, Blacksburg, VA, USA*

ANUM AZAM GLASGOW · *Department of Bioengineering and Therapeutic Sciences, UC San Francisco, San Francisco, CA, USA*

DOMINIC J. GLOVER · *School of Biotechnology and Biomolecular Sciences, The University of New South Wales, Sydney, Australia*

TIJANA Z. GROVE · *Department of Chemistry, Virginia Tech, Hahn Hall South, Blacksburg, VA, USA*

JAN C. M. VAN HEST · *Eindhoven University of Technology, Eindhoven, The Netherlands*

DONALD HILVERT · *Laboratory of Organic Chemistry, ETH Zurich, Zurich, Switzerland*

NOLAN B. HOLLAND · *Cleveland State University, Cleveland, OH, USA*

PRIYA KATYAL · *Department of Chemical and Biomolecular Engineering, Tandon School of Engineering, New York University, Brooklyn, NY, USA*

VALERIE A. KICKHOEFER · *Department of Biological Chemistry, University of California, Los Angeles, CA, USA*
JAYEETA LAHIRI · *Department of Physics, University of Hyderabad, Hyderabad, Telangana, India*
SETH A. LIEBLICH · *Division of Chemistry and Chemical Engineering, California Institute of Technology, Pasadena, CA, USA*
SHAOBO LI · *Department of Chemistry and Biochemistry, University of Texas at Dallas, Richardson, TX, USA*
AIJIE LIU · *Laboratory of Biomolecular Nanotechnology, MESA + Institute for Nanotechnology, University of Twente, Enschede, The Netherlands*
STAN J. MAASSEN · *Laboratory of Biomolecular Nanotechnology, MESA + Institute for Nanotechnology, University of Twente, Enschede, The Netherlands*
SHAILY MAHENDRA · *Department of Civil and Environmental Engineering, University of California, Los Angeles, CA, USA*
JIN KIM MONTCLARE · *Department of Chemical and Biomolecular Engineering, Tandon School of Engineering, New York University, Brooklyn, NY, USA; Department of Biochemistry, SUNY Downstate Medical Center, Brooklyn, NY, USA; Department of Biomaterials, New York University College of Dentistry, New York, NY, USA; Department of Chemistry, New York University, New York, NY, USA*
LISA D. MUIZNIEKS · *Molecular Medicine Program, Research Institute, Hospital for Sick Children, Toronto, ON, Canada*
CHAWITA NETIROJJANAKUL · *Department of Chemistry, University of California, Berkeley, CA, USA*
AVANISH SINGH PARMAR · *Department of Physics, Indian Institute of Technology (BHU) Varanasi, Varanasi, Uttar Pradesh, India*
DUSTIN P. PATTERSON · *Department of Chemistry and Biochemistry, The University of Texas at Tyler, Tyler, TX, USA*
STANISLAV PRESOLSKI · *Yale-NUS College, Singapore, Singapore*
SHEFAH QAZI · *Department of Molecular and Cellular Biochemistry, Indiana University, Bloomington, IN, USA*
JULIA DEL RE · *Department of Chemistry and Center for Self-Assembled Chemical Structures, McGill University, Montreal, QC, Canada*
LEONARD H. ROME · *Department of Biological Chemistry, University of California, Los Angeles, CA, USA*
MARK V. DE RUITER · *Laboratory of Biomolecular Nanotechnology, MESA + Institute for Nanotechnology, University of Twente, Enschede, The Netherlands*
LISE SCHOONEN · *Institute for Molecules and Materials, Radboud University, Nijmegen, The Netherlands*
DEVYANI SHUKLA · *Department of Physics, Indian Institute of Technology (BHU) Varanasi, Varanasi, Uttar Pradesh, India*
KATARZYNA SLOWINSKA · *Department of Chemistry and Biochemistry, California State University Long Beach, Long Beach, CA, USA*
DAVID A. TIRRELL · *Division of Chemistry and Chemical Engineering, California Institute of Technology, Pasadena, CA, USA*

MATTHEW V. TIRRELL · *Institute for Molecular Engineering, University of Chicago, Chicago, IL, USA*
DANIELLE TULLMAN-ERCEK · *Department of Chemical and Biological Engineering, Northwestern University, Evanston, IL, USA*
EDWARD M. TURK · *Gilmour Academy, Gates Mills, OH, USA*
MENG WANG · *Department of Civil and Environmental Engineering, University of California, Los Angeles, CA, USA*
OMAR ZAHR · *Department of Chemistry and Center for Self-Assembled Chemical Structures, McGill University, Montreal, QC, Canada*

Chapter 1

Nanoreactors via Encapsulation of Catalytic Gold Nanoparticles within Cowpea Chlorotic Mottle Virus Protein Cages

Aijie Liu, Mark V. de Ruiter, Stan J. Maassen, and Jeroen J. L. M. Cornelissen

Abstract

Viral protein cage-based nanoreactors can be generated by encapsulation of catalytic metal nanoparticles within the capsid structure. In this method, coat proteins of the cowpea chlorotic mottle virus (CCMV) are used to sequester gold nanoparticles (Au NPs) in buffered solutions at neutral pH to form CCMV-Au hybrid nanoparticles. This chapter describes detailed methods for the encapsulation of Au NPs into CCMV protein cages. Protocols for the reduction of nitroarenes by using CCMV-Au NPs as catalyst are described as an example for the catalytic activity of Au NPs in the protein cages.

Key words CCMV, Protein cage, Gold nanoparticle, Encapsulation, Catalysis, Nanoreactor

1 Introduction

Transition metals such as palladium and platinum are used frequently in heterogeneous catalysis while gold, macro or micro structured, is considered to be inert [1]. However, decreasing the size of gold colloids into the nanometer regime increases their catalytic activity due to the increased surface area. These gold nanoparticles (Au NPs) enable mild reaction conditions and high chemoselectivity in reduction reactions, which make them attractive catalysts [2]. However, a disadvantage is that their stability in solution decreases along with their size. Therefore, surface modifications have to be carried out, or carrier systems have to be designed to enable efficient catalysis and elongated reaction times [3]. Besides synthetic carrier systems, natural assemblies such as protein cages have promising features and recently have gained increased attention in materials science [4–7]. The main improvement compared to artificial carriers is

Andrew K. Udit (ed.), *Protein Scaffolds: Design, Synthesis, and Applications*, Methods in Molecular Biology, vol. 1798, https://doi.org/10.1007/978-1-4939-7893-9_1,

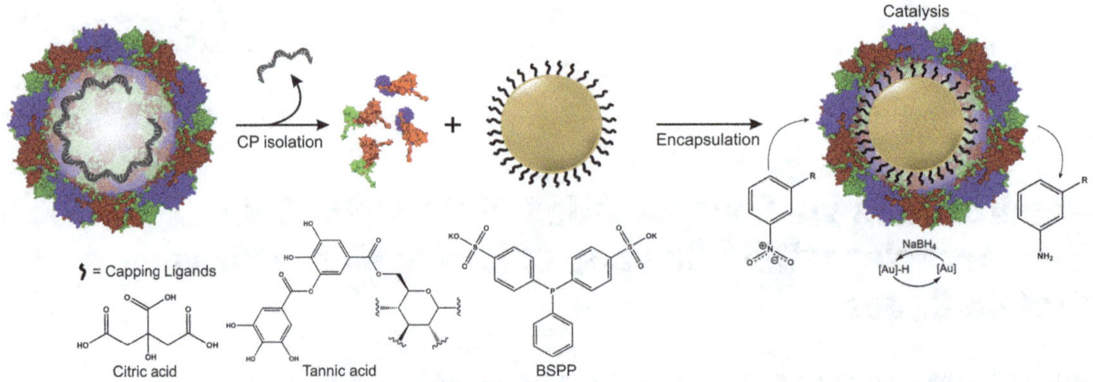

Fig. 1 Encapsulation of gold nanoparticles in the CCMV protein cage. The assembly selectively catalyzes reactions with charged nitroaromatic molecules

their uniformity with respect to size and shape as well as control of catalytic and substrate selectivity [8, 9].

To increase the stability of gold nanoparticles, the capsid of the cowpea chlorotic mottle virus (CCMV) can be used. CCMV is an icosahedral $T = 3$ virus with an average diameter of 28 nm which infects the cowpea plant [10, 11]. The RNA cargo of the virus can be removed by disassembly at neutral pH in the presence of high Ca^{2+} concentration (>0.4 M), resulting in coat protein (CP) dimers in solution [12]. These dimers can self-assemble back into their native cage-like structures when the pH is adjusted to pH 5 at high ionic strength. Alternatively, reassembly can be triggered at neutral pH into different shapes by introducing a negatively charged cargo. CCMV is well suited for this purpose because its N-terminal arginine-rich motif (ARM), which interacts with negatively charged cargo (e.g., RNA), is exposed on the inner surface of the capsid [4, 13, 14].

In this chapter we describe a procedure for encapsulating Au NPs within the CCMV capsid [15]; this is followed by a procedure to use the encapsulated gold nanoparticles as catalysts (Fig. 1) [16]. The viral capsid displays selectivity toward different substrates and enhances the stability of the Au NPs in solution, resulting in better recyclability of the catalytic particles.

2 Materials

1. RNA buffer: add 6.06 g of Tris, 75 g of $CaCl_2$, and 0.154 g of dithiothreitol (DTT) to a 1 L graduated cylinder, add Milli-Q water to a volume of 950 mL, mix and adjust the pH to 7.5 with 5 M HCl, fill up to 1 L with Milli-Q water to yield 50 mM Tris, 500 mM $CaCl_2$, 1 mM DTT.
2. Cleaning buffer: add 6.06 g of Tris, and 29.22 g of NaCl to a 1 L graduated cylinder, add Milli-Q water to a volume of 950 mL, mix and adjust the pH to 7.5 with 5 M HCl, fill up to 1 L with Milli-Q water to yield 50 mM Tris, 0.5 M NaCl.

3. CP storage buffer: add 8.20 g of sodium acetate, 29.22 g of NaCl, and 0.065 g of sodium azide to a 1 L graduated cylinder, add water to a volume of 950 mL. Mix and adjust to pH 5 with 5 M acetic acid, fill up to 1 L with Milli-Q water to yield 100 mM sodium acetate, 500 mM NaCl, and 1 mM sodium azide.
4. Virus storage buffer: add 8.20 g of sodium acetate, 0.372 g of ethylenediaminetetraacetic acid (EDTA), and 0.065 g of sodium azide to a 1 L graduated cylinder, add water to a volume of 950 mL, mix and adjust to pH 5 with 5 M acetic acid, fill up to 1 L with Milli-Q water to yield 100 mM sodium acetate, 1 mM EDTA and 1 mM sodium azide.
5. CCMV: 15–20 mg/mL of wild type CCMV in virus storage buffer (*see* **Note 1**).
6. 12 kDa–14 kDa MWCO dialysis membrane.
7. Centrifuge filters of 10 kDa MWCO with 500 μL, 5 mL, 15 mL capacities, respectively; and 30 kDa MWCO with 5 mL, 15 mL capacities, respectively.
8. Commercially available citrate-stabilized Au NPs, 7 nm diameter (0.05 mg/mL, 1.7×10^{13} particles/mL) (*see* **Note 2**).
9. 10 mM BSPP: dissolve 53.46 mg of bis(p-sulfonatophenyl) phenylphosphine (BSPP) in 1 mL of Milli-Q water, transfer to a 10 mL volumetric flask and add Milli-Q water to 10 mL.
10. 2 mM 3NTA: dissolve 46.13 mg of N,N,N-trimethyl-1-(3-nitrophenyl)methanaminium (3NTA) in 10 mL of Milli-Q water, transfer to a 100 mL volumetric flask and add Milli-Q water to 100 mL (*see* **Note 3**).
11. 2 mM 2M4NP: dissolve 30.63 mg of 2-methyl-4-nitrophenol (2M4NP) in 10 mL of Milli-Q water, transfer to a 100 mL volumetric flask and add Milli-Q water to 100 mL.
12. 30 mM $NaBH_4$: dissolve 11.35 mg of sodium borohydride in 10 mL of Milli-Q water (*see* **Note 4**).
13. Standard HPLC syringe filter (pore size = 0.2 μm).
14. TEM with Formvar/carbon-coated 200 mesh copper grid.
15. 1% w/v uranyl acetate solution in Milli-Q water.
16. Ultracentrifuge (*see* **Note 5**).
17. UV-Vis spectrophotometer with quartz cuvette (*see* **Note 6**).
18. Zeta potential instrument (*see* **Note 7**).
19. Dynamic light scattering instrument (*see* **Note 8**).
20. Fast-protein liquid chromatography system (FPLC) with UV-Vis detectors for size-exclusion chromatography (*see* **Note 9**).
21. 24 mL Superose 6 PC 3.2/30 column (GE Healthcare).
22. High-performance liquid chromatography system (*see* **Note 10**).

3 Methods

3.1 CCMV Disassembly

1. Dialyze 1 mL of CCMV (15–20 mg/mL) in virus storage buffer against 300 mL of RNA buffer. Change the RNA buffer every 3 h and repeat this 3 times (*see* **Note 11**).
2. Isolate the protein using an ultracentrifuge at 40,000 rpm (179,200 × *g*) for 2 h.
3. Collect the supernatant and dialyze against cleaning buffer for a total of 12 h, and refresh the buffer every 3 h (*see* **Note 12**).
4. Determine the final concentration of CP using UV-Vis spectroscopy at λ = 280 nm with extinction coefficient of 1.22 mL mg^{-1} cm^{-1} (*see* **Notes 13** and **14**).

3.2 Au NP Ligand Exchange

Exchanging the citrate ligand on the Au NPs is required to improve encapsulation efficiency (*see* **Note 15**). The Au NP ligand exchange method is performed according to a modified literature procedure [11].

1. Transfer 10 mL of citrate-stabilized 7 nm diameter Au NP solution to a round-bottom flask with a stir bar.
2. Add 750 μL of 10 mM BSPP solution into the above Au NP solution while stirring. Stirring is continued overnight at room temperature.
3. Dilute 6 mL of 10 mM BSPP tenfold with Milli-Q water to yield 60 mL of 1 mM of BSPP.
4. Transfer the mixed solution from **step 2** to two 15 mL 10 kDa MWCO centrifuge filters, 5 mL in each tube, add 1 mM BSPP to 15 mL, and then centrifuge at 2800 rpm (1300 × *g*) for 2 min (*see* **Note 16**). Repeat this washing process three times.
5. Remove the excess BSPP from **step 4** by washing with Milli-Q water using a 15 mL 10 kDa MWCO centrifuge filter. Transfer the solutions from **step 4** to two new centrifuge filters, 5 mL in each tube, add Milli-Q water to 15 mL, and then centrifuge at 2500 rpm (1050 × *g*) for 2 min for each run (*see* **Note 16**). Repeat this washing process two times.
6. Perform dynamic light scattering and Zeta potential measurements to determine particle size and surface potential (*see* **Note 17**).
7. Determine the final concentration of BSPP-capped Au NPs by UV-Vis spectroscopy (*see* **Note 18**).

3.3 Au NP Encapsulation

1. Concentrate Au NPs to 55 nM using a 10 kDa MWCO centrifuge filter (*see* **Note 19**).
2. Prepare fresh 100 μL of CP in cleaning buffer with a concentration of 20 mg/mL by concentrating with a 5 mL of 10 kDa MWCO centrifuge filter.

3. Add 400 μL of 55 nM Au NP solution to 100 μL of 20 mg/mL CCMV CP in cleaning buffer (*see* **Note 20**).
4. Allow the mixture to incubate for 5 min at room temperature.
5. Dialyze the mixture against cleaning buffer overnight (about 17 h) (*see* **Note 21**).
6. Purify the particles using FPLC with a Superose 6 PC 3.2/30 column and a flow rate of 0.5 mL/min, with UV-Vis detectors set to λ = 280 nm and λ = 260 nm for the coat protein, and λ = 520 nm for the Au NPs. Collect the particles using a fraction collector (*see* **Notes 22** and **23**).
7. Concentrate the purified CCMV-Au NPs using 5 mL 30 kDa MWCO centrifuge filters, and centrifuge at 2800 rpm (1300 × *g*) and 4 °C to the desired concentration.

3.4 Characterization of CCMV-Au NPs

3.4.1 UV-Vis Analysis

1. Fill a quartz glass UV-Vis cuvette with 500 μL of CCMV-Au NPs solution.
2. Look for an increase in absorption at 280 nm compared to naked Au NPs, which indicates assembly of CCMV around the gold.

3.4.2 TEM Analysis

1. Cast a droplet of the sample on a 200 mesh copper grid for 2 min.
2. Blot away excess solvent using sterile filter paper.
3. Negatively stain the samples by applying 5 μL of 1% (w/v) uranyl acetate onto the grid for 30 s and then blot away the excess solution using sterile filter paper (*see* **Note 24**).
4. Image the sample.

3.4.3 DLS Analysis

1. Perform dynamic light scattering of the samples.
2. Calculate the size and standard deviation of the nanoparticles by taking an average of at least 5 measurements.

3.5 Catalytic Study of CCMV-Au NPs

1. Dilute 1 mL of CCMV-Au NPs with Milli-Q water to 15 mL, place in a 15 mL 30 kDa MWCO centrifuge filter and centrifuge at 2800 rpm (1300 × *g*) for 2 min. Repeat this wash step three times.
2. Concentrate the CCMV-Au NP solution to 63 μM (based on Au atom concentration) using a 5 mL 30 kDa MWCO centrifuge filter. Determine the concentrations by UV-Vis spectroscopy (*see* **Note 25**).
3. Mix 170 μL of above CCMV-Au NP solution with 30 μL of 2 mM 3NTA nitroarene substrate solution in a quartz cuvette (*see* **Note 26**).
4. Start UV-Vis analysis immediately after adding 150 μL of 30 mM of sodium borohydride. Record the absorption over a wavelength range of 200–350 nm (*see* **Note 27**).

5. Confirm the catalytic conversion of 3-nitroarenes with HPLC using 2M4NP as an internal standard molecule:
   (a) Remove CCMV-Au NPs from the reaction using a 500 μL 10 kDa MWCO centrifuge filter. Collect the filtrate (the solution in the bottom part of the centrifuge tube).
   (b) Mix 174 μL of the filtrate with 3 μL of 2 mM 2M4NP solution.
   (c) Filter the sample using a standard HPLC syringe filter (pore size of 200 nm) to remove possible aggregates and then inject into HPLC for analysis.
   (d) Determine the amount of nonreacted nitroaromatic component by integrating the eluting peaks (*see* **Note 28**).

4 Notes

1. Wild type CCMV can be obtained according to literature procedures [4]. PDB ID: 1CWP.
2. Au NPs with different sizes and capping ligands can be used for studying the Au NPs encapsulation efficiency based on different sizes and capping ligands; for details see [15]. 7 nm Au NPs are chosen for this work, their extinction coefficient (ε) is $2.79 \times 10^7\ M^{-1}\ cm^{-1}$.
3. The electrostatic potential of the capsid can affect catalysis of charged substrates. To investigate this effect, in addition to the positively charged 3NTA the following substrates can be used: negatively charged 3-nitrobenzenesulfonate (3NBS), and noncharged 1-methyl-3-nitrobenzene (3NTT) and nitrobenzene (NB).
4. Sodium borohydride can self-hydrolyze in solution; therefore, it should be freshly prepared for each reduction reaction.
5. We used a Sorvall WX80 equipped with a Fiberlite F50L-24 × 1.5 rotor.
6. We used a Perkin Elmer Lambda 850 UV-Vis spectrometer.
7. We used a Zetasizer Nano ZS ZEN3600 instrument (Malvern Instruments) at 25 °C with 633 nm laser.
8. We used a nanotrac wave setup by Anaspec operating with Microtrac FLEX Operating Software at 25 °C, using a laser wavelength of 780 nm with measurement angle of 180°.
9. A fast protein liquid chromatography (FPLC) system was used for size-exclusion chromatography in this work. We used Amersham Ettan LC system, equipped with Superose 6 PC 3.2/30 analytical 24 mL column from GE Life sciences. Other size exclusion techniques can be used (e.g., Asymmetric flow field flow fractionation (AF4) and gradient centrifugation), but the appropriate elution volumes need to be determined.

10. We used a Waters assembly consisting of a 2535 Quaternary Gradient Module with Flex Inject setup and a 2998 Photodiode Array Detector.
11. Alternatively, the dialysis can be performed by dialyzing against 1 L of RNA buffer overnight at 4 °C.
12. For coat proteins used for nitroarene reduction by CCMV-Au NPs, DTT from the RNA buffer must be completely removed before encapsulation of Au NPs. DTT can occupy Au NP active sites and decrease the catalytic activity of Au NPs. DTT can be removed by washing against cleaning buffer using a centrifuge filter. To test whether DTT is completely removed from the coat protein solution, take 10 μL of filtrate from the bottom part of the centrifuge tube, add 20 μL of $Ni(II)SO_3$ (100 μM) to this filtrate: if a transparent solution results it means DTT is fully removed. Otherwise, the coat protein needs to be further washed by cleaning buffer.
13. In this step, we used a NanoDrop setup to determine the concentration of CP, as well as the purity of the protein. Free CP with Abs 280/260 $\geq$ 1.65 is used for encapsulation. If the ratio < 1.65, a second run with the ultracentrifuge may help to get rid of contaminants.
14. Always store free CP in capsid storage buffer for later use. The CP of CCMV can be stored for a maximum of 1 month in the sodium acetate buffer pH 5 and 1 week in Tris–HCl buffer pH 7.5. Without interaction with a negative cargo the capsid proteins gradually lose their ARM. This process is accelerated at elevated pH.
15. BSPP has a higher binding affinity with Au NPs than citrate.
16. The centrifuge time depends on the centrifuge tube volume. In principle, the first three runs are to remove citric acid (and/or excess BSPP capping ligand) but not to concentrate the Au NP solution. An appropriate running time and centrifugation rate of 1050–1300 × *g* are required to avoid NP aggregation during washing.
17. After ligand exchange from citrate to BSPP, the zeta-potential of the Au NP solution decreases from −6 mV to −25 mV (± 5 mV), indicating successful ligand exchange. BSPP-stabilized Au NPs can be stored in the fridge at 4 °C for at least 6 months without precipitation, and the encapsulation efficiency remains 99%.
18. The absorption of the Au NPs can shift upon ligand exchange and after assembly in CCMV, due to a change in the dielectric constant of the surrounding environment. However, the change in extinction coefficient is negligible, therefore, the extinction coefficient of Au NPs $\varepsilon = 2.79 \times 10^7\ M^{-1}cm^{-1}$ for all 7 nm Au NPs can be used.
19. Centrifuge speeds in the range of 1050–1300 × *g* can be used without causing Au NP aggregation. Au NP concentration was calculated by: $C_{Au\ NPs}$ = (Au NPs particles/mL)/N_A, N_A is Avogadro constant.

20. This step encapsulates Au NPs into viral cages by electrostatic interaction at neutral pH. Buffer salt concentrations should be minimal when working with gold nanoparticles because precipitation occurs due to decreased electrostatic repulsion. After encapsulation in CCMV the particles are stable in salt up to at least 0.5 M NaCl.
21. Leaving the mixture on a rotator at 4 °C overnight rather than dialyzing to the cleaning buffer yields similar results. The encapsulation described here is at neutral pH, but a similar procedure is also possible at more acidic pH (pH 4–5) by dialyzing to CP storage buffer after mixing. However, a disadvantage of this approach is that it yields more "empty" virus particles. For increased long-term stability of the particles, final concentrations of 1 mM sodium azide and 5 mM $MgCl_2$ can be added to the buffer.
22. Under these conditions FPLC runs of each purified solution elute fractions at approximately 7–8, 10–12, and 18 mL corresponding to aggregated particles and larger assemblies, hybrid particles and empty capsids, and coat protein dimers respectively.
23. The encapsulation efficiency can be determined using the following formula: (Y_{yield}) = # monodispersed Au particles/total # Au particle = $Abs_{520\ nm-11.5\ mL}/(Abs_{520\ nm-10\ mL} + Abs_{520\ nm-8\ mL})$, where $Abs_{520\ nm-8\ mL}$ is the absorption maximum at $\lambda = 520$ nm of the peak eluting at 7–8 mL. $Abs_{520\ nm-11.5\ mL}$ is the absorption maximum at $\lambda = 520$ nm of the peak eluting at 10–12 mL. Due to the overlap of peaks at elution volume of 7–8 mL and 10–12 mL, instead of integrating these two peaks, absorption maxima at $\lambda = 520$ nm of these two peaks can be taken for the calculation.
24. Appropriate staining time results in good TEM images, though this is substrate dependent.
25. Other concentration determination techniques are possible, for example SDS-PAGE.
26. The ratio between Au NPs to substrate used in this work is 1:5.6; for this pseudo first-order kinetic reaction, the apparent rate constant is decreased exponentially. Therefore to study the catalytic activity of Au NPs, a ratio between Au NP to substrate at around 1:5 to 1:15 is suggested [17, 18].
27. The absorption at $\lambda = 260$ nm indicates the initial nitroarene substrate, the absorbance at $\lambda = 286$ nm indicates the intermediate of the reduction reaction and the absorption at $\lambda = 238$ nm indicates the product amine.
28. The reaction conversion can be checked using an analytical HPLC (Waters 2535 quaternary gradient module XBridge C18 5 μM column with 4.6 × 250 mm dimensions). We observed HPLC retention times following analytical HPLC with a solvent gradient of 10–100% acetonitrile for 10 min of initial stage (solvent: 90% water + 10% acetonitrile); 30 min gradient, and in total over 90 min timescale with flow rate of 0.5 mL/min in all

timescale. The internal standard molecule 2M4NP elutes at 40.1–40.5 min with UV-vis absorption 325 nm; the substrate 3NBS elutes at around 13 min with UV-vis absorption 260 nm; 3NTA elutes at around 18.5 min with UV-vis absorption 260 nm; 3NTT elutes at around 44.8 min with UV-Vis absorption 260 nm; NB elutes at around 41.8 min with UV-Vis absorption 260 nm.

Acknowledgments

We acknowledge financial support from the research program of the Dutch Polymer Institute (DPI), Project #777t (DPI, P.O. Box 902, 5600 AX Eindhoven, the Netherlands) and ERC Consolidator Grant (Protcage#616907).

References

1. Takale BS, Bao M, Yamamoto Y (2014) Gold nanoparticle (AuNPs) and gold nanopore (AuNPore) catalysts in organic synthesis. Org Biomol Chem 12:2005–2027
2. Chen Y, Qiu J, Wang X et al (2006) Preparation and application of highly dispersed gold nanoparticles supported on silica for catalytic hydrogenation of aromatic nitro compounds. J Catal 242:227–230
3. Wu S, Dzubiella J, Kaiser J et al (2012) Thermosensitive Au-PNIPA Yolk–Shell nanoparticles with tunable selectivity for catalysis. Angew Chem Int Ed 51:2229–2233
4. Comellas-Aragones M, Engelkamp H, Claessen VI et al (2007) A virus-based single-enzyme nanoreactor. Nat Nanotechnol 2:635–639
5. Bode SA, Minten IJ, Nolte RJM et al (2011) Reactions inside nanoscale protein cages. Nanoscale 3:2376–2389
6. Douglas T, Young M (1998) Host-guest encapsulation of materials by assembled virus protein cages. Nature 393:152–155
7. Wang T, Zhang Z, Gao D et al (2011) Encapsulation of gold nanoparticles by simian virus 40 capsids. Nanoscale 3:4275–4282
8. Jutz G, van Rijn P, Santos Miranda B et al (2015) Ferritin: a versatile building block for bionanotechnology. Chem Rev 115:1653–1701
9. Culver JN, Brown AD, Zang F et al (2015) Plant virus directed fabrication of nanoscale materials and devices. Virology 479–480:200–212
10. Bancroft JB, Hiebert E (1967) Formation of an infectious nucleoprotein from protein and nucleic acid isolated from a small spherical virus. Virology 32:354–356
11. Bancroft JB, Hills GJ, Markham R (1967) A study of the self-assembly process in a small spherical virus formation of organized structures from protein subunits in vitro. Virology 31:354–379
12. Speir JA, Munshi S, Wang G et al (1995) Structures of the native and swollen forms of cowpea chlorotic mottle virus determined by X-ray crystallography and cryo-electron microscopy. Structure 3:63–78
13. Kwak M, Minten IJ, Anaya D-M et al (2010) Virus-like particles templated by DNA micelles: a general method for loading virus nanocarriers. J Am Chem Soc 132:7834–7835
14. Brasch M, de la Escosura A, Ma Y et al (2011) Encapsulation of phthalocyanine supramolecular stacks into virus-like particles. J Am Chem Soc 133:6878–6881
15. Liu A, Verwegen M, de Ruiter MV et al (2016) Protein cages as containers for gold nanoparticles. J Phys Chem B 120:6352–6357
16. Liu A, Traulsen CHH, Cornelissen JJLM (2016) Nitroarene reduction by a virus protein cage based Nanoreactor. ACS Catal 6:3084–3091
17. Biondi I, Laurenczy G, Dyson PJ (2011) Synthesis of gold nanoparticle catalysts based on a new water-soluble ionic polymer. Inorg Chem 50:8038–8045
18. Gangula A, Podila R, M R, Karanam L et al (2011) Catalytic reduction of 4-Nitrophenol using biogenic gold and silver nanoparticles derived from Breynia rhamnoides. Langmuir 27:15268–15274

Chapter 2

Encapsulation of Active Enzymes within Bacteriophage P22 Virus-Like Particles

Dustin P. Patterson

Abstract

Virus-like particles (VLPs) are nonpathogenic protein cage structures derived from viral coat proteins that have found utility in the area of biomaterials and nanotechnology. VLPs have been exploited as containers for the sequestration and encapsulation of a wide range of guest molecules in their hollow interiors. The robust nature of VLPs lend them as versatile scaffolds that can be exploited to provide protection to encapsulated guest molecules, such as enzymes which are often susceptible to inactivation and degradation, and for organization and construction of new nanomaterials incorporating the chemical properties of the guest molecules. In this chapter a background and methodology for the encapsulation of enzymes on the interior of the bacteriophage P22 derived VLP is described.

Key words Enzyme, Virus-like particle (VLP), Protein cage, Encapsulation, Protein Engineering, Hierarchical Assembly, Nanoreactor, Nanomaterials, Bioinspired, Biomaterials

1 Introduction

Protein cages are a unique class of structures found in nature that are comprised of a protein coat, composed of multiple protein subunits, surrounding a hollow interior space. The assembly of protein cages is the result of programmed hierarchical self-assembly of protein subunits, directed by noncovalent interactions between protein subunits comprising the protein coat. Functions of protein cages include metal sequestration and storage in the cell, organization of enzymes to coordinate metabolic processes, encapsulation and protection of viral genomes, and as containers that perform chaperoning of misfolded protein [1–4]. Investigation of these structures and their function has been of intense interest and provided greater understanding in the areas of protein self-assembly, cellular metabolisms, and human health and disease. In addition, protein cages have become a major source of inspiration in the area of bioinspired materials, where researchers look to adapt and engineer these structures directly or to synthetically mimic their structural

Andrew K. Udit (ed.), *Protein Scaffolds: Design, Synthesis, and Applications*, Methods in Molecular Biology, vol. 1798, https://doi.org/10.1007/978-1-4939-7893-9_2,

properties and functions for applications in materials and biochemical sciences [5, 6]. Of the protein classes described thus far, virus-like particles have found great utility in the field of nanomaterials and the discussion going forward will focus specifically on this class of protein cage structures.

Virus-like particles, or VLPs for short, are protein cage structures derived from the capsid (protein coat) forming proteins of a virus. Typically, VLPs are comprised of nonpathogenic components and are devoid of viral genomes, tail spikes, and other pathogenic proteins. VLPs have several unique characteristics that make them especially useful as nanomaterials. The assembly of VLPs is often well characterized due to the association of viruses with health, and research performed by virologists provides extensive details about viral organization and assembly dynamics. In addition, the structures of VLPs are often known to the molecular level, with either high resolution crystal structures or molecular models available. Knowing the molecular arrangement of proteins in VLPs allows rational modification and redesign of the VLPs with molecular precision to an extent not seen in synthetic, nonbiological based systems. Not only do VLPs have precisely defined compositions and structures, they are typically quite stable, owed to their function as containers for harboring and protecting the viral genome, which makes them robust scaffolds from which to construct materials. Another unique feature of VLPs for materials construction is that they can be produced genetically by exploiting heterologous recombinant protein expression in various hosts, including *E. coli*, yeast, and plant systems, allowing one to harness and exploit organisms as biofactories for nanomaterials production. Standard approaches in genetic modification enable rapid generation of new VLP constructs with tailored properties as desired, with changes ensured by the genetic coding.

To date, a large array of VLPs have been described, providing a library of VLPs with diverse properties, including external and internal diameters (ranging from 10 to 200 nm), physical/chemical stabilities, protein compositions, assembly pathways, and biological origins, to name a few [5, 7]. VLPs can be treated as scaffolds that have an outside, interface, and inside that can be modified for various applications [5, 8, 9]. The outside surface can be modified to control external chemical/physical properties and be used to tune the properties of association and interaction of the VLP with other molecular species (i.e., solvating solution, receptors and other proteins, etc.). Modification of the interface allows the association between subunits to be modified and manipulated. The interior provides both a molecular surface that can be chemically modified and a hollow space that can be inhabited by guest molecules. The discussion here will focus on exploiting the hollow interior for encapsulation of guest molecules and further discussion of outside and interface modification can be found in the referenced articles and elsewhere [5, 6, 8]. Research in modifying the inside of VLPs has attracted much attention and has been used to sequester a large variety of guest molecules. Methodology has been developed to encapsulate drug molecules [10, 11], exploit the VLP as a size constrained reaction container for synthesizing both inorganic and organic nanoparticles [12], incorporate MRI contrast agents [13], encapsulate preformed nanoparticles [9, 14], and encapsulate a range of peptides and proteins with various biological functions [15, 16]. The aim of this chapter is to describe methodology for the encapsulation of proteins, specifically catalytically active enzymes, on the interior of VLPs. Enzymes are often sensitive to denaturation, both heat and chemical denaturation, and are susceptible to proteolysis, degradation, and inactivation upon attachment to solid supports. VLPs provide a robust container for sequestering enzymes, providing a shell that can protect the enzymes from degradation and providing an intermediary sacrificial scaffold that can be modified to construct higher order materials or be attached to surfaces, eliminating direct modification of the enzymes encapsulated that could lead to inactivation.

Two approaches have been described for the encapsulation of enzymes inside of VLPs, which can be described as either: (1) passive encapsulation or (2) directed encapsulation. Passive encapsulation is performed in vitro and involves mixing of disassembled capsid coat proteins (CPs) with high concentrations of an enzyme cargo, then stimulating reassembly of the CPs which leads to encapsulation of some of the enzyme cargoes in solution that get trapped based on their concentration. This mode of enzyme encapsulation has been described for cow pea chlorotic mottle virus (CCMV), where changes in pH leads to disassembly or assembly of the CCMV CPs. Jeroen Cornelissen and coworkers showed that by mixing enzymes with disassembled CP at 7.5 pH and then lowering the pH to 5.0, enzymes could be entrapped on the inside without loss in activity of the enzymes [17, 18]. However, this strategy does not to be ensure encapsulation of the enzymes, some VLPs will be empty, and careful consideration must be taken as to the enzyme that is to be encapsulated since changes in pH or use of denaturants that are often used to disassemble VLPs will lead to irreversible inactivation of the enzymes that one wishes to encapsulate. Directed encapsulation employs the use of an affinity tag that is added to the enzyme guest of interest, via genetic modification, that targets and associates to part of the CP that forms the interior surface of the VLP. Co-mixing of the tagged-guest enzymes with disassembled CPs, and subsequent assembly of the CPs leads to formation of VLPs with enzymes on the interior due to their noncovalent association with the CP prior to assembly. Often directed encapsulation exploits peptide tags that are found on accessory proteins that are normally encapsulated on the interiors of the viruses and encode/direct their integration on the viral interior. This approach enables encapsulation of enzymes inside VLPs in vivo, by genetic coexpression of tagged-guest enzymes with CP, or encapsulation in vitro from purified components in some cases. A number of examples of directed encapsulation have been described that utilize various targeting tags including engineered peptide dimerization domains genetically added to the CP and the guest protein (CCMV VLP) [19], RNA aptamers that bind to both the CP and guest enzyme (Qβ VLP) [20], fusion of guest enzymes to minor CP subunits (SV40 VLP) [21], and attachment of guest enzymes to scaffolding proteins that direct and catalyze the assembly of the CP into the VLP (P22 VLP) [22]. Among systems exploiting directed encapsulation, the P22 VLP has been found to be particularly versatile due to its general encapsulation strategy and the robust nature of its capsid structure. The remaining text will focus on the P22 VLP and methods for encapsulating enzymes within the P22 VLP to produce catalytically active nanoparticles.

The P22 VLP is derived from the bacteriophage P22, whose native host is *Salmonella typhimurium*. Studies investigating the assembly and maturation of bacteriophage P22 identified two main components required for proper assembly and eventual maturation of the bacteriophage P22, a 47 kDa coat protein (CP) that forms the capsid shell and a 33.6 kDa scaffolding protein (SP) that templates and directs the assembly of CPs into the procapsid structure. In the absence of proteins forming the portal complex, DNA, tail spike proteins and other processing proteins, mixing of the P22 CP and SP leads to the spontaneous self-assembly of a $T = 7$, 58 nm nanoparticle known as the procapsid (PC), which is comprised of 420 copies of the CP and approximately 100–300 copies of the SP [23]. In addition to the PC structure, other morphologies were found to be produced upon heating at temperatures at or above 60 °C, a more angular enlarged structure closely mimicking the final virus structure called the expanded shell form (EX) and a porous form of EX missing the CP pentons at the fivefold vertices called the wiffleball form (WB), both of which have an increased external dimension of 60 nm and an internal volume almost twice that of PC [24–28]. As noted the SP templates assembly of the CP into the PC structure, and interaction of the SP to the CP leads to trafficking of the SPs to the interior of the final PC P22 VLP. Further investigations of the assembly pathway found that the N-terminus of SP can be extensively modified and severely truncated, only requiring the last 18 amino acids on the C-terminus, with no deleterious effects on assembly of the PC structure [29, 30]. This feature of the P22 system has enabled the truncated SP to be exploited as a vehicle for directing the encapsulation of a wide array of protein cargoes on the interior of the assembled P22 VLP, whereby truncation of N-terminal amino acids yields empty space, previously inhabited by portions of the SP, in the interior of the P22 VLP that can be occupied by guest proteins and enzymes [22, 31–39]. For encapsulation of guest proteins, genes encoding desired guest proteins are fused to the gene encoding the truncated SP, which upon expression leads to the production of a guest protein-SP fusion protein. Coproduction of the guest protein-SP fusion protein with CP in the cell leads to in vivo encapsulation of the guest protein-SP on the interior of the P22 VLP. Enzyme substrates are purportedly freely diffusible into the VLP interior through 2 nm pores found in the VLP capsid shell. This method has proved to be very robust and general, with the ability to encapsulate fusion proteins up to 180 kDa in size reported thus far, without compromise to the P22 VLP structure [31]. P22 VLPs can also be used to encapsulate proteins that form inclusion bodies or are typically difficult to produce in *E. coli* by rapidly sequestering them on the inside of the VLP interior and insulating them from aggregation or degradation [35]. In addition, enzymes that require integration of special cofactors or have unique maturation requirements can be encapsulated using a temporal controlled expression strategy, whereby the guest enzyme-SP can be expressed first, allowed to mature for some time, then the CP can be induced to express at a later time leading to encapsulation of a mature, active enzyme, as has been previously reported [34, 40].

The sections below outline the materials and methodology for encapsulating enzymes within the P22 VLP. The methodology focuses on a single plasmid approach for coexpression of a guest enzyme-SP with CP for encapsulation, however the notes section also provides instruction for modifying to a two plasmid approach allowing temporal control over the expression of the guest enzyme-SP in relation to the CP. These techniques allow for rapid production of catalytic protein nanoparticles through in vivo heterologous expression in *E. coli*.

2 Materials

1. pRSFDuet-1 plasmid (Novagen) containing the genes encoding the bacteriophage P22 CP and truncated SP (amino acids 141–303). The P22 CP and genes can be purchased from companies providing gene synthesis (Genescript, etc.) from known sequences and subcloned into the pRSFDuet-1 plasmid [32]. The gene for truncated SP should incorporate BamHI (N-terminal end of gene) and SacI (C-terminal end of gene) restriction sites for insertion into multiple cloning site 1 and the CP gene should incorporate NdeI (N-terminal end of gene) and XhoI restriction sites (C-terminal end of gene) for insertion into multiple cloning site 2 of pRSFDuet-1. The pRSFDuet-1 plasmid contains NcoI and BamHI sites allowing insertion of enzyme guest genes upstream of the truncated SP. *See* **Note 1** for information about plasmid construction and use of alternative plasmids.
2. Gene encoding the guest enzyme of choice with NcoI restriction enzyme site on the N-terminal end of the gene and a BamHI restriction site on the C-terminal end of the gene (*see*

Note 2). Make sure to omit the stop codon (C-terminal end) from the guest enzyme gene.

3. Restriction enzyme NcoI with proper buffer.
4. Restriction enzyme BamHI with proper buffer.
5. T4 DNA ligase with proper buffer.
6. Reaction cleanup kit (*see* **Note 3**).
7. XL1-Blue electrocompetent or chemical competent cells.
8. Miniprep DNA purification kit.
9. SOC medium: 2% tryptone, 0.5% yeast extract, 10 mM sodium chloride, 2.5 mM potassium chloride, 10 mM magnesium chloride, 10 mM magnesium sulfate in water and sterilized by autoclaving, supplement sterilized solution with sterile filtered D-glucose to a final concentration of 20 mM.
10. Sterile petri dishes (4–6 dishes).
11. LB agar: 1.5 g agar, 1 g tryptone, 0.5 g yeast extract, and 1 g sodium chloride suspended in water to a final volume of 100 mL, autoclave to sterilize and fully dissolve solids. Before solidifying kanamycin should be added to LB agar to 50 μg/mL, mixed, then poured into petri dishes (4–6 dishes per 100 mL of LB agar) to cool and solidify.
12. LB broth: 10 g tryptone, 5 g yeast extract, and 10 g sodium chloride suspended in water to a total volume of 1 L, autoclave to sterilize and fully dissolve solids.
13. Kanamycin stock solution: 50 mg/mL kanamycin prepared in sterile water, can be stored at −20 °C until use.
14. An autoclave for sterilization of media.
15. *E. coli* BL21(DE3) electrocompetent or chemical competent cells.
16. IPTG solution: 500 mM molecular biology grade isopropyl-β-D-thiogalactopyranoside in sterile water.
17. Incubator/shaker with temperature and speed control capabilities outfitted with clamps for 2–2.8 L Erlenmeyer or Fernbach flasks.
18. Spectrophotometer with absorbance setting of 600 nm.
19. Hydrochloric acid: 1 M in water.
20. Sodium hydroxide: 1 M in water.
21. Lysis Buffer: 50 mM sodium phosphate, 100 mM sodium chloride, pH 7.6. Use hydrochloric acid and sodium hydroxide to adjust pH.
22. Lysozyme (lyophilized powder).
23. RNase (lyophilized powder).
24. Benchtop centrifuge (e.g., Beckman Allegra 25R Centrifuge).
25. Centrifuge tubes.

26. Sucrose 35% (w/v) in Lysis Buffer.
27. Ultracentrifuge tubes (e.g., Thermo Scientific F50L-8×39)
28. Ultracentrifuge.
29. Ultracentrifuge rotor (e.g., Thermo Scientific TH-641 rotor or equivalent)
30. Phosphate Buffered Saline (PBS): 50 mM sodium phosphate, 25 mM sodium chloride, pH 7.0. Use hydrochloric acid and sodium hydroxide to adjust pH.
31. Syringe filter (0.2 micron).
32. 5 mL syringe.
33. S-500 Sephadex size exclusion column.
34. An FPLC for column chromatography with UV-Vis detector.
35. 4× SDS-PAGE loading buffer: 2.5 mM Tris–HCl pH 6.8, 25% glycerol, 2% SDS, 2% β-mercaptoethanol, and 0.01% Bromophenol Blue.
36. SDS-PAGE gel: 4% acrylamide stacking gel and a 16% acrylamide running or separating gel.
37. Gel electrophoresis apparatus.
38. Destain solution: 50% methanol, 10% acetic acid, 40% water solution.
39. Coomassie Blue Stain: 0.5 g Coomassie Blue in 0.5 L Destain solution.
40. Dynamic light scattering (DLS) instrument (e.g., Brookhaven 90Plus particle size analyzer).
41. Transmission electron microscope (TEM) access.
42. TEM stain: 1% uranyl acetate in water.
43. Formvar coated TEM grids.

3 Methods

3.1 Construction of Plasmids

1. Digest the pRSFDuet-1 plasmid containing P22 truncated SP and CP and the gene encoding the guest enzyme, separately, with NcoI and BamHI restriction endonucleases.
2. Column-purify the digests using an enzyme reaction cleanup kit to obtain purified DNA devoid of restriction enzymes. *See* **Note 4** for additional preparatory methods to prevent religation of the plasmid.
3. Perform ligation reaction using T4 DNA ligase according to the manufacturer's specifications by mixing the enzyme gene and plasmid products previously digested with the same restriction sites.

4. Transform ligation reactions into XL1-Blue *E.coli* strain or equivalent (*see* **Note 5**) and incubate after addition of SOC media according to the transformation requirements for the strain used and plate on LB agar plates containing 50 μg/mL kanamycin for selection of plasmid incorporation.
5. Allow growth overnight at 37 °C for visualization of colonies.
6. Grow up 5 mL LB overnight cultures inoculated with individual colonies at 37 °C with shaking at 250 rpm. *See* **Note 6** for prescreening colonies by PCR.
7. Extract DNA using a miniprep extraction kit to obtain plasmid DNA.
8. Send DNA samples out for sequencing to verify insertion of proper sequence of the desired gene(s).
9. Store purified DNA at −20 °C.

3.2 Recombinant Protein Expression

1. Transform pRSFDuet-1 plasmid containing the desired Enzyme-SP and CP genes into BL21(DE3) cells, plate on LB agar plates containing kanamycin (50 μg/mL) and grow overnight as described above in Subheading 3.1.
2. From the LB agar plates, inoculate 5 mL cultures of LB containing antibiotics with *E. coli* strains harboring plasmids for the coexpression of Enzyme-SP/CP at 37 °C with shaking at 250 rpm and incubate overnight. *See* **Note 7**.
3. Transfer a 5 mL overnight culture showing good bacterial growth (cloudy solution appearance) to 1 L of LB containing proper antibiotic to maintain selection for the plasmid and grow at 37 °C with shaking at 250 rpm.
4. Monitor the optical density at 600 nm on a UV-Vis spectrophotometer to determine growth (typically start to monitor after 2–3 h of growth).
5. Induce expression of proteins by addition of IPTG to a final concentration of 0.5 mM once the cells reach mid log phase (OD_{600} = 0.8).
6. Grow cultures for 4 h after addition of IPTG (*see* **Note 8**).
7. Harvest cells by centrifugation at 4000 × *g* and store cell pellets at −20 °C overnight.
8. Resuspend the cell pellet after thawing, on ice, in Lysis Buffer supplemented with lysozyme (15 mg/mL) and RNase (30 mg/mL) added (5 mL of Lysis buffer per gram of cell pellet) and incubate on ice for 30 min.
9. Lyse cell suspension by using mild/moderate duty cycle sonication (*see* **Note 9**).
10. Remove cell debris by centrifugation at 12,000 × *g* for 45 min at 4 °C and decant resulting supernatant (soluble fraction) into a clean centrifuge tube (*see* **Note 10**).

11. Place the supernatant into an ultracentrifuge tube leaving enough room for ultracentrifugation over a 35% (w/v) sucrose cushion in Lysis Buffer applied below the supernatant by syringe (5 mL of 35% sucrose) and spin tubes at 148,790 × *g* (38,000 rpm) for 50 min.
12. After completion, immediately remove centrifuge tubes from the ultracentrifuge rotor, decant the supernatant making sure not to disturb the pellet, and resuspended the VLP pellet by addition of a suitable amount (typically 1–5 mL for 1 L growth) of PBS (50 mM sodium phosphate, 25 mM sodium chloride, pH 7.0) with rocking overnight at 4 °C.
13. Centrifuge or filter (0.2 micron) the resuspended VLP pellet to remove solid precipitates.
14. Purify the solution over an S-500 Sephadex size exclusion column using a suitable FLPC. Flow rate for size exclusion chromatography purification is typically performed at 0.5–1 mL/min with PBS (*see* **Note 11**).
15. Concentrate fractions taken from size exclusion chromatography containing P22 by ultracentrifugation of combined fractions and subsequently resuspend the resulting VLP pellet in an adequate volume of PBS to give a VLP solution of 1–5 mg/mL protein concentration.
16. Store Enzyme-VLP product at 4 °C.

3.3 Enzyme-P22 VLP Characterization

3.3.1 SDS-PAGE

1. Mix protein samples with 4× loading buffer and heat in a boiling water bath for 10 min, remove, and spin down in a microcentrifuge.
2. Run samples on a gel containing a 4% acrylamide stacking gel and a 16% acrylamide running or separating gel using a constant current of 38 milliamperes for approximately 1–1.5 h or until the loading dye has travelled a sufficient distance (3/4 gel length).
3. Stain gel with Coomassie Blue stain, rinse with water, and destain.
4. Take images with a suitable imaging system (e.g., UVP MultDoc-IT Digital Imaging System) or camera.
5. Perform densitometry on SDS-PAGE gels to determine the approximate ratio of enzyme-SP to CP to determine amount of enzymes packaged per VLP particle (420 copies of CP per VLP).

3.3.2 Light Scattering

1. Spin down VLP samples at 17,000 × *g* for 10 min and filter the supernatant through a 0.2 micron syringe filter prior to taking measurements.
2. Take measurements on a DLS instrument and fit the correlation function found for the sample to a nonnegatively constrained least squares analysis or other suitable analysis to yield the hydrodynamic radius (R_H) for the particles (*see* **Note 12**).

3.3.3 Activity

1. Evaluate the kinetic activity of the enzyme encapsulated inside the P22 VLP according to methodology previously described for the enzyme (*see* **Note 13**).

3.3.4 Microscopy

1. Apply samples (10 μL, 0.1 mg/mL protein) to glow-discharged formvar coated grids and incubate for 30 s and wick away excess liquid with filter paper. Apply 10 μL of distilled water to the grid to wash and wick away liquid shortly after addition with filter paper. Stain the sample by addition of 5 μL 1% uranyl acetate to the grid and wick away excess stain with filter paper. Image VLP particles on a transmission electron microscope at accelerating voltage of 80–100 kV.

4 Notes

1. The construction of the P22 VLP assembler plasmid (pRSF-Duet-1 vector described) can be performed according to the same methodology described for insertion of the enzyme guest gene of interest; digestion with the proper restriction enzymes and ligation of the plasmid and gene products. For the pRSF-Duet-1 vector and other Duet systems, so-called because they contain two multiple cloning sites, the CP has generally been incorporated with NdeI/XhoI restriction endonuclease sites into multiple cloning site 2 and the truncated SP into multiple cloning site 1 with BamHI/SacI restriction endonuclease sites. Other Duet plasmid systems include pETDuet, pACYCDuet, and pCDFDuet, which could also be utilized, but require alternative antibiotics for selection.
2. The gene encoding the enzyme of interest can either be purchased from a gene synthesis company with restriction enzyme sequences incorporated, typically inserted into a vector, or alternatively the gene can be amplified from genomic DNA via polymerase chain reaction (PCR) using primers incorporating the NcoI and BamHI restriction sites with sufficient nucleotide overhangs for efficient restriction endonuclease cutting. For the PCR approach two primers are required, a forward primer comprised of several base overhangs (5–7 bases) flanked by the restriction enzyme DNA sequence followed by the first 15–20 bases found in the enzyme gene of interest and a reverse primer set up in a similar manner having several base overhangs, the restriction enzyme site sequence, but comprised of the first 15–20 bases found at the 5′-end of the complementary (noncoding) strand. PCR amplification will result in the gene product of the enzyme of interest flanked at each end by restriction enzyme sites that allows directional insertion of the gene into the cut plasmids. Resources for PCR cloning are available from

numerous sources (New England Biolabs, Agilent, etc.) and examples of primers utilized for enzyme incorporation into the P22 VLP system can be found in the noted references [35, 36].

3. The author utilizes the Qiagen MinElute Reaction Cleanup kit, although equivalent kits are available from other manufacturers.
4. The pRSFDuet-1 plasmid may be dephosphorylated after digest with Antarctic phosphatase to prevent religation, in the event of single cutting, and subsequently heat inactivated according to the manufacturer's specifications before mixing with gene insert and performing ligation.
5. Other DNA propagation *E. coli* strains include the DH1, DH5α, and 10G (Lucigen) strains.
6. Colonies of transformed *E. coli* can be screened prior to growing up cultures to determine the presence of the enzyme gene of interest by PCR. Make a PCR reaction mixture containing primers for the enzyme gene of interest and make 10 μL aliquots of the solution into PCR tubes. Scratch a small bit of cells from a colony of interest with a small pipet tip and place it in the PCR tube containing the solution. Rock the PCR tube(s) for 5–10 min, then perform PCR with an initial denaturation step of 10 min to lyse the cells and perform all subsequent steps as normal. Run PCR reactions out on an agarose gel by electrophoresis according to standard methods and look for bands corresponding to the molecular size of the enzyme gene of interest when compared to a DNA ladder. Grow up any colonies showing hits for the gene of the desired enzyme.
7. Colonies of *E. coli* BL21(DE3) can be used to make glycerol stocks of the cells for long term storage. For this, inoculate a 5 mL culture of LB media with BL21(DE3) containing the plasmid and incubate overnight at 37 °C with shaking at 250 rpm. Place 0.3 mL of glycerol in a cryo-vial of 1.5–2.0 mL volume and add 0.7 mL of overnight culture showing good growth to it, mix with glycerol by capping and inverting the vial, then flash freeze with liquid nitrogen and store at −80 °C. Future cultures can be obtained by quickly taking the glycerol stock, stabbing it with a sterile loop or pipet tip, and subsequently inoculating an LB solution to start growth. Return the glycerol stock to −80 °C before allowing it to thaw if possible. Multiple freeze thaw cycles will eventually lead to loss of the vitality of the *E. coli* glycerol stock and thus new glycerol stocks should be made periodically.
8. Optimal protein expression levels can be determined empirically by evaluating expression levels of proteins at different expression times after induction by SDS-PAGE. For this, induce the cells to express protein by addition of IPTG, then take samples from the culture at different time points (1 h, 2 h, 4 h, etc.), pellet cells of each sample by centrifugation, then

add SDS-PAGE loading buffer and boil to lyse cells. Run an SDS-PAGE, stain then destain the gel, and evaluate the intensity levels of the protein bands corresponding to the enzyme-SP and CP to determine best incubation times.

9. The freeze thaw method of lysis is a simple method that can be utilized if sonication is unavailable. Place resuspended cells in a disposable centrifugation tube. Plunge the centrifuge tube into liquid nitrogen until the cell suspension is frozen, then remove the tube from the liquid nitrogen and plunge it into a beaker of warm water. Perform this liquid nitrogen to warm water cycle several times to lyse the cells then carry on with the purification as described above.
10. In the event that lipid or solids are not fully removed after one centrifugation spin, perform an additional 45 min centrifugation on the supernatant at 12,000 × *g* to remove any excess cellular debris.
11. Ultracentrifugation of P22 VLPs through a cesium chloride gradient can be utilized in conjunction with or in lieu of size exclusion chromatography. For poorly expressing enzyme guest-SP, CP aggregates may form that have similar shape/size as the mature enzyme-P22 VLP leading to co-elution by SEC. Applying P22 VLP containing sample to the top of a cesium chloride gradient (0.26–0.54 g CsCl/mL PBS) and subsequent centrifugation on an ultracentrifuge at 36,000 rpm (96,500 × *g*) for 2 h produces a band of the properly formed P22 VLP with encapsulated guest proteins. Bands can be removed by syringe or pipette and composition determined by SDS-PAGE [33].
12. An additional method for characterizing enzyme guest-VLPs is to utilize size exclusion chromatography coupled with Multiangle Light Scattering and Refractive Index Detection. For this samples are separated over a size exclusion column utilizing an HPLC (e.g., WTC-0200S column (Wyatt Technologies) on an Agilent 1200 HPLC) to apply and maintain a 0.7 mL/min flow rate. Typically samples of 25 μL (1–2 mg/mL P22 VLP concentration) are injected onto the column and run for a total time of 30 min using a 50 mM phosphate, pH 7.2 buffer containing 100 mM sodium chloride and 200 ppm sodium azide (to prevent microbial growth) as an elution buffer. Samples are detected using a UV-Vis detector, a Multi-Angle Laser Light Scattering (e.g., Wyatt HELEOS) detector, and a differential refractometer (e.g., Wyatt Optilab rEX). The number average molecular weight, Mn, are then calculated with user software (e.g., Wyatt Technology Corporation Astra 5.3.14 software) based on the molecular weight distribution. These provide accurate molecular weights for the P22 VLP, which can be used to calculate the number of enzymes packaged as previously described [22].

13. In the event that an enzyme is found to be inactive upon encapsulation with the P22 VLP using the co-induction approach described above, a simple alteration to the encapsulation strategy can often produce active particles. Improved catalytic properties have been observed for enzymes having special maturation requirements, such as incorporation of heme or other unique cofactors required for activity, by expressing the enzyme guest-SP fusion protein temporally before the CP. For this approach transfer the enzyme guest-SP gene to the pBAD plasmid, which if incorporated into the pRSFDuet-1 plasmid as described above can be removed by double digest with NcoI/SacI and transferred into an NcoI/SacI cut pBAD vector according to the methods described above. For sequential expression, co-transform the pBAD plasmid containing the enzyme guest-SP with a pRSFDuet-1 vector void of SP, but containing the CP and utilize media containing both ampicillin and kanamycin to select for *E. coli* containing both plasmids. It is important that the plasmid containing CP does not also contain an expressible SP to ensure VLPs only form with the enzyme guest-SP. Perform growth as described above and induce expression of enzyme guest-SP fusion protein contained on pBAD by addition of L-arabinose to 0.2% (w/v) upon reaching mid log phase (OD600 = 0.6–0.8), allow expression of the enzyme guest-SP for 4 h to overnight and then add IPTG to induce CP expression, leading to encapsulation of the mature enzyme guest-SP fusion protein [40]. For optimal activity, evaluate enzyme guest-SP and CP expression for different durations.

References

1. Sutter M, Boehringer D, Gutmann S et al (2008) Structural basis of enzyme encapsulation into a bacterial nanocompartment. Nat Struct Mol Biol 15(9):939–947
2. Bode SA, Minten IJ, Nolte RJ et al (2011) Reactions inside nanoscale protein cages. Nanoscale 3(6):2376–2389
3. Corchero JL, Cedano J (2011) Self-assembling, protein-based intracellular bacterial organelles: emerging vehicles for encapsulating, targeting and delivering therapeutical cargoes. Microb Cell Factories 10(1):92
4. Giessen TW (2016) Encapsulins: microbial nanocompartments with applications in biomedicine, nanobiotechnology and materials science. Curr Opin Chem Biol 34:1–10
5. Uchida M, Klem MT, Allen M et al (2007) Biological containers: protein cages as multifunctional nanoplatforms. Adv Mater 19(8):1025–1042
6. Flenniken M, Uchida M, Liepold L et al (2009) A library of protein cage architectures as nanomaterials. Curr Top Microbiol Immunol 327:71–93
7. Pushko P, Pumpens P, Grens E (2013) Development of virus-like particle technology from small highly symmetric to large complex virus-like particle structures. Intervirology 56(3):141–165
8. Smith MT, Hawes AK, Bundy BC (2013) Reengineering viruses and virus-like particles through chemical functionalization strategies. Curr Opin Biotechnol 24(4):620–626
9. Dixit SK, Goicochea NL, Daniel M-C et al (2006) Quantum dot encapsulation in viral capsids. Nano Lett 6(9):1993–1999
10. Rother M, Nussbaumer MG, Renggli K et al (2016) Protein cages and synthetic polymers: a fruitful symbiosis for drug delivery applications, bionanotechnology and materials science. Chem Soc Rev 45(22):6213–6249
11. Ma Y, Nolte RJ, Cornelissen JJ (2012) Virus-based nanocarriers for drug delivery. Adv Drug Deliv Rev 64(9):811–825
12. Maity B, Fujita K, Ueno T (2015) Use of the confined spaces of apo-ferritin and virus capsids as nanoreactors for catalytic reactions. Curr Opin Chem Biol 25:88–97

13. Cormode DP, Jarzyna PA, Mulder WJ et al (2010) Modified natural nanoparticles as contrast agents for medical imaging. Adv Drug Deliv Rev 62(3):329–338
14. Aniagyei SE, DuFort C, Kao CC et al (2008) Self-assembly approaches to nanomaterial encapsulation in viral protein cages. J Mater Chem 18(32):3763–3774
15. Al-Barwani F, Donaldson B, Pelham SJ et al (2014) Antigen delivery by virus-like particles for immunotherapeutic vaccination. Ther Deliv 5(11):1223–1240
16. Frietze KM, Peabody DS, Chackerian B (2016) Engineering virus-like particles as vaccine platforms. Curr Opin Virol 18:44–49
17. Minten IJ, Claessen VI, Blank K et al (2011) Catalytic capsids: the art of confinement. Chem Sci 2(2):358–362
18. Comellas-Aragonès M, Engelkamp H, Claessen VI et al (2007) A virus-based single-enzyme nanoreactor. Nat Nanotechnol 2(10):635–639
19. Minten IJ, Hendriks LJ, Nolte RJ et al (2009) Controlled encapsulation of multiple proteins in virus capsids. J Am Chem Soc 131(49):17771–17773
20. Fiedler JD, Brown SD, Lau JL et al (2010) RNA-directed packaging of enzymes within virus-like particles. Angew Chem Int Ed 49(50):9648–9651
21. Inoue T, M-a K, Takahashi R-u et al (2008) Engineering of SV40-based nano-capsules for delivery of heterologous proteins as fusions with the minor capsid proteins VP2/3. J Biotechnol 134(1):181–192
22. Patterson DP, Prevelige PE, Douglas T (2012) Nanoreactors by programmed enzyme encapsulation inside the capsid of the bacteriophage P22. ACS Nano 6(6):5000–5009
23. Botstein D, Waddell CH, King J (1973) Mechanism of head assembly and DNA encapsulation in salmonella phage P22: I. Genes, proteins, structures and DNA maturation. J Mol Biol 80(4):697–731
24. Teschke CM, McGough A, Thuman-Commike PA (2003) Penton release from P22 heat-expanded capsids suggests importance of stabilizing penton-hexon interactions during capsid maturation. Biophys J 84(4):2585–2592
25. Kang S, Uchida M, O'Neil A et al (2010) Implementation of p22 viral capsids as nanoplatforms. Biomacromolecules 11(10):2804–2809
26. Parent KN, Khayat R, Tu LH et al (2010) P22 coat protein structures reveal a novel mechanism for capsid maturation: stability without auxiliary proteins or chemical crosslinks. Structure 18(3):390–401
27. Casjens S, Adams M, Hall C et al (1985) Assembly-controlled autogenous modulation of bacteriophage P22 scaffolding protein gene expression. J Virol 53(1):174–179
28. Schwarz B, Madden P, Avera J et al (2015) Symmetry controlled, genetic presentation of bioactive proteins on the P22 virus-like particle using an external decoration protein. ACS Nano 9(9):9134
29. Parker MH, Casjens S, Prevelige PE (1998) Functional domains of bacteriophage P22 scaffolding protein. J Mol Biol 281(1):69–79
30. Weigele PR, Sampson L, Winn-Stapley D et al (2005) Molecular genetics of bacteriophage P22 scaffolding protein's functional domains. J Mol Biol 348(4):831–844
31. Patterson DP, Schwarz B, Waters RS et al (2013) Encapsulation of an enzyme cascade within the bacteriophage P22 virus-like particle. ACS Chem Biol 9(2):359–365
32. Patterson DP, Schwarz B, El-Boubbou K et al (2012) Virus-like particle nanoreactors: programmed encapsulation of the thermostable CelB glycosidase inside the P22 capsid. Soft Matter 8(39):10158–10166
33. Patterson DP, Rynda-Apple A, Harmsen AL et al (2013) Biomimetic antigenic nanoparticles elicit controlled protective immune response to influenza. ACS Nano 7(4):3036–3044
34. Jordan PC, Patterson DP, Saboda KN et al (2015) Self-assembling biomolecular catalysts for hydrogen production. Nat Chem 8(2):179–185
35. Patterson DP, LaFrance B, Douglas T (2013) Rescuing recombinant proteins by sequestration into the P22 VLP. Chem Commun 49(88):10412–10414
36. Patterson DP, McCoy K, Fijen C et al (2014) Constructing catalytic antimicrobial nanoparticles by encapsulation of hydrogen peroxide producing enzyme inside the P22 VLP. J Mater Chem B 2(36):5948–5951
37. O'Neil A, Prevelige PE, Basu G et al (2012) Coconfinement of fluorescent proteins: spatially enforced communication of GFP and mCherry encapsulated within the P22 capsid. Biomacromolecules 13(12):3902–3907
38. O'Neil A, Prevelige PE, Douglas T (2013) Stabilizing viral nano-reactors for nerve-agent degradation. Biomater Sci 1(8):881–886
39. Qazi S, Miettinen HM, Wilkinson RA et al (2016) Programmed self-assembly of an active P22-Cas9 nanocarrier system. Mol Pharm 13(3):1191–1196
40. Sánchez-Sánchez L, Tapia-Moreno A, Juarez-Moreno K et al (2015) Design of a VLP-nanovehicle for CYP450 enzymatic activity delivery. J Nanobiotechnol 13(1):66

Chapter 3

Encapsulation of Exogenous Proteins in Vault Nanoparticles

Meng Wang, Danny Abad, Valerie A. Kickhoefer, Leonard H. Rome, and Shaily Mahendra

Abstract

Natural vault nanoparticles are ribonucleoprotein particles with a mass of 13 MDa that have been found in a wide variety of eukaryotes. Empty recombinant vaults are assembled from heterologously expressed Major Vault Protein (MVP), forming the barrel-shaped vault shell. These structures are morphologically indistinguishable from natural vault particles. Here, we describe the packaging and purification of exogenous proteins into these recombinant vault particles by mixing with proteins attached to the INT domain that binds to MVP.

Key words Nanostructures, Enzyme immobilization, Baculovirus expression, Nanocages, Protein scaffold

1 Introduction

Vault nanoparticles, which are classified as ribonucleoprotein particles, are commonly found in most eukaryotic organisms with dimensions of 41 × 41 × 72.5 nm and have a highly conserved barrel-like shape [1–3]. Naturally occurring vaults are composed of three different proteins and one or more copies of a small untranslated vault RNA [4–8]. The Major Vault Protein (MVP) forms the outer vault shell and makes up over 70% of the total vault mass. Inside of natural vaults are multiple copies of vault poly(ADP-ribose) polymerase (VPARP, also called PARP4), which contains a poly-(ADP-ribosylation) catalytic domain and a domain that tightly binds to the inside of the vault shell (INT), and multiple copies of the telomerase-associated protein-1 (TEP1) that binds to the untranslated vault RNAs. Seventy-eight copies of heterologously expressed MVP are assembled on a polyribosome to build a recombinant vault in insect cells, which are morphologically indistinguishable from natural vault particles (Fig. 1a, b) [9, 10]. Each recombinant vault particle has a hollow, barrel-shaped structure, and consists of two identical cup-like

Andrew K. Udit (ed.), *Protein Scaffolds: Design, Synthesis, and Applications*, Methods in Molecular Biology, vol. 1798, https://doi.org/10.1007/978-1-4939-7893-9_3, © Springer Science+Business Media, LLC, part of Springer Nature 2018

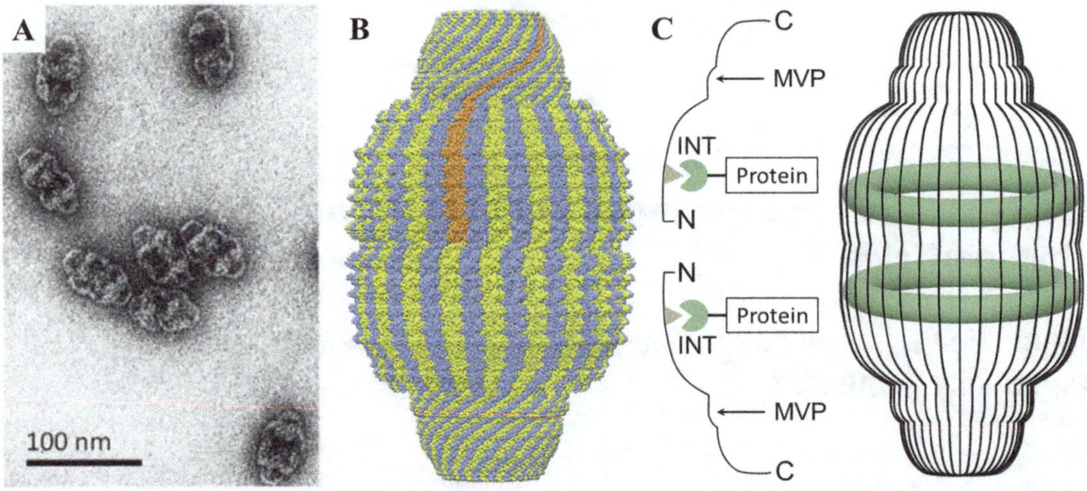

Fig. 1 Structure of vault particles. (**a**) TEM image of vaults purified from rat liver. (**b**) A side view showing the arrangement of MVP chains forming vault shell. (Reprinted from Rome et al. (2013) with permission from the American Chemical Society.) (**c**) Vault packaging strategy using INT domain attachment *(12)*. (Adapted from Han et al. (2011) with permission from the American Chemical Society)

half-vaults where the N-termini of all MVP chains meet together at the waist. The inner core of recombinant vaults provides ample space to enable vaults to be plausible vehicles for carrying small chemical molecules and/or large macromolecules.

Here, we describe a protocol to package proteins into recombinant vaults using INT domain. The INT domain is defined as the 162-amino acid region at the C-terminus of VPARP, and has been identified as the smallest region enabling interaction with the MVP shell [11, 12]. It serves as a packaging signal and directs the INT attached components, such as enzymes [11], chemokines [13], and lipoproteins [14], to the inner core of vault particles (Fig. 1c). In addition, due to the dynamic nature of the vault shell, exterior components can be taken up by preassembled vaults [15, 16], thus vault-packaging can be achieved via mixing vault particles with INT attached components. In this protocol, we first outline how to construct the specialized vectors used in synthesizing vault particles and INT fusion proteins. The expression, vault-packaging, and purification procedures are then described in the following sections.

2 Materials

2.1 Solutions and Reagents

1. 10× Tris–borate–EDTA buffer solution (TBE): 890 mM Tris, 890 mM boric acid, 20 mM EDTA, adjust pH to 7.6 with HCl. Store at room temperature.
2. Dithiothreitol (DTT) solution: 1 M DTT in deionized water. Store at −20 °C.

3. Phenylmethylsulfonyl fluoride (PMSF) solution: 200 mM PMSF in ethanol. Store at room temperature.
4. Streptomycin sulfate solution: 200 g/L streptomycin sulfate in deionized water. Store at −20 °C.
5. Protease Inhibitor (PI): Protease Inhibitor Cocktail for use in purification of Histidine-tagged proteins (Sigma-Aldrich, St. Louis, MO), or any equivalent protease inhibitors. Store at −20 °C.
6. Trypan Blue solution: 137 mM NaCl, 2.7 mM KCl, 10 mM Na_2HPO_4, 1.8 mM KH_2PO_4, 2% Trypan Blue, adjust pH to 7.4 with HCl. Store at room temperature.
7. Uranyl acetate solution: 1% uranyl acetate in deionized water. Store at room temperature.
8. Glucose stock solution: 1 M glucose in water. Dissolve 180 g of glucose in 700 mL of deionized water and bring volume to 1 L. Sterilize by filtration (0.2 μm filter). Store at 4 °C.
9. Buffer A: 50 mM Tris, 75 mM NaCl, 0.5 mM $MgCl_2$, pH to 7.4 with HCl. Sterilize by filtration (0.2 μm filter). Store at 4 °C.
10. Buffer A containing 1% v/v Triton X-100 (Buffer A w/ Triton). Store at 4 °C.
11. Lysis buffer: Buffer A w/Triton containing 1 mM DTT, 1 mM PMSF and PI (50 μL/gram cell pellet). Prepare fresh.
12. Buffer B: Buffer A containing 5 M NaCl. Store at 4 °C.
13. Sucrose gradient solutions: 20%, 30%, 40%, 45%, 50%, and 60% sucrose in Buffer A. Sterilize by filtration (0.2 μm filter). Store at 4 °C.
14. 100× Antibiotic-Antimycotic (Anti/Anti): 6000 μg/mL of penicillin, 10,000 μg/mL of streptomycin, and 25 μg/mL of amphotericin B. Store at −20 °C.
15. Cellfectin II Transfection Reagent (Life Technologies, Grand Island, NY). Store at 4 °C.
16. DNA Agarose Gel Recovery kit.
17. Plasmid DNA purification kit.
18. *Taq* DNA polymerase
19. *Taq* DNA polymerase 10× reaction buffer: 100 mM Tris–HCl, 500 mM KCl, 15 mM $MgCl_2$, pH 8.3. Store at −20 °C.
20. dNTP mix: dATP, dCTP, dGTP, dTTP, each at a final concentration of 10 mM. Store at −20 °C.
21. T4 DNA ligase
22. T4 DNA ligase 10× reaction buffer: 500 mM Tris–HCl, 100 mM $MgCl_2$, 10 mM ATP, 100 mM DTT, pH 7.5. Store at −20 °C.

2.2 Media

1. LB medium: 1% tryptone, 0.5% yeast extract, 1% NaCl, adjust pH to 7.5 with NaOH. Autoclave 20 min on liquid cycle, cool to about 55 °C and add 100 μg/mL ampicillin (Amp), or all of 50 μg/mL kanamycin (Km), 7 μg/mL gentamicin (Gen), 10 μg/mL tetracycline (Tet) if needed (*see* **Note 1**). Store at 4 °C.
2. LB-Amp plates: 1% tryptone, 0.5% yeast extract, 1% NaCl, 20 g/L agar, adjust pH to 7.5 with NaOH. Autoclave for 20 min on liquid cycle, cool to about 55 °C and add 100 μg/mL Amp (*see* **Note 1**). Store at 4 °C.
3. Bacmid selection plates: 1% tryptone, 0.5% yeast extract, 1% NaCl, 20 g/L agar, adjust pH to 7.5 with NaOH. Autoclave for 20 min on liquid cycle, cool to about 55 °C and add 50 μg/mL Km, 7 μg/mL Gen, and 10 μg/mL Tet, 200 μg/mL Bluo-Gal, and 80 μg/mL IPTG (*see* **Note 1**). Store at 4 °C.
4. S.O.C. medium. Store at 4 °C (*see* **Note 2**).
5. Sf-900 II SFM medium (Life Technologies, Grand Island, NY) supplemented with a 1/100 dilution of Anti/Anti unless otherwise specified. Store at 4 °C.

2.3 Plasmids and Cells

1. pFastBac1 vector (Invitrogen, Carlsbad, CA).
2. Chemically competent *E. coli* cells.
3. MAX Efficiency DH10Bac Competent *E. coli* (Invitrogen, Carlsbad, CA).
4. *Spodoptera frugiperda Sf9* cells.

2.4 Equipment and Supplies

1. Glass Dounce homogenizer.
2. Hemacytometer.
3. Inverted microscope with a 10× ocular and 10× objective.
4. Centrifuge that can go up to 100,000 × *g*, and has acceleration/deceleration function.
5. Fixed angle rotor and swinging bucket rotor.
6. Biosafety cabinet.
7. 27 °C and 37 °C stationary and shaking incubator.
8. 42 °C water bath.
9. 15 mL round-bottom tube.
10. 6-well tissue culture plate.
11. 125 mL and 500 mL flasks with vented caps.
12. 5 mL, 10 mL, and 25 mL serological pipettes.
13. Apparatus for agarose gel electrophoresis.

3 Methods

3.1 Expression of INT Fused Protein and Recombinant Vaults in Sf9 Cells

Sf9 insect cells are chosen as the host organism in this protocol (*see* **Notes 3** and **4**).

3.1.1 Isolation of Gene of Interest (GOI) from Source Organisms

Coding sequences of INT domain derived from human VPARP (GenBank No. AF158255, amino acids 1563–1724) [7, 11] and rat MVP (GenBank No. U09870) [17] can be accessed at GenBank. MVP from other species, such as human and *Dictyostelium discoideum*, can also be used to synthesize recombinant vault particles (*see* **Note 5**). GOIs encoding MVP, INT domain and protein of interest (POI) can be isolated from their source organisms and cloned into vectors using the method of user's choice, or can be chemically synthesized.

3.1.2 Construction of Baculovirus Expression Vectors

1. To construct the POI-INT expression cassette (*see* **Notes 6** and **7**):
   (a) PCR amplify POI and INT coding fragments with primers containing appropriate restriction sites.
   (b) Digest amplified fragments and pFastBac1 vectors to generate complementary sticky ends (*see* **Note 8**).
   (c) Verify the digestion by agarose-gel electrophoresis in TBE buffer, and extract the digested products using a commercial gel recovery kit (*see* **Note 9**).
   (d) Ligate purified fragments together using T4 DNA ligase to construct recombinant POI-INT-pFastBac1 plasmid (*see* **Note 10**).
2. To construct the MVP vector:
   (a) Digest PCR-amplified MVP coding fragment and pFastBac1 vector with the same restriction enzymes, leaving complementary sticky ends.
   (b) Verify, purify and ligate digested fragments to construct recombinant MVP-pFastBac1 plasmid as described above (**steps 1c–d**).
3. Add 50 μL of chemically competent *E. coli* to 1 ng of each ligation product in a 15 mL round-bottom tube, mix gently and incubate on ice for 30 min.
4. Heat-shock for 30–45 s in a 42 °C water bath. Place the tube on ice for 1–2 min.
5. Add 450 μL of prewarmed S.O.C. medium. Shake horizontally at 37 °C, 200 rpm for 1 h, and then spread transformation mixture onto LB-Amp plates. Incubate overnight at 37 °C.

6. For each transformation, pick 4–10 colonies, grow in LB-Amp medium at 37 °C and 200 rpm overnight.
7. Pellet the cells and purify plasmid DNAs with a commercial plasmid DNA purification kit. Sequence to verify the insertion of target sequences into pFastBac1.
8. Mix 10 μL of the DH10Bac Competent *E. coli* cells with 100 ng of each recombinant pFastBac1 plasmid in separate 15 mL round-bottom tubes, and incubate on ice for 30 min.
9. Heat-shock the cells for 45 s in a 42 °C water bath, then immediately transfer the tubes to ice and chill for 2 min, and then add 90 μL of prewarmed S.O.C. medium. Shake tubes at 37 °C at 200 rpm for 4 h.
10. Spread 150 μL of 100-fold and 1000-fold diluted mixture on bacmid selection plates. Incubate plates at 37 °C for 48 h.
11. For each transformation, pick and restreak 5 white colonies on bacmid selection plates to confirm the colonies are truly white (*see* **Note 11**).
12. Inoculate a white colony from the restreaked plates in 2 mL of LB medium overnight (12–24 h), and purify bacmid DNAs with a commercial plasmid DNA purification kit.

3.1.3 Production of Recombinant Baculovirus

Baculoviruses are produced by transfection of *Sf9* cells with bacmid DNA following the Bac-to-Bac Expression System Manual, described below (*see* **Note 12**).

1. Establish a *Sf9* cell stock growing in unsupplemented Sf-900 II SFM medium.
2. Determine the *Sf9* cell count:
   (a) Transfer 5 μL of the *Sf9* cell stock into a test tube, add 85 μL of room temperature Sf-900 II SFM medium and 10 μL of 2% Trypan Blue solution, and gently mix.
   (b) Add 10 μL of the cell–Trypan Blue mixture to a hemacytometer and count cells under an inverted microscope with a 10× ocular and 10× objective.
   (c) Verify the cells are in log phase (1.5–2.5 × 10^6 cells/mL) with greater than 95% viability (*see* **Note 13**).
3. Make a master mix of the *Sf9* cells at 4 × 10^5 cell/mL in unsupplemented Sf-900 II SFM medium. Add 2 mL of this mix into each well of a 6-well tissue culture plate for a final concentration of 8 × 10^5 cells per well.
4. Allow the cells to attach for 15 min at room temperature in the biosafety cabinet, then remove the medium and replace with 2 mL of fresh unsupplemented Sf-900 II SFM medium.
5. Dilute 8 μL of Cellfectin II reagent into 100 μL of unsupplemented Sf-900 II SFM media, and mix gently. Dilute 5 μL (or 1–2 μg) of bacmid DNA in 100 μL of unsupplemented Sf-900

II SFM media, and mix gently. Combine the diluted Cellfectin II with diluted DNA. Mix gently and incubate for 30 min at room temperature to allow lipid–DNA complexes to form.

6. Slowly add all the DNA–Cellfectin transfection mixture onto the *Sf9* cells attached on the 6-well plate. Incubate at 27 °C for 3–5 h without shaking.
7. Remove the transfection mixture and medium without disturbing attached cells and replace with 2 mL of fresh Sf-900 II SFM medium, and incubate cells at 27 °C for 96 h without shaking.
8. Collect the medium containing the virus from each well and centrifuge at 500 × *g* for 5 min to remove cells (*see* **Note 14**). The clarified supernatant will be the P1 baculoviral stock.
9. Count cells from *Sf9* stocks and verify the cells are in normal shape and with greater than 95% viability as described in **step 2** (*see* **Note 15**). Seed 50 mL of *Sf9* culture (2 × 10^6 cells/mL) in Sf-900 II SFM medium in a 150 mL flask with a vented cap and add 500 μL of P1 viral stock. Incubate the cells at 27 °C at 120 rpm.
10. Three days post-infection, transfer the culture to sterile 50 mL conical tubes and centrifuge the tubes at 3000 × *g* for 5 min at 4 °C to remove cells and large debris (*see* **Note 14**). The clarified supernatant will be the P2 baculoviral stock. P2 stock can be stored at 4 °C for a few months, or can be aliquoted into small tubes and long-term stored at −80 °C.
11. Infect 50 mL *Sf9* cultures (2 × 10^6 cells/mL) in Sf-900 II SFM medium with P2 viral stock at different volumes (1–500 μL). Collect 1 mL of each infected culture at 24, 48, 72, and 96 h, and analyze each for protein expression to determine the appropriate infection volume and time (*see* **Note 16**).

3.1.4 Production of Recombinant Vault Particles

1. Count and verify *Sf9* cell viability as described in Subheading 3.1.3, **step 2**.
2. Seed 150 mL of *Sf9* culture (2 × 10^6 cells/mL) in Sf-900 II SFM medium in a 500 mL flask with a vented cap, and infect with MVP baculoviral P2 stock prepared in Subheading 3.1.3, **step 10** (*see* **Note 17**). Incubate the culture at 27 °C at 120 rpm for about 96 h.
3. Transfer the culture into a preweighed conical tube and centrifuge for 5 min at 3000 × *g* at 4 °C. Discard the supernatant and determine the cell pellet weight (*see* **Note 18**). Use immediately or store at −80 °C (*see* **Note 19**).

3.1.5 Production of INT Fused POI

1. Count and verify *Sf9* cell viability as described in Subheading 3.1.3, **step 2**.
2. Seed 50 mL of *Sf9* culture (2 × 10^6 cells/mL) in Sf-900 II SFM medium in a 150 mL flask with a vented cap, and infect with POI-INT baculoviral P2 stock (*see* **Note 17**). Incubate the culture at 27 °C and 120 rpm for 72 h.

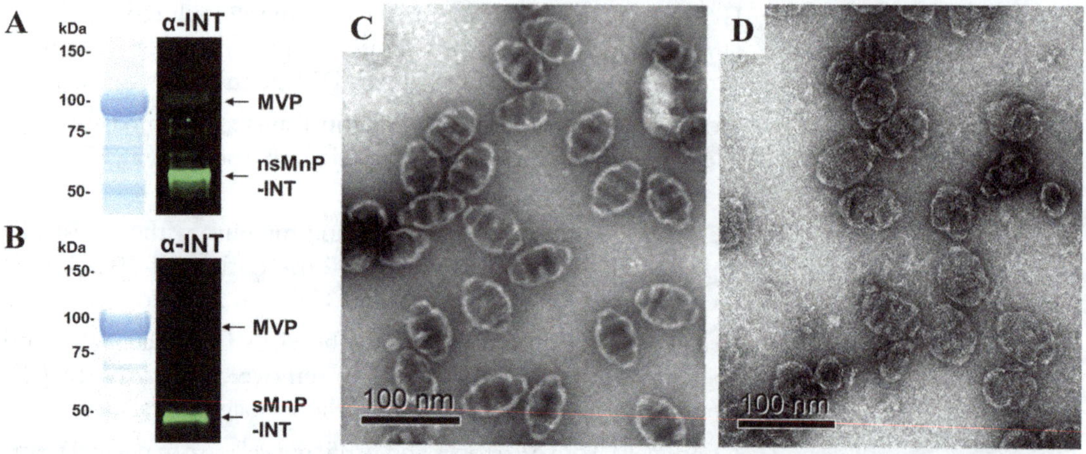

Fig. 2 Vault Particles Packaged with Manganese Peroxidase (MnP). (**a**, **b**) Coomassie stains (left panels) of purified vaults containing intracellularly expressed MnP-INT (nsMnP-INT, **a**) and secreted MnP-INT (sMnP-INT, **b**). The MVP and (ns or s-)-MnP-INT bands are indicated by arrows. Western blot analysis (right panels) confirmed the presence of (ns or s-)-MnP-INT in the purified vaults. (**c**, **d**) Negative-stained TEM images of purified nsMnP-INT vaults (**c**) and sMnP-INT vaults (**d**). (Reprinted from Wang et al. (2015) with permission from the American Chemical Society)

3. Depending on the POI, expression may be intracellular or extracellular:
   (a) For intracellularly expressed POI-INT, collect cell pellets by centrifugation for 5 min at 3000 × *g* at 4 °C in pre-weighted conical tubes, and determine the pellet weight. Use immediately or freeze at −80 °C.
   (b) For extracellularly expressed POI-INT, centrifuge infected culture for 5 min at 3000 × *g* at 4 °C to remove cells and large debris and obtain clarified supernatant, followed by spinning the supernatant at 100,000 × *g* for 1 h at 4 °C to remove baculovirus particles. Use immediately or store at 4 °C (*see* **Note 20**).

3.2 Packaging INT Fused Proteins into Vaults

INT domain maintains its vault-binding affinity in both intracellular and extracellular expression (Fig. 2). However, the secretion process may affect posttranslational modification on the fused proteins, which may alter their bioactivity and functionality (*see* **Note 21**).

All procedures and centrifugation are carried out at 4 °C or on ice.

3.2.1 Packaging of Intracellularly Expressed POI-INT

1. For each gram of infected cell pellet, resuspend in 5 mL of Lysis Buffer (*see* **Note 22**). Mix lysates from MVP cells and POI-INT cells and homogenize with 10–20 strokes in a glass Dounce homogenizer. Incubate on ice for 30 min (*see* **Notes 23** and **24**).
2. Centrifuge the mixture for 20 min at 20,000 × *g*, collect the supernatant and further centrifuge at 100,000 × *g* for 1 h to separate packaged vault particles from the lysate.

3. Resuspend the pellet (P100) by douncing 20 times in 1 mL of Buffer A with 1 mM DTT and 0.1% PI/gram of cell pellet.
4. Add streptomycin sulfate, which will precipitate contaminating ribosomes, to the resuspended P100 to 2% (w/v) final concentration, and nutate at room temperature for 10 min.
5. Centrifuge the mixture at 13,000 × *g* for 10 min at room temperature.
6. Load the supernatant onto a sucrose gradient composed of equal volumes of 20, 30, 40, 45, 50, 60% (w/v) sucrose layers, and spin at 78,000 × *g* for 16 h using a swinging bucket rotor at slow acceleration and deceleration.
7. Collect 40, 45, and 50% sucrose layers (*see* **Note 25**). Dilute each fraction in 5 volumes of Buffer A, and then spin for 2 h at 100,000 × *g*.
8. Resuspend the pellets in 0.5–1 mL of Buffer A by douncing 10–20 times (*see* **Notes 26** and **27**).
9. Packaged vaults can be stored at 4 °C for a few months, or can be lyophilized and long-term stored at −20 °C.

3.2.2 Packaging of Secreted POI-INT

1. For each gram of MVP infected cell pellet, resuspend in 5 volumes of Lysis Buffer.
2. Lyse the cells with 10–20 strokes in a glass Dounce homogenizer. Incubate on ice for 30 min and mix every 5 min.
3. Centrifuge the crude lysate for 20 min at 20,000 × *g*, discard cell debris and transfer the supernatant to 50 mL conical tubes.
4. Estimate the supernatant volume using serological pipettes and add 0.25 volume of Buffer B to a final concentration of 1 M NaCl. Incubate on ice for 30 min and shake every 5 min (*see* **Note 28**).
5. Centrifuge the mixture for 10 min at 13,000 × *g*. Carefully take out the centrifuge tubes, and decant the supernatant.
6. Resuspend the pellets (P13) in 6 mL of Buffer A/gram of MVP infected cell pellet by douncing 10–20 times, mix with virus-free culture supernatant containing POI-INT, and rotate at 4 °C for 1 h (*see* **Notes 29** and **30**).
7. Centrifuge the mixture at 100,000 × *g* for 1 h, and collect and resuspend the pellets in 1 mL of Buffer A/gram of MVP infected cell pellet using a glass douncer. Remove insoluble clumps by centrifugation at 13,000 × *g* for 10 min.
8. Transfer the aqueous packaged vaults suspension to sterile tubes, and store at 4 °C (*see* **Notes 26** and **27**).
9. Packaged vaults can be stored at 4 °C for a few months, or can be lyophilized and long-term stored at −20 °C.

4 Notes

1. The shelf life of antibiotics in LB medium or agar is 3–4 weeks, which is much shorter than the life of LB. For usage of LB medium with antibiotics, they can be added right before use.
2. Sterile solutions can be obtained from commercial sources. The medium can also be made according to the following composition: 2% tryptone, 0.5% yeast extract, 20 mM glucose, 10 mM NaCl, 2.5 mM KCl, 10 mM $MgCl_2$, and 10 mM $MgSO_4$.
3. INT domain fused protein can be expressed in other organisms, such as *E. coli*. Both prokaryotic and eukaryotic cells can synthesize and fold INT domain correctly that has vault binding affinity.
4. Polyribosome is involved in the synthesis and assembly of intact vault particles [18], which suggests that eukaryotic cells are required for assembling vaults. Since indigenous vault particles have been found in many eukaryotes, to simplify the purification of recombinant vaults, organisms that do not naturally express vaults are preferred hosts for synthesizing vaults. These organisms include yeasts, nematodes, insects, and plants.
5. Vault structures are highly conserved among eukaryotic species. Rat and human vaults that are assembled in *Sf9* cells expressing rat or human MVP have similar morphology and both show INT domain binding ability, suggesting vaults from other eukaryotes may also be assembled by expressing corresponding MVPs.
6. Primers containing restriction sites for amplifying INT and target protein coding sequences depend on specific proteins that need to be packaged into vaults and the location of INT domain. INT can be fused to either the N-terminal or C-terminal of the target proteins. If needed, linkers can be added in between to increase the functionality, stability and bioactivity of fusion proteins [19].
7. The POI-INT expression cassette can be constructed with or without a secretion signal sequence. INT domain works properly and has vault binding affinity in both intracellularly expressed and secreted fusion proteins [20].
8. Digestion that generates blunt ends can also be used. However the ligation efficiency is lower and it may cause incorrect fragment orientation. Overlap PCR with multiple primer sets can also be used to fuse POI and INT coding sequences.
9. To increase the purity of DNA recovered from the gel, incubate spin columns with wash buffer for 20 min at room temperature. To maximize the amount of recovered DNA, incubate spin columns with nuclease-free water for 10 min at room temperature.

10. To lower the possibility of vector self-ligation, use a high molar ratio of insert to vector (3:1 to 5:1 is typical).
11. Blue and white colonies appear on bacmid selection agar surface. The successful insertion of GOI into bacmid disrupt the expression of *lacZ* gene, which encodes β-galactosidase that hydrolyzing Bluo-Gal to produce a blue precipitate. Blue colonies contain bacmids and pFastBac1 plasmids, but no recombinant bacmids with GOI inserted.
12. The manual for the Bac-to-Bac TOPO Expression System has more details and common troubleshooting. The manual can be accessed at ThermoFisher's website. (https://tools.thermofisher.com/content/sfs/manuals/bactobac_topo_exp_system_man.pdf).
13. Healthy *Sf9* cells are regular in size and spherical in shape. Viable cells are unstained, while dead cells are blue.
14. Virus production causes cell lysis, releasing the virus into the medium. The cell pellet from centrifugation can be lysed and analyzed for expression of POI or MVP protein before amplifying P1 viral stock or testing P2 viral stock.
15. *Sf9* cell stocks are maintained in Sf-900 II SFM medium supplemented with Anti/Anti. Passage cells into fresh medium at a seeding density of 2×10^6 viable cells/mL every 3 days. Cell density reaches about $1.2–1.6 \times 10^7$ cells/mL in 3 days.
16. Depending on the quality of P2 viral stock, the infection viral stock volume per 50 mL culture ranges from a few microliters up to a few hundred microliters.
17. The infection volume is determined in Subheading 3.1.3, **step 11**.
18. Generally, 1 g wet cell pellet can be obtained from each 50 mL infection.
19. Recombinant vault particles in *Sf9* cells can be stored for more than 1 year at −80 °C.
20. When being stored at 4 °C, fusion proteins POI-INT easily lose their bioactivity and may be degraded over time. Frozen at −80 °C is not convenient for a large volume of supernatant and results in precipitation after thawing. Consequently, for extracellularly expressed POI-INT, freshly prepared fusion proteins are preferred for vault-packaging.
21. We find manganese peroxidase (MnP), which is naturally synthesized as a secreted protein in fungal cells, only maintains its activity as secreted MnP-INT, but not as intracellularly expressed MnP-INT [20]. Thus, secretion needs to be taken into consideration when expressing INT fused proteins. We suggest extracellular expression of POI-INT for POI that is naturally secreted. If the POI is naturally intracellularly expressed, both extracellular and intracellular expression can be tried.

22. To facilitate cell lysis, pipet up and down. For frozen pellets, break pellets up with a spatula.
23. The ratio of POI-INT cells to MVP cells can be optimized based on the expression level of each protein. Depending on the POI-INT protein size, each vault particle can hold up to about 80 protein copies. If POI-INT is expressed at a relatively low level, a higher POI-INT cells/MVP cells ratio can be used to fully use vault internal space.
24. Thirty minutes is sufficient for POI-INT uptake and binding to the vault interior. Longer incubation on ice does not seem to increase packaging efficiency. For purifying recombinant vault particles only, POI-INT cell pellets can be omitted.
25. Empty vaults are usually found in the 40 and 45% sucrose layers, while vaults containing INT fusion proteins are usually found in the 40, 45, and 50% sucrose fractions.
26. The final vault-packaged POI-INT may now be visualized by negative staining TEM to confirm correct vault assembly and structural intactness. Samples are absorbed on carbon-coated copper EM grids by floating the grids on 20 μL of vault containing samples for 5 min. Afterward, the grids are floated for 5 min on 1 mL of 1% uranyl acetate solution and dried on filter paper prior to be viewed in a TEM.
27. The final vault-packaged POI-INT is ready for further analysis of packaged protein. However, note that the suspension has roughly 90% particle purity. To further remove impurities, anion-exchange chromatography can be used.
28. The clear supernatant turns opaque and cloudy after adding NaCl, which is an indication of the presence of vault particles.
29. P13 is not as tight as cell debris. Thus carefully decant the supernatant, and avoid disturbing and resuspending pellets on the bottom.
30. Concentrating POI-INT using molecular weight cutoff filters may improve vault-packaging efficiency. Although, for extracellularly expressed MnP-INT we find that the improvement is very limited.

Acknowledgments

This work was supported by Strategic Environmental Research and Development Program (SERDP) Award ER-2422 and UCLA Department of Civil and Environmental Engineering. S.M. received the Paul L. Busch Award from the Water Environment and Reuse Foundation.

References

1. Kedersha NL, Heuser JE, Chugani DC et al (1991) Vaults. III. Vault ribonucleoprotein particles open into flower-like structures with octagonal symmetry. J Cell Biol 112:225–235
2. Mikyas Y, Makabi M, Raval-Fernandes S et al (2004) Cryoelectron microscopy imaging of recombinant and tissue derived vaults: localization of the MVP N termini and VPARP. J Mol Biol 344:91–105
3. Tanaka H, Kato K, Yamashita E et al (2009) The structure of rat liver vault at 3.5 angstrom resolution. Science 323:384–388
4. Kedersha NL, Rome LH (1986) Isolation and characterization of a novel Ribonucleoprotein particle: large structures contain a single species of small RNA. J Cell Biol 103:699–709
5. Kickhoefer VA, Poderycki MJ, Chan EKL et al (2002) The La RNA-binding protein interacts with the vault RNA and is a vault-associated protein. J Biol Chem 277:41282–41286
6. Kickhoefer VA, Stephen AG, Harrington L et al (1999) Vaults and telomerase share a common subunit, TEP1. J Biol Chem 274: 32712–32717
7. Kickhoefer VA, Siva AC, Kedersha NL et al (1999) The 193-kD vault protein, VPARP, is a novel poly(ADP-ribose) polymerase. J Cell Biol 146:917–928
8. Kickhoefer VA, Searles RP, Kedersha NL et al (1993) Vault ribonucleoprotein-particles from rat and bullfrog contain a related small RNA that is transcribed by RNA polymerase-III. J Biol Chem 268:7868–7873
9. Stephen AG, Raval-Fernandes S, Huynh T et al (2001) Assembly of vault-like particles in insect cells expressing only the major vault protein. J Biol Chem 276:23217–23220
10. Rome LH, Kickhoefer VA (2013) Development of the vault particle as a platform technology. ACS Nano 7:889–902
11. Kickhoefer VA, Garcia Y, Mikyas Y et al (2005) Engineering of vault Nanocapsules with enzymatic and fluorescent properties. Proc Natl Acad Sci U S A 102:4348–4352
12. Han M, Kickhoefer VA, Nemerow GR et al (2011) Targeted vault nanoparticles engineered with an endosomolytic peptide deliver biomolecules to the cytoplasm. ACS Nano 5:6128–6137
13. Kar UK, Srivastava MK, Andersson A et al (2011) Novel CCL21-vault nanocapsule intratumoral delivery inhibits lung cancer growth. PLoS One 6:e18758
14. Buehler DC, Toso DB, Kickhoefer VA et al (2011) Vaults engineered for hydrophobic drug delivery. Small 7:1432–1439
15. Poderycki MJ, Kickhoefer VA, Kaddis CS et al (2006) The vault exterior shell is a dynamic structure that allows incorporation of vault-associated proteins into its interior. Biochemistry 45:12184–12193
16. Yang JA, Kickhoefer VA, Ng BC et al (2010) Vaults are dynamically unconstrained cytoplasmic nanoparticles capable of half vault exchange. ACS Nano 4:7229–7240
17. Kickhoefer VA, Rome LH (1994) The sequence of a Cdna-encoding the major vault protein from Rattus-Norvegicus. Gene 151:257–260
18. Mrazek J, Toso D, Ryazantsev S et al (2014) Polyribosomes are molecular 3D nanoprinters that orchestrate the assembly of vault particles. ACS Nano 8:11552–11559
19. Chen XY, Zaro JL, Shen WC (2013) Fusion protein linkers: property, design and functionality. Adv Drug Deliv Rev 65:1357–1369
20. Wang M, Abad D, Kickhoefer VA et al (2015) Vault nanoparticles packaged with enzymes as an efficient pollutant biodegradation technology. ACS Nano 9:10931–10940

Chapter 4

Enzyme Encapsulation in an Engineered Lumazine Synthase Protein Cage

Yusuke Azuma and Donald Hilvert

Abstract

The packaging of active enzymes in protein cages is a powerful strategy to control catalytic activity. Using a positively supercharged variant of green fluorescent protein, GFP(+36), as a genetically programmable tag, enzymes can be rapidly and quantitatively loaded into an engineered variant of the *Aquifex aeolicus* cage-forming protein lumazine synthase (AaLS-13) that possesses a negatively charged lumen. The cargo is spontaneously localized within AaLS-13 cages by simply mixing the components in aqueous solution. This chapter describes a detailed protocol for the preparation of AaLS-13 cages and GFP(+36)-enzyme fusions, as well as characterization of the inclusion complexes. Suitable conditions for encapsulation and enzyme kinetic assays are also discussed.

Key words Lumazine synthase, Supercharged protein, Enzyme encapsulation, Nanoreactors, Synthetic biology

1 Introduction

Biology uses spatial organization of enzymes in segregated compartments to optimize metabolic or biosynthetic reactions in complex intracellular environments. Lipid bilayers serve as physical barriers in eukaryotic organisms, while many bacteria produce cage-forming proteins that confine enzymes that promote short metabolic pathways [1, 2]. Compartmentalization leads to an enhancement of pathway flux by sequestering intermediates, as well as minimizing off-target reactions by limiting the uptake of undesired substrates [1–3]. Inspired by these natural systems, several strategies have been devised to package enzymes in liposomes [4, 5], polymers [6, 7], and natural and artificial protein cages [8–14] toward biotechnological applications.

We have established a general packaging system based on the cage-forming protein lumazine synthase from *Aquifex aeolicus* (AaLS). Wild-type AaLS spontaneously assembles as a $T = 1$ icosahedral cage structure that contains 60 identical subunits organized

Andrew K. Udit (ed.), *Protein Scaffolds: Design, Synthesis, and Applications*, Methods in Molecular Biology, vol. 1798,
https://doi.org/10.1007/978-1-4939-7893-9_4, © Springer Science+Business Media, LLC, part of Springer Nature 2018

as 12 pentameric capsomers [15] (Fig. 1a, d). In nature, this bacterial cage encapsulates riboflavin synthase, which is the next enzyme in the vitamin B2 biosynthetic pathway [16]. In order to package foreign guest molecules, we introduced negatively charged residues on the lumenal face of AaLS-wt, affording a variant called AaLS-neg that encapsulates proteins bearing high positive charge (Fig. 1b [17]. Subsequent optimization of AaLS-neg using directed evolution yielded the variant AaLS-13, which has seven additional mutations that further increase the net negative charge of the lumenal cavity and enhance its loading capacity (Fig. 1c) [18]. The changes introduced by design and evolution caused significant expansion of the protein shell (Fig. 1d–f) [19]. Whereas AaLS-wt has a diameter of ~16 nm, AaLS-neg and AaLS-13 are ~30 nm and ~40 nm wide, reflecting their assembly from 180 and 360 subunits, respectively [19]. Moreover, the expanded structures contain large, keyhole-shaped pores in the shell, not present in the wild-type cage, which enable efficient uptake of a wide variety of positively charged guest molecules (Fig. 1e, f) [20–24]. For example, GFP(+36), a positively supercharged variant of green fluorescent protein [25], is spontaneously loaded into empty AaLS-13 cages in quantitative yield at diffusion-limited rates by simply mixing the guests with the cages in aqueous solution [21, 24].

Exploiting such simple and efficient host–guest interactions, active enzymes can be loaded into AaLS-13 assemblies by genetic fusion to GFP(+36), which acts as a localization tag (Fig. 2) [22]. Like GFP(+36) itself, simply mixing the components in aqueous solution leads to internalization of the fusion proteins. In the case of a computationally designed retro-aldolase, loading was found to be quantitative with a capacity for up to ~80 guests per 360mer AaLS-13 [19], allowing tight regulation of guest density in the lumenal space. This simple encapsulation strategy works for monomeric and dimeric enzymes possessing moderate net positive charge [4, 11]. The utility of such AaLS-13-based nanoreactors has been illustrated by the templated synthesis of polymers [26] and mimicry of the carboxysome [27], a natural bacterial microcompartment.

2 Materials

All buffers were prepared using doubly distilled water (~17 MΩ/cm at 25 °C) and commercially supplied reagents without further purification. Media, agar plates, and all buffers were autoclaved prior to use. All running buffers for size-exclusion and ion-exchange chromatography were filtered through a 0.22 μm-pore polyethersulfone membrane and degassed in vacuo.

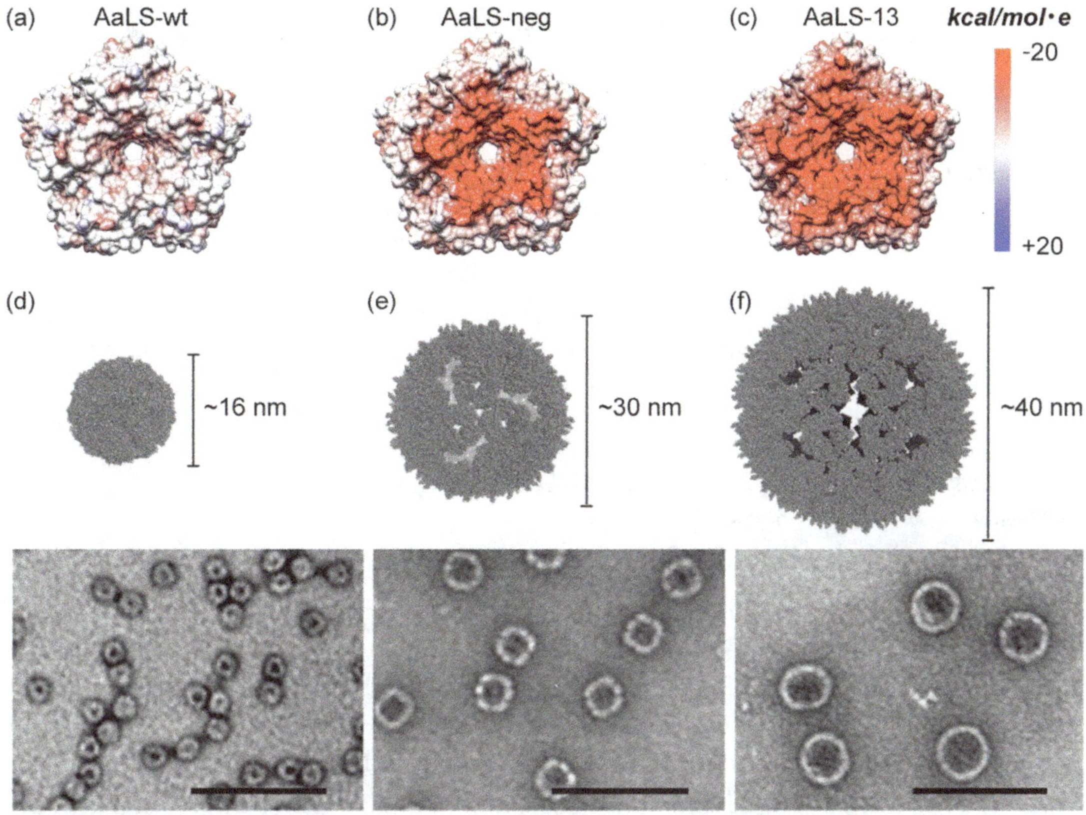

Fig. 1 Structural characteristics of AaLS variants. (**a**–**c**) Lumenal coulombic surface potentials of the AaLS-wt (**a**), AaLS-neg (**b**), and AaLS-13 (**c**) pentamers. The models for engineered variants were constructed from the crystal structure of the AaLS-wt pentamer (PDB: 1HQK) [15]. (**d**–**f**) Cryo-electron microscopy single-particle reconstructions (upper panel) and negative-staining transmission electron microscopic images (lower panel) of AaLS-wt (**d**), AaLS-neg (**e**), and AaLS-13 (**f**) assemblies, respectively (PDB ID: 5MPP, AaLS-wt; 5MQ3, AaLS-neg; 5MQ7, AaLS-13) [19]. Scale bar = 100 nm

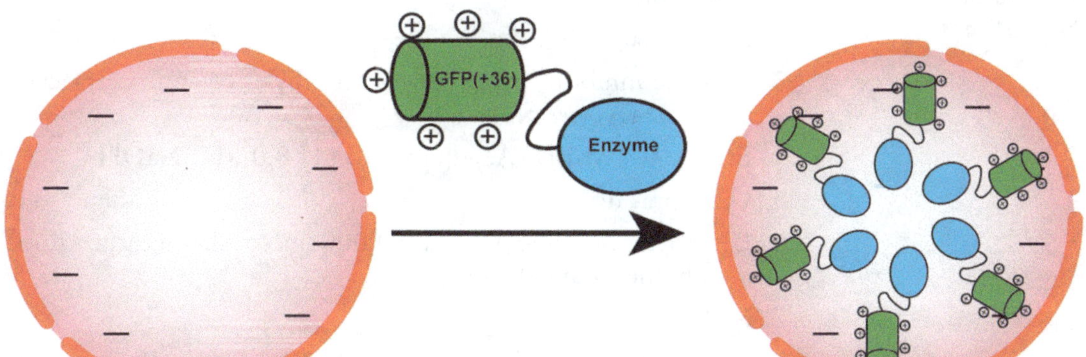

Fig. 2 Schematic representation of enzyme encapsulation in AaLS-13 cages using GFP(+36) as a pilot tag

Fig. 3 The design of typical GFP(+36)-fusion enzymes

2.1 Plasmids

1. pMG_AaLS-13 [18]: a pMG-based plasmid containing a pBR322 origin of replication, genes encoding β-lactamase, the *nah* regulator, the *lac* repressor, and AaLS-13 with a C-terminal hexahistidine tag (*see* **Note 1**).
2. pACYC_GFP(+36)-enzyme [22]: a pACYC-based plasmid with a p15A origin of replication, genes encoding chloramphenicol acetyl transferase, the *nah* regulator, and a GFP(+36)-enzyme fusion (Fig. 3) (*see* **Notes 2–4**).

2.2 Protein Expression

1. *E. coli* strain BL21 (DE3).
2. Incubator.
3. Shaker with a temperature controller.
4. LB medium: 10 g of yeast extract, 5 g of tryptone, and 10 g of NaCl per liter.
5. LB-agar plates: 15 g of agar per liter of LB medium containing 100 μg/mL ampicillin or 25 μg/mL chloramphenicol.
6. Ampicillin solution: 100 mg/mL aqueous stock solution.
7. Chloramphenicol solution: 25 mg/mL stock solution in aqueous 70% ethanol.
8. Isopropyl β-D-1-thiogalactopyranoside (IPTG) solution: 0.1 M aqueous stock solution.
9. Centrifuge capable of ≥9500 × *g* relative centrifugal force.
10. PBS: 10 mM sodium phosphate buffer pH 7.4 containing 150 mM NaCl.

2.3 Protein Purification

1. Lysozyme.
2. Benzonase: 250 U/mL.
3. Protease inhibitor cocktail without ethylenediaminetetraacetic acid (EDTA) (*see* **Note 5**).
4. EDTA: 0.4 M EDTA, adjusted to pH 8.0 with NaOH.
5. Ultrasonicator.
6. Ultrafiltration tube with a 30,000 molecular weight cutoff membrane (4 and 15 mL).
7. Ni-NTA agarose resin.
8. Gravity column (1.5 × 30 cm).
9. Bradford reagent.
10. Fast protein liquid chromatography (FPLC) pump system equipped with a UV-Vis absorbance detector.

11. HiPrep 16/60 Sephacryl S-400 HR size-exclusion column (GE Healthcare), or similar column.
12. Superdex 200 10/300 GL size-exclusion column (GE Healthcare), or similar column.
13. Superose 6 10/300 GL size-exclusion column (GE Healthcare), or similar column.
14. UnoS1 (Bio-Rad) or Mono S HR 16/10 cation-exchange column (GE Healthcare), or similar column.
15. Polyacrylamide gel (12.5 and 20%).
16. Equipment for SDS-PAGE.
17. Glycerol.
18. UV–visible absorbance spectrometer.

2.4 Buffers for AaLS-13 Purification

1. Lysis buffer A: 100 mM sodium phosphate buffer, pH 8.0, containing 300 mM NaCl and 10 mM imidazole (*see* **Note 6**).
2. Ni-NTA wash buffer A1: 50 mM sodium phosphate buffer, pH 8.0, containing 800 mM NaCl and 20 mM imidazole (*see* **Note 7**).
3. Ni-NTA wash buffer A2: 50 mM sodium phosphate buffer, pH 8.0, containing 800 mM NaCl and 80 mM imidazole (*see* **Note 7**).
4. Ni-NTA elution buffer A: 50 mM sodium phosphate buffer, pH 8.0, containing 300 mM NaCl and 500 mM imidazole.
5. Size exclusion chromatography (SEC) buffer A: 50 mM sodium phosphate buffer, pH 8.0, containing 200 mM NaCl and 5 mM EDTA (*see* **Note 8**).

2.5 Buffers for Purification of GFP(+36)-Enzymes

1. PBS-HS: 20 mM sodium phosphate buffer, pH 7.4, containing 2 M NaCl.
2. Ni-NTA wash buffer B: 20 mM sodium phosphate buffer, pH 7.4, containing 2 M NaCl and 20 mM imidazole.
3. Ni-NTA elution buffer B: 20 mM sodium phosphate buffer, pH 7.4, containing 2 M NaCl and 250 mM imidazole.
4. Ion exchange chromatography (IEX) buffer A: 20 mM sodium phosphate buffer, pH 7.4, containing 300 mM NaCl.
5. IEX buffer B: 20 mM sodium phosphate buffer, pH 7.4, containing 2 M NaCl.
6. Storage buffer: 50 mM sodium phosphate buffer, pH 7.4, containing 1 M NaCl.

2.6 Characterization of Cage Complexes

1. 96-well plate (black, flat bottom).
2. Fluorescence microplate reader.

3. 2% uranyl acetate aqueous solution (filtered through a 0.22 μm membrane).
4. Copper grids: 400 mesh, covered with formvar and carbon (Ted Pella, 01822-F).
5. Transmission electron microscope.

3 Methods

3.1 Expression and Purification of AaLS-13 Cages

A method to purify AaLS-13 cages was previously reported [21, 22]. Here we describe a modified procedure with additional remarks.

1. Transform the *E. coli* BL21 (DE3) strain with plasmid pMG-AaLS-13, and plate the cells on LB-agar plates with 100 μg/mL of ampicillin. Incubate overnight at 37 °C.
2. Inoculate 5 mL of LB medium containing 100 μg/mL ampicillin with an isolated colony. Culture the cells with a shaker at 37 °C overnight.
3. Inoculate 400 mL of LB medium containing 100 μg/mL ampicillin with 2 mL of the overnight culture. Grow the cells in a shaker at 37 °C until an OD_{600} of ~0.7 is reached, then induce protein expression with IPTG (0.1 mM final).
4. Shake the cell culture at 180 rpm at 25 °C for 18 h.
5. Harvest the cells by centrifugation at 4 °C and 5000 × *g* for 10 min. Resuspend the cell pellet in PBS and transfer to a 50 mL tube. Repeat this step once.
6. Store the cell pellet at −20 °C until purification.
7. Resuspend the cell pellet in 20 mL of lysis buffer A, and add lysozyme (0.1 mg/mL), benzonase (250 U), and protease inhibitor cocktail (*see* **Note 5**). Incubate at room temperature for 1 h.
8. Lyse the cells by sonication using a 50% duty cycle and 80% amplitude setting for 2 min on ice, followed by cooling on ice for 2 min. Repeat 5 times.
9. Remove the insoluble fraction by centrifugation at 9500 × *g* at room temperature for 25 min.
10. Add the supernatant to 4 mL of Ni-NTA agarose resin (50% v/v) preequilibrated with lysis buffer A in a gravity column. Gently mix and incubate at room temperature for 10 min (*see* **Note 9**).
11. Wash the resin with Ni-NTA wash buffer A1 (40 mL), followed by Ni-NTA wash buffer A2 (40 mL).

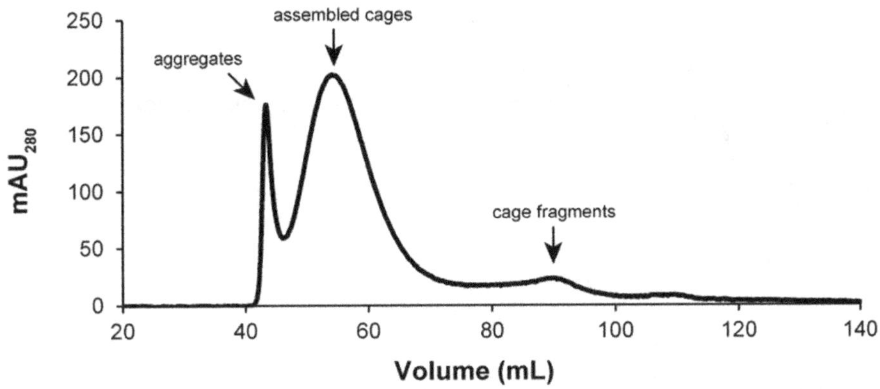

Fig. 4 Size-exclusion chromatogram of AaLS-13. The sample was purified by Ni-NTA affinity chromatography and subjected to high salt treatment for 3 days prior to SEC

12. Elute protein with Ni-NTA elution buffer A while monitoring aliquots of eluent with Bradford reagent. Immediately add EDTA (5 mM final concentration) to the eluate.
13. Exchange the buffer to SEC buffer A by ultrafiltration at room temperature. Measure protein concentration by absorbance (ε_{280} = 13,980 M^{-1} cm^{-1}), and then adjust the protein concentration to 800–1200 μM.
14. Add 10% by volume of 5 M NaCl (*see* **Note 10**), and then incubate at room temperature for 3–5 days to favor cage assembly. If precipitation is observed, sonicate the sample (*see* **Note 11**).
15. Centrifuge the sample at 25 °C and 15,000 × *g* for 15 min to remove aggregates. Load the supernatant onto the FPLC with a HiPrep 16/60 Sephacryl S-400 HR, and elute with SEC buffer A at a flow rate of 1.0 mL/min at room temperature while monitoring UV absorbance at 280 nm.
16. Pool the fractions corresponding to assembled AaLS-13 cage (Fig. 4), and concentrate the protein solution by ultrafiltration at room temperature to a final concentration of ~300 μM.
17. Check the purity by SDS-PAGE using a 20% polyacrylamide gel.
18. Store the purified protein at room temperature.
19. Determine the concentration by absorbance at 280 nm, correcting for light scattering (Fig. 5) (*see* **Note 12**).

3.2 Expression and Purification of GFP(+36)-Fusion Enzymes

Here we describe the purification of a GFP(+36)-enzyme construct based on the published protocol by Thompson et al. [28] for purifying supercharged proteins. We typically isolate the fusion proteins using Ni-NTA chromatography, followed by size-exclusion chromatography, a sequence that usually affords >90% purity (Fig. 6).

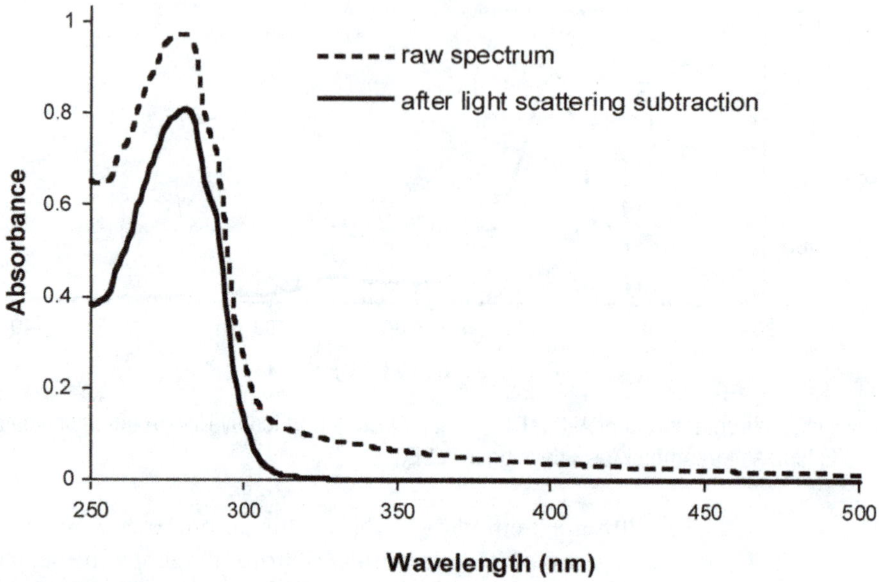

Fig. 5 Absorbance spectra of AaLS-13 cages

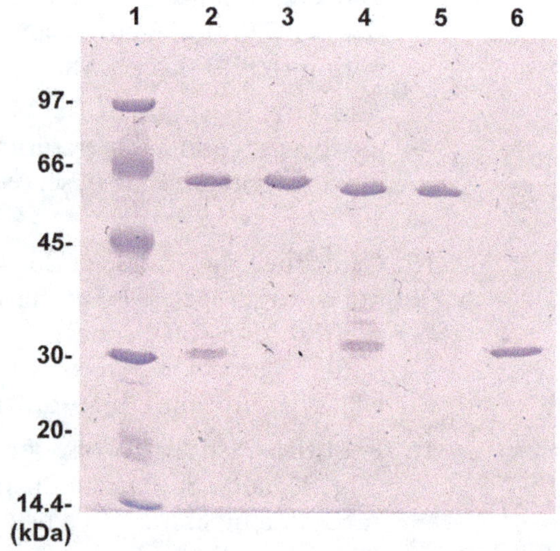

Fig. 6 SDS-PAGE analysis of purified fusion proteins. Lane 1, low weight molecular marker; lanes 2 and 3, GFP(+36)-fused Kemp eliminase variant HG3.17 (GFP(+36)-KE, 64.4 kDa) after Ni-NTA (2) and size-exclusion chromatography (3); lanes 4 and 5, GFP(+36)-fused β-lactamase (GFP(+36)-βLac, 59.5 kDa) after Ni-NTA (4) and size-exclusion chromatography (5); lane 6: GFP + 36 (29.5 kDa). Electrophoresis was performed on a 12.5% acrylamide gel using a Phast system (GE healthcare); the gel was stained with Coomassie R350 (GE healthcare)

1. Transform the pACYC plasmid encoding a GFP(+36) fusion protein into *E. coli* BL21 (DE3), and plate the cells on an LB-agar plate with 25 μg/mL chloramphenicol. Incubate the sample overnight at 37 °C.
2. Inoculate 5 mL of LB medium containing 25 μg/mL chloramphenicol with an isolated colony. Culture the cells with a shaker at 37 °C overnight.
3. Inoculate 400 mL of LB medium containing 25 μg/mL chloramphenicol with 2 mL of the overnight culture. Culture the cells with a shaker at 37 °C until an OD600 of ~0.6 is reached, at which point add IPTG to a final concentration of 0.1 mM.
4. Shake the cell culture at 180 rpm at 25 °C for 18 h, or at 18 °C for 36 h.
5. Harvest the cells by centrifugation at 4 °C and 5000 × *g* for 10 min. Resuspend the cell pellet in PBS and transfer to a 50 mL tube. Repeat this step once.
6. At this point the cell pellet can be stored at −20 °C until ready to proceed (below).
7. Resuspend the cell pellets in 20 mL PBS-HS, and add lysozyme (0.1 mg/mL), benzonase (250 U), and protease inhibitor cocktail (*see* **Note 5**). Incubate at room temperature for 1 h.
8. Lyse the cells by sonication using a 50% duty cycle and 80% amplitude setting on ice for 2 min, followed by cooling on ice for 2 min. Repeat the process 5 times.
9. Remove the insoluble fraction by centrifugation at room temperature and 9500 × *g* for 25 min.
10. Load the supernatant onto 4 mL of Ni-NTA sepharose resin (50% v/v) pre-equilibrated with lysis buffer in a gravity column. Gently mix and incubate at room temperature for 10 min.
11. Wash the resin with 100 mL of PBS-HS and then with 40 mL of Ni-NTA wash buffer B.
12. Elute the protein with Ni-NTA elution buffer B and immediately add EDTA to a final concentration of 5 mM.
13. Concentrate and exchange the buffer to PBS-HS by ultrafiltration at room temperature.
14. Load onto the FPLC with a Superdex 200 10/300 GL and run with PBS-HS at a flow rate of 0.5 mL/min at room temperature while monitoring UV-Vis absorbance signals at 280 nm and 488 nm.
15. Pool the fractions corresponding to GFP(+36) fusion protein (*see* **Notes 13** and **14**).

16. Check the purity by SDS-PAGE using a 12.5% polyacrylamide gel (Fig. 6).
17. Buffer exchange to storage buffer by ultrafiltration at room temperature, and then add glycerol to a final concentration of 50% and store at −20 °C.
18. Determine the concentration by absorbance at 488 nm prior to use (ε_{488} = 36,600 M^{-1} cm^{-1}).

3.3 Cation-Exchange Chromatography of GFP(+36)-Fusion Enzymes

This section is optional. If purification using Ni-NTA and size-exclusion chromatography does not yield a sufficiently pure fusion enzyme, cation-exchange chromatography can be used for further purification. Since GFP(+36) is highly positively charged, most of the impurities from *E. coli* cells are removed by this method.

1. Dilute protein samples purified by Ni-NTA and size-exclusion chromatography with PBS to a final NaCl concentration < 0.8 M.
2. Load the sample onto the FPLC with a UnoS1 column pre-equilibrated with 20% IEX buffer B in IEX buffer A.
3. Wash the column with 10 mL 20% IEX buffer B at a flow rate of 1.0 mL/min at room temperature.
4. Elute the protein using a linear gradient from 20 to 80% IEX buffer B in IEX buffer A over 30 mL at a flow rate of 1.0 mL/min at room temperature while monitoring UV-Vis absorbance signals at 280 and 488 nm (*see* **Note 15**).
5. Pool the fractions corresponding to the GFP(+36)-fusion protein.
6. Buffer exchange to storage buffer using ultrafiltration at room temperature. Add glycerol to a final concentration of 50%, and store at −20 °C.
7. Determine the concentration by the absorbance at 488 nm prior to use (ε_{488} = 36,600 M^{-1} cm^{-1}).

3.4 Enzyme Encapsulation in AaLS-13 Cages

1. Mix purified empty AaLS-13 cages with a GFP(+36)-enzyme fusion in aqueous buffer at a final AaLS-13 monomer concentration of <100 μM (*see* **Note 16**). GFP(+36)-enzyme concentrations can be varied, but the highest concentration should typically not exceed 1/5 of the AaLS-13 monomer concentration (*see* **Note 17**). Keep the final NaCl concentration at ~200 mM [24] (*see* **Note 18**).
2. Incubate the mixture at room temperature for 1 h (*see* **Note 19**).

3.5 Size-Exclusion Chromatography of Enzyme–AaLS-13 Complexes

Complex formation is confirmed using size-exclusion chromatography (*see* **Note 20**).

1. Load the enzyme–AaLS-13 mixture onto the FPLC with a Superose 6 10/300 GL column and elute with SEC buffer A at a flow rate of 0.5 mL/min at room temperature while monitoring UV-Vis absorbance at 280 and 488 nm. Collect 0.5 mL fractions.
2. Transfer 200 μL of each fraction to a 96-well plate (black, flat bottom) and detect GFP(+36) by measuring fluorescence in a microplate reader (excitation at 488 nm; emission at 509 nm).
3. Combine the fractions containing assembled AaLS-13 cages (*see* **Note 21**), and concentrate by ultrafiltration.
4. Measure the absorbance spectra of the samples (*see* **Note 12**) to determine the loading efficiency using the following equation:

$$\text{Loading efficiency} = [\text{E}] / ([\text{cage}] / N)$$
$$[\text{E}] = A_{488} / \epsilon_{\text{GE488}}$$
$$[\text{cage}] = (A_{280} - \epsilon_{\text{GE280}} \times [\text{E}]) / \epsilon_{\text{A280}}$$

where [E] and [cage] are the concentrations of enzyme and cage monomer, respectively, and N is the number of subunits per cage (for AaLS-13, N = 360 [19]). A_{280} and A_{488} indicate observed absorbance at 280 nm and 488 nm. $\varepsilon_{\text{GE488}}$ and $\varepsilon_{\text{A280}}$ are extinction coefficients for the fusion enzyme at 488 nm (for GFP(+36) fusion enzymes, $\varepsilon_{\text{GE488}}$ = 36,600 $\text{M}^{-1}\ \text{cm}^{-1}$) and cage at 280 nm (for AaLS-13, $\varepsilon_{\text{A280}}$ = 13,980 $\text{M}^{-1}\ \text{cm}^{-1}$). Extinction coefficients for the fusion enzyme at 280 nm ($\varepsilon_{\text{GE280}}$) are determined by the spectral ratio (A_{280}/A_{488}) for the purified fusion enzyme using the following equation;

$$\epsilon_{\text{GE280}} = (A_{280} / A_{488}) \times \epsilon_{\text{GE488}}$$

3.6 Transmission Electron Microscopy of Enzyme/AaLS-13 Complexes

The simple mixing protocol usually results in nearly 100% association of guests with AaLS-13. However, very negatively charged enzymes may also partially bind the exterior of AaLS-13 assemblies (Fig. 7) [22]. Thus, it is recommended to characterize enzyme/AaLS-13 complexes by transmission electron microscopy (TEM). Here, we describe our typical protocol to prepare TEM samples and stain them with uranyl acetate.

1. Prepare copper grids by glow discharging in the negative mode at 20 mA for 30 s.
2. Adjust AaLS-13 concentration to 3–5 μM (with respect to monomer), and add 4 μL of the solution on a discharged grid.

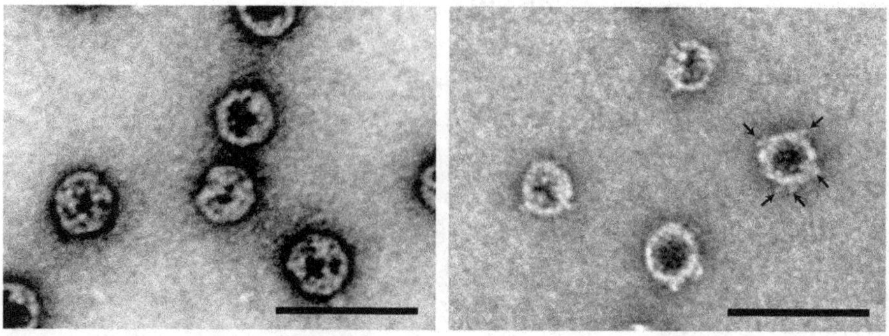

Fig. 7 Transmission electron microscopic images of enzyme/AaLS-13 complexes. The guests fused to GFP(+36) are (**a**) a Kemp eliminase (GFP(+36)-KE) and (**b**) ***Rhodobacter capsulatus*** B10 catalase-peroxidase (GFP(+36)-KatG). Native KE is a monomeric enzyme with a net charge of −4.3 at pH 8.0, whereas KatG forms a dimer with a total net charge of −22.9. Arrows indicate objects on the exterior of the AaLS-13 cage. Scale bar = 100 nm

3. Allow proteins to adsorb for 1 min, and then remove excess solution with filter paper.
4. Wash the grid surface by positioning it on a droplet of SEC buffer A, then blot it dry with filter paper. Repeat the washing step twice.
5. Quickly rinse the grid surface with 4 μL of 2% aqueous uranyl acetate solution, blot with filter paper, and then stain the sample with the same solution for 20 s.
6. Remove excess solution with filter paper and dry the grid at room temperature.
7. Visualize the sample by transmission electron microscopy (*see* **Note 22**).

3.7 Measuring the Activity of Encapsulated Enzymes

Encapsulated enzymes should be analyzed using the same activity assays one would use for the free enzyme, with some considerations for the AaLS-13 and GFP proteins (*see* **Notes 23–27**).

4 Notes

1. This variant contains 11 mutations (D28G, R52C, R83E, T86E, T112S, V115D, A118D, T120E, Q123E, R127C, and K131E), which differentiate it from wild-type AaLS (WP_010880027.1; NP_213089.1) [18]. Transcription of the AaLS-13 gene is regulated by both a salicylate promoter and a T7 promoter coupled to the *lac* operon.
2. We obtained the GFP(+36) construct from the lab of David R. Liu (Department of Chemistry and Chemical Biology,

Harvard University). The supercharged variant of GFP has the following 37 mutations compared to the wild-type protein: E6R, T9R, V11K, D19K, S30R, E32K, E34K, Y39R, F64L, S65T, D76K, Q80R, E90K, F99S, D102K, N105K, D117R, E124K, I128R, D133K, E142R, Y145F, N148K, M153T, Q157R, V163A, N164K, I171V, E172K, D190R, D197R, Q204R, A206K, N212K, T230K, M233R, and L236R [25, 29].

3. We typically use fusion enzymes composed of an N-terminal hexahistidine tag for purification, followed by GFP(+36), and ending with the target enzyme (Fig. 3). For some applications, it may be preferable to have the hexahistidine tag at the C-terminus or to switch the order of the target enzyme and GFP(+36) [28]. In our hands, the N-terminal histidine-tagged $(His)_6$-GFP(+36) generally gives high yields of the fusion protein in *E. coli*. However, genetic fusion to the N-terminus of enzymes can potentially decrease the catalytic activity. For example, attachment of GFP(+36) to the N-terminus of a retro-aldolase resulted in a tenfold decrease in activity compared to the unmodified enzyme [22]. In such cases, affinity and localization tags could be fused to the enzyme C-terminus. Additionally, five glycine-glycine-serine repeats were introduced between GFP(+36) and the guest enzymes to ensure active site accessibility irrespective of the binding orientation of the targeting protein on the lumenal surface of the AaLS-13 cages. However, this long and flexible linker sometimes results in significant proteolytic cleavage of the fusion construct in the cytosol of *E. coli* cells (Fig. 6). Shorter and/or more rigid linkers, such as poly-prolines, can be alternatively used to minimize undesired truncation.
4. Transcription of the gene encoding the fusion enzyme is regulated by a T7 promoter coupled to the *lac* operon.
5. For example, we use EDTA-free Sigma*FAST* protease inhibitor cocktail tablets (Sigma, S8830) in the recommended concentration, i.e., 1 tablet per 100 mL cell suspension, containing 2 mM 4-(2-aminoethyl)benzenesulfonyl fluoride hydrochloride (AEBSF), 1 μM phosphoramidon disodium salt, 130 μM bestatin hydrochloride, 14 μM *N*-(trans-epoxysuccinyl)-L-leucine 4-guanidinobutylamide (E-64), 1 μM leupeptin, 0.2 μM aprotinin, and 10 μM pepstatin A.
6. Optionally, up to 1 mM EDTA can be added to the lysis buffer, which helps to prevent aggregation of AaLS-13 without causing problems in the Ni-NTA affinity chromatography step.
7. Optionally, 0.05% Tween-20 can be added to prevent hydrophobic interactions of some cellular components with the resin.

8. Buffer systems other than Tris–HCl (pH 7–9) can also be used. At least 1 mM EDTA is necessary, however, to minimize AaLS-13 aggregation and sticking to the size-exclusion chromatography columns.
9. AaLS-13 may aggregate and settle on the Ni-NTA resin. Although lowering the ionic strength of lysis and wash buffers to 0.24 M minimizes aggregation and increases yields of isolated materials, almost no assembled AaLS-13 cages are observed under these conditions [23]. The aggregates can be eluted with buffer containing 500 mM imidazole as a cloudy suspension and cleared by the addition of EDTA. Due to Ni-triggered aggregation, chromatography columns fitted with fine-pored frits at the bottom are not recommended. We use Econo chromatography columns (Bio-Rad).
10. AaLS-13 usually elutes as a mixture of assembled cages and fragments, typically in a ratio of approximately 1:2 [21]. Incubation at high salt concentrations promotes the assembly of fragments into intact cages, probably by reducing charge repulsion between fragments. Lower pH (pH 7.4) can also be used to facilitate the assembly process. After 3–5 days, more than 90% of the protein assembles into cage structures (Fig. 4).
11. The sonication step helps to solubilize aggregated and precipitated proteins. Once assembled, AaLS-13 cages are stable against sonication. For example, after sonication using a 50% duty cycle and 60% amplitude setting on ice for 10 min, no disassembly of AaLS-13 cage was observed.
12. Light scattering contributes substantially (10–20%) to the total 280 nm absorbance of AaLS-13 samples (Fig. 5). This background can be subtracted with the help of the UV-Vis-IR Spectral Software a|e (FluorTools.com, http://www.fluortools.com/software/ae), which employs the following equation:

$$A_{\text{scat}} = \log\left(1 / \left(1 - a \times \lambda^{-4}\right)\right) + b$$

where A_{scat}, λ, a, and b represent Rayleigh scattering, the wavelength, the theoretical scattering at 1 nm, and a spectral offset, respectively. The unknown parameters a and b can be estimated by curve fitting using spectral data recorded in a range where the chromophores do not have any absorbance.

13. For reference, GFP(+36) itself elutes at ~16 mL.
14. Before the size-exclusion chromatography step, GFP(+36)-fusion enzymes often precipitate if stored on ice or in lower ionic strength buffer, presumably by aggregating with negatively charged molecules from *E. coli* cells still present in the sample [28]. Thus, it is recommended to store the sample in

buffer containing 2 M NaCl at room temperature. Once purified, however, fusion enzymes remain soluble at lower ionic strength at 4 °C.

15. For reference, GFP(+36) itself elutes at ~1.2 M NaCl.
16. Overloading AaLS-13 with GFP(+36) causes coprecipitation of both proteins [21, 22]. Maximum loading of GFP(+36) alone is ~70 per AaLS-13 cage, corresponding to ~1 guest per pentamer unit [22]. For fusion enzymes, the maximum number of guests depends on the characteristics of the cargo enzymes. In the case of dimeric NADH oxidase from *Thermus thermophilus* HB8 (net charge of +78.2 at pH 8.0 as GFP(+36) fusion), maximum loading was ~30 guests per cage [22].
17. At the moment of mixing, the sample often turns cloudy due to protein aggregation [24]. However, it usually clarifies within a few seconds.
18. If the enzyme is positively charged, a lower enzyme concentration or a higher NaCl concentration may be needed. For example, the optimal NaCl concentration for encapsulation of positively supercharged ferritin in AaLS-13 is 600 mM [20].
19. Loading of GFP(+36) into AaLS-13 occurs at diffusion-limited rates [24].
20. Since quantitative loading was observed for most fusion enzymes [22], separation of AaLS-13 cages from free enzyme is usually unnecessary once complex formation has been confirmed by size-exclusion chromatogram.
21. Assembled AaLS-13 cages elute at ~9 mL in our setup. Since empty and filled cages are not distinguishable by SEC, we simply pool all the fractions corresponding to AaLS-13 cages.
22. We use an FEI Morgagni 268 transmission electron microscope, operated at 100 kV.
23. The dissociation constant for the binding between AaLS-13 and GFP(+36) is ~0.2 nM at pH 8, I = 0.35 (M), and 25 °C when 56 guests are loaded per cage [24]. Lowering the ionic strength increases the affinity, and vice versa.
24. AaLS-13 cages precipitate at low pH. They remain soluble between pH 7–9 [30].
25. Since the encapsulation system relies on electrostatic interactions, ionic strength is an important factor that influences binding affinity. At $I > 0.6$ M, GFP(+36) leaks out of AaLS-13 cages [24].
26. AaLS-13 is very sensitive to the presence of metal ions. Many divalent metal ions including Ni(II), Zn(II), and Co(II) cause AaLS-13 to precipitate. In contrast, in the presence of 5 mM Mg(II) it remains soluble. Thus, it is recommended to

add 1 mM EDTA to the assay buffer to ensure the absence of metal ions.

27. AaLS-13 cages are thermostable. Temperature can be increased to at least ~70 °C without dissociation or precipitation of the assemblies. However, elevated temperatures may result in leakage of guests from the cages.

Acknowledgments

This work was generously supported by ETH Zurich and the European Research Council (Advanced ERC Grant ERC-AdG-2012-321295 to D.H.). Y.A. is grateful for an Uehara Memorial Foundation Research Fellowship and an ETH Zurich Postdoctoral Fellowship (cofunded by the Marie Curie Actions program).

References

1. Yeates TO, Kerfeld CA, Heinhorst S et al (2008) Protein-based organelles in bacteria: carboxysomes and related microcompartments. Nat Rev Microbiol 6:681–691
2. Yeates TO, Thompson MC, Bobik TA (2011) The protein shells of bacterial microcompartment organelles. Curr Opin Struct Biol 21:223–231
3. Chen AH, Silver PA (2012) Designing biological compartmentalization. Trends Cell Biol 22:662–670
4. Walde P, Ichikawa S (2001) Enzymes inside lipid vesicles: preparation, reactivity and applications. Biomol Eng 18:143–177
5. Chaize B, Colletier J-P, Winterhalter M, Fournier D (2004) Encapsulation of enzymes in liposomes: high encapsulation efficiency and control of substrate permeability. Artif Cells Blood Substit Immobil Biotechnol 32:67–75
6. Marguet M, Bonduelle C, Lecommandoux S (2012) Multicompartmentalized polymeric systems: towards biomimetic cellular structure and function. Chem Soc Rev 42:512–529
7. Peters RJRW, Marguet M, Marais S et al (2014) Cascade reactions in multicompartmentalized polymersomes. Angew Chem Int Ed 53:146–150
8. Comellas-Aragonès M, Engelkamp H, Claessen VI et al (2007) A virus-based single-enzyme nanoreactor. Nat Nanotechnol 2:635–639
9. Minten IJ, Claessen VI, Blank K et al (2011) Catalytic capsids: the art of confinement. Chem Sci 2:358–362
10. Glasgow JE, Capehart SL, Francis MB, Tullman-Ercek D (2012) Osmolyte-mediated encapsulation of proteins inside MS2 viral capsids. ACS Nano 6:8658–8664
11. Fiedler JD, Brown SD, Lau JL, Finn MG (2010) RNA-directed packaging of enzymes within virus-like particles. Angew Chem Int Ed 49:9648–9651
12. O'Neil A, Reichhardt C, Johnson B et al (2011) Genetically programmed in vivo packaging of protein cargo and its controlled release from bacteriophage P22. Angew Chem Int Ed 50:7425–7428
13. Patterson DP, Schwarz B, El-Boubbou K et al (2012) Virus-like particle nanoreactors: programmed encapsulation of the thermostable CelB glycosidase inside the P22 capsid. Soft Matter 8:10158–10166
14. Patterson DP, Prevelige PE, Douglas T (2012) Nanoreactors by programmed enzyme encapsulation inside the capsid of the bacteriophage P22. ACS Nano 6:5000–5009
15. Zhang X, Meining W, Fischer M et al (2001) X-ray structure analysis and crystallographic refinement of lumazine synthase from the hyperthermophile *Aquifex aeolicus* at 1.6 Å resolution: determinants of thermostability revealed from structural comparisons. J Mol Biol 306:1099–1114
16. Azuma Y, Zschoche R, Hilvert D (2017) The C-terminal peptide of *Aquifex aeolicus* riboflavin synthase directs encapsulation of native and foreign guests by a cage-forming lumazine synthase. J Biol Chem 292:10321–10327

17. Seebeck FP, Woycechowsky KJ, Zhuang W et al (2006) A simple tagging system for protein encapsulation. J Am Chem Soc 128:4516–4517
18. Wörsdörfer B, Woycechowsky KJ, Hilvert D (2011) Directed evolution of a protein container. Science 331:589–592
19. Sasaki E, Böhringer D, van de Waterbeemd M et al (2017) Structure and assembly of scalable porous protein cages. Nat Commun 8:14663
20. Beck T, Tetter S, Künzle M, Hilvert D (2015) Construction of matryoshka-type structures from supercharged protein nanocages. Angew Chem Int Ed 54:937–940
21. Wörsdörfer B, Pianowski Z, Hilvert D (2012) Efficient in vitro encapsulation of protein cargo by an engineered protein container. J Am Chem Soc 134:909–911
22. Azuma Y, Zschoche R, Tinzl M, Hilvert D (2016) Quantitative packaging of active enzymes into a protein cage. Angew Chem Int Ed 55:1531–1534
23. Sasaki E, Hilvert D (2016) Self-assembly of Proteinaceous multishell structures mediated by a supercharged protein. J Phys Chem B 120:6089–6095
24. Zschoche R, Hilvert D (2015) Diffusion-limited cargo loading of an engineered protein container. J Am Chem Soc 137:16121–16132
25. Lawrence MS, Phillips KJ, Liu DR (2007) Supercharging proteins can impart unusual resilience. J Am Chem Soc 129:10110–10112
26. Frey R, Hayashi T, Hilvert D (2016) Enzyme-mediated polymerization inside engineered protein cages. Chem Commun 52:10423–10426
27. Frey R, Mantri S, Rocca M, Hilvert D (2016) Bottom-up construction of a primordial carboxysome mimic. J Am Chem Soc 138:10072–10075
28. Thompson DB, Cronican JJ, Liu DR (2012) Chapter twelve—Engineering and identifying supercharged proteins for macromolecule delivery into mammalian cells. In: Wittrup KD, Verdine GL (eds) Methods enzymol. Academic Press, Cambridge, MA, pp 293–319
29. Pédelacq JD, Cabantous S, Tran T et al (2006) Engineering and characterization of a superfolder green fluorescent protein. Nat Biotechnol 24:79–88
30. Wörsdörfer B (2010) Functionalizing capsid forming proteins. PhD thesis, ETH Zurich

Chapter 5

Metal Ion-Induced Self-Assembly and Packaging of CCMV Nanocapsules

Lise Schoonen, Mark B. van Eldijk, and Jan C. M. van Hest

Abstract

The controlled self-assembly of protein cages is vital for the use of these nanocompartments in biomedical and nanotechnological applications. Recently, we showed that by combining different structural peptide elements, it is possible to assemble viral capsid proteins in distinct well-defined morphologies. In this chapter, a triblock copolypeptide is discussed, consisting of a metal ion-coordinating hexahistidine tag, a stimulus-responsive elastin-like polypeptide and a pH-responsive self-assembling viral capsid protein. This protein is able to form two different types of capsids, depending on the assembly pathway that is followed. Here, we focus on the metal ion-induced assembly process and describe the relevant experimental procedures to induce and utilize this assembly behavior.

Key words Affinity chromatography, Cowpea chlorotic mottle virus, Elastin-like polypeptide, Hexahistidine, His-tag, Metal ion, Self-assembly, Viral capsid

1 Introduction

Self-assembled nanocompartments have become a major player in nanotechnology [1]. They are, for instance, being applied in drug delivery to improve blood circulation times and targeted delivery to the tissue of interest, and as nanoreactors to mimic cellular architectures, i.e., as artificial organelles [2, 3]. The nanocompartments that are used, e.g., micelles, liposomes, polymersomes, and layer-by-layer capsules, are often synthetic. The architectures of these particles, however, often have a relatively high size distribution and can only be controlled to a certain extent. This makes protein-based nanocages, and in particular viral capsids, especially interesting for nanotechnological applications, due to their uniform size and shape, biocompatibility, and robustness [4]. Furthermore, the primary structure of the building blocks of the cages, i.e., the sequence of the capsid proteins, can easily be manipulated, resulting in changes on the tertiary and quaternary levels.

Andrew K. Udit (ed.), *Protein Scaffolds: Design, Synthesis, and Applications*, Methods in Molecular Biology, vol. 1798, https://doi.org/10.1007/978-1-4939-7893-9_5, © Springer Science+Business Media, LLC, part of Springer Nature 2018

The cowpea chlorotic mottle virus (CCMV) is an excellent example of a virus whose capsid proteins can function as a nanocompartment platform in different applications. This is mainly due to its ability to reversibly assemble and disassemble upon a pH shift, even in the absence of its viral RNA [5]. At pH 5, viral capsids exist with $T = 3$ icosahedral symmetry and an outer diameter of 28 nm [6, 7]. Upon a pH increase to pH 7.5, the capsids fall apart into 90 protein dimers. Recently, we equipped the CCMV capsid protein with an elastin-like polypeptide (ELP), a stimulus responsive polymer [8]. ELPs consist of repeating VPGXG pentapeptides (where X is any amino acid, except proline). These peptides exhibit lower critical solution temperature (LCST) behavior; they can be reversibly switched between a water-soluble state to a collapsed, hydrophobic state by changes in the environment, such as temperature or salt concentration. ELPs were attached to the N-termini of CCMV capsid proteins, which are located in close proximity to each other inside the viral capsids after assembly. In ELP-CCMV, the properties of both the stimulus-responsive ELP and the self-assembling CCMV capsid protein were combined towards a system with two types of assembly behavior. Firstly, the endogenous assembly behavior was retained: pH-induced assembly resulted in viral capsids with $T = 3$ icosahedral symmetry. Additionally, the ELPs enabled salt or temperature induced assembly. The latter assembly mechanism yielded 18 nm sized viral capsids with $T = 1$ icosahedral symmetry.

Recently, we reported on another expansion of the self-assembly properties of CCMV [9]. In this case, the ELP-CCMV construct was expanded with an N-terminal hexahistidine-tag, to yield H_6-ELP-CCMV. Hexahistidines are able to coordinate divalent metal ions, a property which is cleverly exploited for affinity protein purification [10]. We demonstrated that the hexahistidine tag can efficiently induce capsid formation upon the addition of divalent ions, such as nickel and cobalt (Fig. 1a). We postulated that upon binding of multiple hexahistidine tags to a metal ion, the local ELP concentration

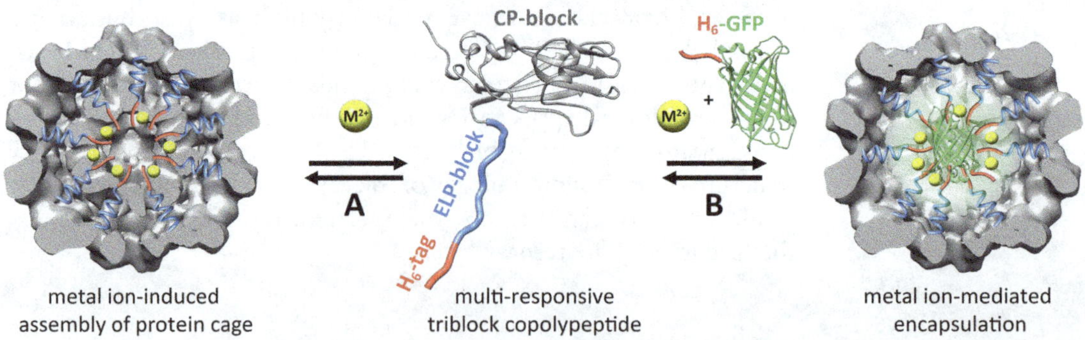

Fig. 1 (**a**) Schematic overview of metal ion-induced assembly of H_6-ELP-CCMV into capsids with $T = 1$ icosahedral symmetry. (**b**) Encapsulation of hexahistidine-tagged green fluorescent protein using metal-ion induced assembly. Reproduced from van Eldijk et al. 2016 with permission from John Wiley and Sons [9]

increases, resulting in aggregation of the ELPs and subsequent assembly of H_6-ELP-CCMV towards $T = 1$ assemblies. Additionally, it was shown that this same procedure can be used to incorporate hexahistidine-tagged proteins (Fig. 1b). Here, we will elaborate on the experimental procedures and techniques that are involved when using metal ion-induced assembly to form H_6-ELP-CCMV capsids.

2 Materials

2.1 Expression and Purification of H_6-ELP-CCMV Capsid Protein

1. *E. coli* BLR(DE3)pLysS competent cells.
2. Expression vector pET-15b-H_6-ELP[$V_4L_4G_1$–9]-CCMV(ΔN26) (*see* **Note 1**).
3. Sterile LB agar plates containing 100 mg/L ampicillin and 50 mg/L chloramphenicol.
4. Autoclaved 250 mL and 5 L Erlenmeyer flasks.
5. Sterile LB medium containing 100 mg/L ampicillin and 50 mg/L chloramphenicol.
6. Sterile 2× TY medium containing 100 mg/L ampicillin.
7. 1 M isopropyl β-D-1-thiogalactopyranoside (IPTG) stock solution in sterile Milli-Q water.
8. Lysis buffer: 50 mM NaH_2PO_4, 1300 mM NaCl, 10 mM imidazole, pH adjusted to 8.0 using NaOH (filtered over a 0.2 μm filter and autoclaved).
9. Wash buffer: 50 mM NaH_2PO_4, 1300 mM NaCl, 20 mM imidazole, pH adjusted to 8.0 using NaOH (filtered over a 0.2 μm filter and autoclaved).
10. Elution buffer: 50 mM NaH_2PO_4, 1300 mM NaCl, 250 mM imidazole, pH adjusted to 8.0 using NaOH (filtered over a 0.2 μm filter and autoclaved).
11. Ni^{2+}-NTA agarose beads.
12. Disposable plastic columns (approximately 15 × 120 mm).
13. pH 7.5 dimer buffer: 50 mM Tris·HCl, 500 mM NaCl, 10 mM $MgCl_2$, 1 mM EDTA, pH adjusted to 7.5 using HCl (filtered over a 0.2 μm filter).
14. pH 5.0 capsid buffer: 50 mM sodium acetate, 500 mM NaCl, 10 mM $MgCl_2$, 1 mM EDTA, pH adjusted to 5.0 using HCl (filtered over a 0.2 μm filter).
15. Dialysis tubing (12–14 kDa molecular weight cutoff) and clamps.

2.2 Metal-Ion Induced Assembly

1. pH 7.5 0.3 M NaCl buffer: 50 mM Tris–HCl, 300 mM NaCl, 10 mM $MgCl_2$, 1 mM EDTA, pH adjusted to 7.5 using HCl (filtered over a 0.2 μm filter).
2. Dialysis tubing (12–14 kDa molecular weight cutoff) and clamps.

3. Fresh 450 μM nickel chloride ($NiCl_2$) stock solution in pH 7.5 0.3 M NaCl buffer.
4. Fresh 4.5 mM cobalt chloride ($CoCl_2$) stock solution in pH 7.5 0.3 M NaCl buffer.

2.3 EDTA-Induced Disassembly

1. pH 7.5 0.3 M NaCl buffer: 50 mM Tris–HCl, 300 mM NaCl, 10 mM $MgCl_2$, 1 mM EDTA, pH adjusted to 7.5 using HCl (filtered over a 0.2 μm filter).
2. 5 mM ethylenediaminetetraacetic acid (EDTA) stock solution in pH 7.5 0.3 M NaCl buffer.

2.4 Metal-Ion Mediated Encapsulation

1. pH 7.5 0.3 M NaCl buffer: 50 mM Tris–HCl, 300 mM NaCl, 10 mM $MgCl_2$, 1 mM EDTA, pH adjusted to 7.5 using HCl (filtered over a 0.2 μm filter).
2. Fresh 100 μM stock solution of H_6-tagged cargo, e.g., H_6-tagged green fluorescent protein (GFP) in pH 7.5 0.3 M NaCl buffer.
3. Fresh 1.8 mM nickel chloride ($NiCl_2$) stock solution in pH 7.5 0.3 M NaCl buffer.
4. Dialysis tubing (12–14 kDa molecular weight cutoff) and clamps.

2.5 Asymmetric-Flow Field-Flow Fractionation Analysis of CCMV Capsules

1. Asymmetric-flow field-flow fractionation equipment (*see* **Note 2**).
2. Regenerated cellulose 10 kDa membrane (Millipore).
3. Bovine serum albumin (BSA).
4. pH 7.5 0.3 M NaCl buffer: 50 mM Tris–HCl, 300 mM NaCl, 10 mM $MgCl_2$, 1 mM EDTA, pH adjusted to 7.5 using HCl (filtered over a 0.2 μm filter).
5. pH 7.5 dimer buffer: 50 mM Tris–HCl, 500 mM NaCl, 10 mM $MgCl_2$, 1 mM EDTA, pH adjusted to 7.5 using HCl (filtered over a 0.2 μm filter).

3 Methods

3.1 Expression and Purification of H_6-ELP-CCMV Capsid Protein

1. Transform the pET-15b-H_6-ELP[$V_4L_4G_1$–9]-CCMV(ΔN26) vector into *E. coli* BLR(DE3)pLysS competent cells and plate onto an LB agar plate containing 100 mg/L ampicillin and 50 mg/L chloramphenicol. Incubate at 37 °C for 16–24 h.
2. Use a single colony to inoculate 50 mL of LB medium containing 100 mg/L ampicillin and 50 mg/L chloramphenicol. Grow at 37 °C for 16–24 h while shaking vigorously (200–250 rpm) (*see* **Note 3**).
3. Dilute the overnight culture in 1 L of 2× TY medium containing 100 mg/L ampicillin and grow at 37 °C while shaking

vigorously (200–250 rpm) until a culture density of OD_{600} of 0.4–0.6 is reached (approximately 1–2 h).

4. Induce protein expression by the addition of IPTG (final concentration 1 mM). Incubate the culture at 30 °C for 4–6 h while shaking vigorously (200–250 rpm).
5. Harvest the cells by centrifugation (4000 × *g*, 4 °C, 15 min) (*see* **Note 4**).
6. Resuspend the cells in 25 mL of cold lysis buffer.
7. Lyse the cells by ultrasonic disruption (3 times 30 s) (*see* **Notes 5** and **6**).
8. Centrifuge the lysate to pellet the cellular debris (15,000 × *g*, 4 °C, 15 min).
9. Add 1.5 mL of Ni^{2+}-NTA agarose beads resuspended in lysis buffer to the supernatant. Incubate the mixture at 4 °C for at least 1 h while mixing gently to allow binding of H_6-tagged ELP-CCMV to the beads.
10. Load the mixture into a column and flow the lysate through the column.
11. Wash the beads 2 times with 10 mL of cold wash buffer.
12. Elute H_6-ELP-CCMV with cold elution buffer (first time with 0.5 mL to remove residual wash buffer, then 7 times with 1.5 mL).
13. Combine the column fractions containing the desired protein (*see* **Note 7**). Dialyze the protein to 100 mL of pH 7.5 dimer buffer at 4 °C to remove the excess imidazole, using a dialysis membrane with a 12–14 kDa molecular weight cutoff. Dialyze for 3 h and refresh the buffer approximately every hour.
14. Continue with dialysis to 100 mL of pH 5.0 capsid buffer at 4 °C. Refresh the buffer after the first and second hours of dialysis. Leave the protein solution at 4 °C for 16–72 h during the last dialysis step (*see* **Note 8**).
15. Verify the purity of the protein by SDS-PAGE analysis. Measure the protein mass by electrospray ionization time-of-flight (ESI-TOF) mass spectrometry to validate the expected molecular mass of H_6-ELP-CCMV. The expected SDS-PAGE and ESI-TOF mass spectrometry results are shown in Fig. 2.
16. The self-assembling properties of the protein can be verified by size exclusion chromatography (SEC) and transmission electron microscopy (TEM) (*see* **Note 9**).
17. Determine the protein yield using the molar extinction coefficient at 280 nm (23,470 M^{-1} cm^{-1}) [11]. A yield of 50–100 mg/L of bacterial culture is expected for H_6-ELP-CCMV under these conditions.
18. Store the protein solution at 4 °C (*see* **Note 10**).

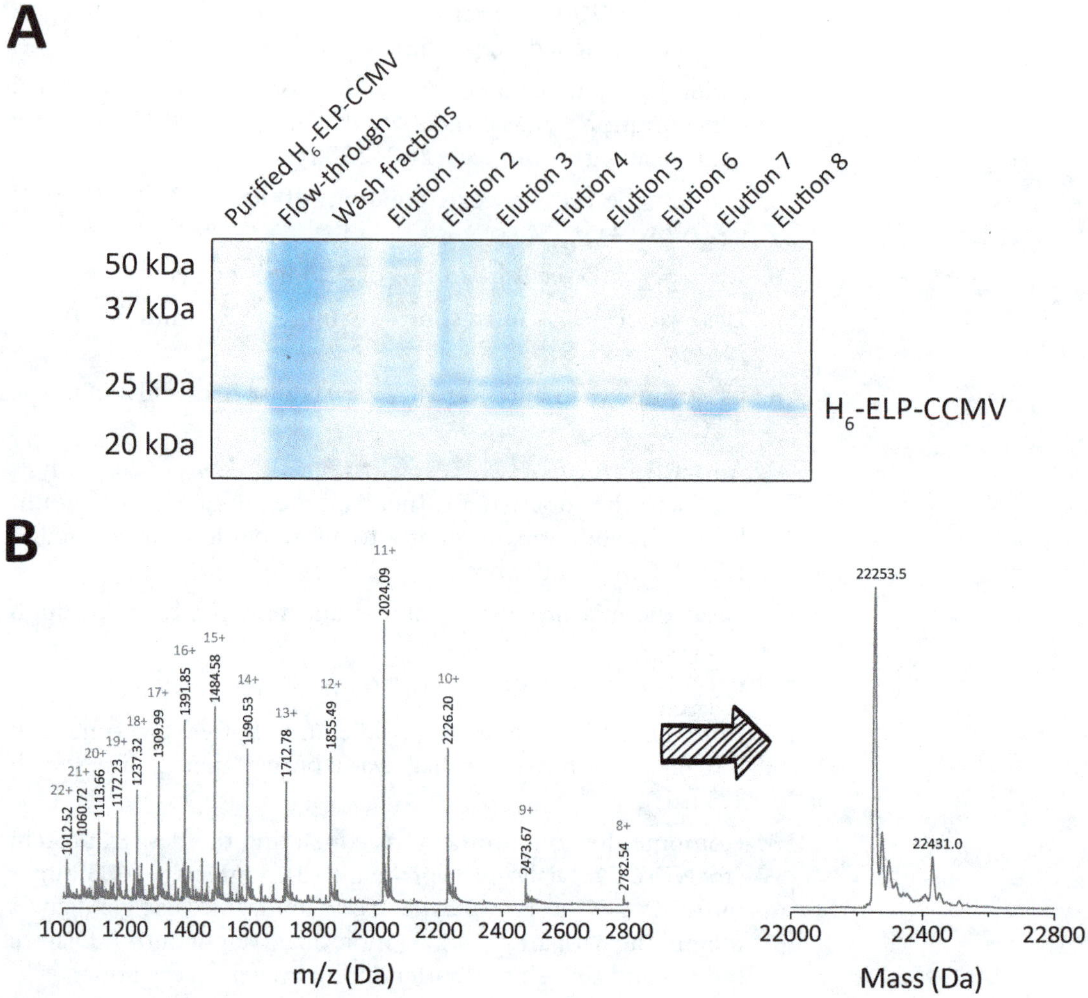

Fig. 2 (**a**) SDS-PAGE of the affinity purification of H_6-ELP-CCMV and purified H_6-ELP-CCMV. (**b**) ESI-TOF mass spectrometry of purified H_6-ELP-CCMV: multiply charged ion series (left) and deconvoluted total mass spectrum (right). The expected molecular weight is 22253.4 Da

3.2 Metal-Ion Induced Assembly

1. Dialyze H_6-ELP-CCMV to pH 7.5 0.3 M NaCl buffer at 4 °C, using a dialysis membrane with a 12–14 kDa molecular weight cutoff. Dialyze for 1.5 h and refresh the buffer approximately every 30 min.
2. Determine the protein concentration using the molar extinction coefficient at 280 nm (23,470 M^{-1} cm^{-1}) [11].
3. Dilute the protein solution to 90 μM (2 mg/mL) with pH 7.5 0.3 M NaCl buffer.
4. Add pH 7.5 0.3 M NaCl buffer (80 μL) and the $NiCl_2$ or $CoCl_2$ stock solution (20 μL) to the H_6-ELP-CCMV solution (100 μL).

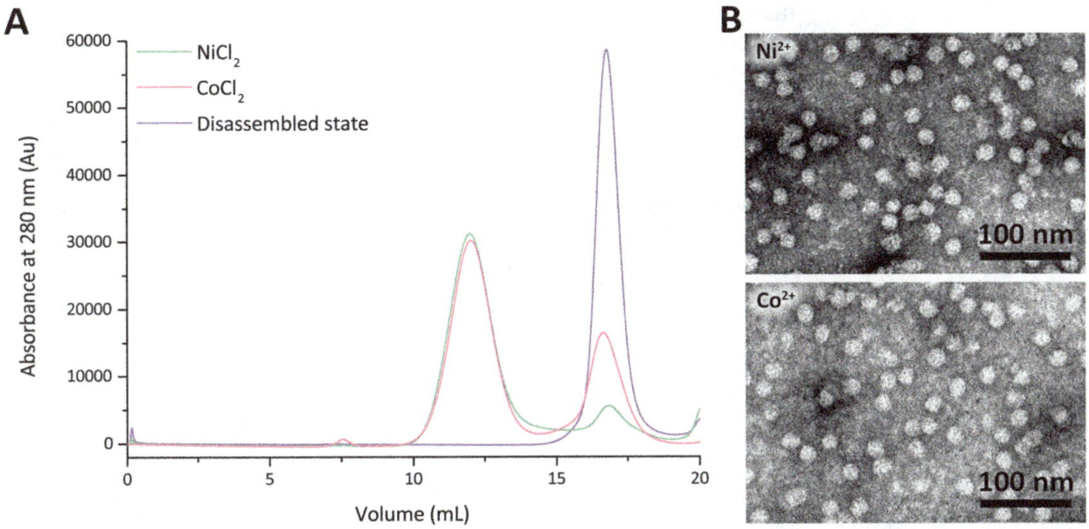

Fig. 3 (**a**) Size exclusion chromatograms of Ni^{2+}- and Co^{2+}-induced assembly of H_6-ELP-CCMV and H_6-ELP-CCMV in the disassembled state as a control. (**b**) Uranyl acetate stained TEM micrographs of Ni^{2+}- and Co^{2+}-induced H_6-ELP-CCMV capsids

5. Incubate the solution at room temperature for at least 15 min (no shaking required).
6. Analyze capsid assembly using SEC and TEM (*see* **Note 9**). The expected SEC chromatograms and representative TEM images of Ni- and Co-induced assemblies are shown in Fig. 3.
7. Further analysis of the resulting capsids can be done using asymmetric-flow field-flow fractionation (AF_4) in combination with multiangle laser light scattering (MALLS) analysis. This will be discussed in more detail in Subheading 3.5.

3.3 EDTA-Induced Disassembly

1. Add the EDTA stock solution (10 μL) to a metal-ion induced capsid solution (90 μL) (*see* Subheading 3.2).
2. Analyze capsid disassembly using SEC. The expected SEC chromatogram of disassembled CCMV capsids is shown in Fig. 3.

3.4 Metal-Ion Mediated Encapsulation

1. Add the $NiCl_2$ stock solution (10 μL) to the cargo stock solution (90 μL).
2. Incubate the solution at room temperature for at least 5 min (no shaking required).
3. Dialyze H_6-ELP-CCMV to pH 7.5 0.3 M NaCl buffer at 4 °C, using a dialysis membrane with a 12–14 kDa molecular weight cutoff. Dialyze for 1.5 h and refresh the buffer approximately every 30 min.
4. Determine the protein concentration using the molar extinction coefficient at 280 nm (23,470 M^{-1} cm^{-1}) [11].

Table 1
Overview of the method used for the AF_4 experiments, mentioned in this chapter

Duration (min)	Mode	Cross flow rate start (mL/min)	Cross flow rate end (mL/min)
2	Elution	2.00	2.00
1	Focus	–	–
2	Focus + inject	–	–
1	Focus	–	–
15	Elution	2.00	0.30
4	Elution	0.00	0.00
4	Elution + inject	0.00	0.00

1.00 mL/min detector flow, 1.50 mL/min focus flow, 0.20 mL/min inject flow

5. Dilute the protein solution to 90 μM (2 mg/mL) with pH 7.5 0.3 M NaCl buffer.
6. Add the preincubated solution of cargo and 55 μL of $NiCl_2$ to 55 μL of the H_6-ELP-CCMV solution.
7. Incubate the solution at room temperature for at least 15 min (no shaking required).
8. If desired, purify the assemblies using preparative SEC (*see* **Note 9**).
9. Analyze the incorporation of the cargo with the appropriate technique, e.g., SEC, fluorescence spectroscopy, SDS-PAGE, or another technique suitable for the type of cargo.

3.5 Asymmetric-Flow Field-Flow Fractionation Analysis of CCMV Capsules

1. Prepare a method, including the desired flow rates, for the asymmetric-flow field-flow fractionation (AF_4) (*see* **Note 11**). For the experiments mentioned in this chapter, the method in Table 1 was used.
2. Preequilibrate the membrane channel with the desired buffer.
3. Inject 5 μL of 3.0 mg/mL BSA in the desired buffer and normalize the detectors.
4. Inject 5 μL of 1.0 mg/mL of the desired sample in the desired buffer.
5. The expected AF_4-MALLS chromatograms after Ni^{2+}-induced assembly and Co^{2+}-induced assembly in pH 7.5 0.3 M NaCl buffer are shown in Fig. 4. As a comparison, the CCMV capsid

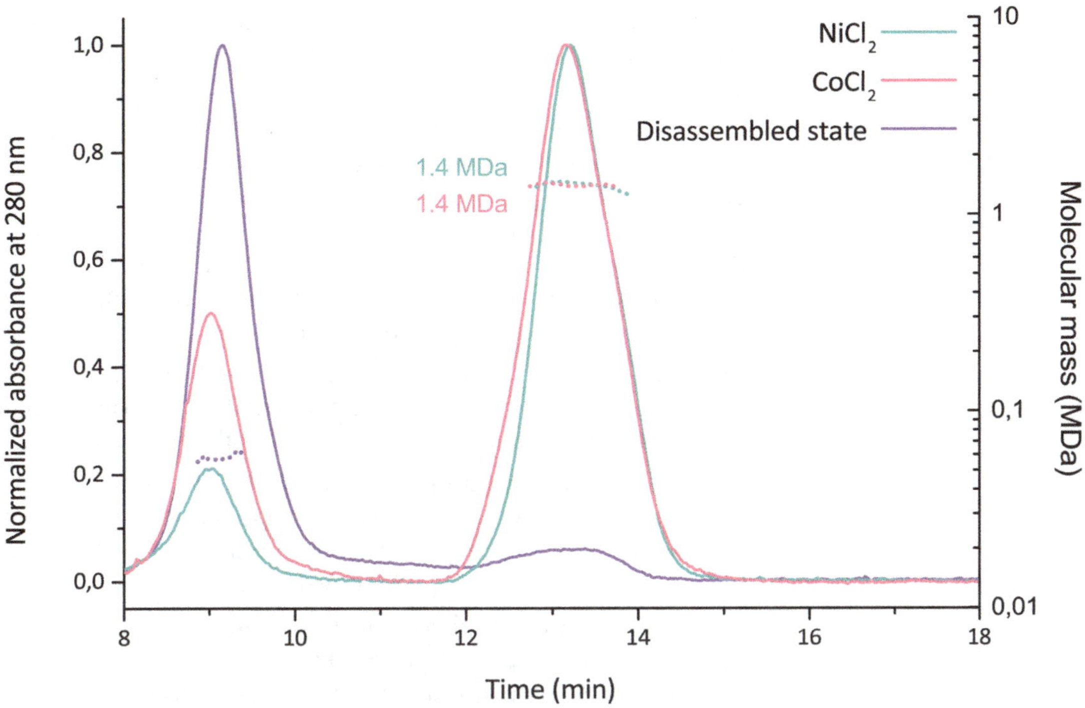

Fig. 4 AF_4-MALLS chromatograms of Ni^{2+}- and Co^{2+}-induced assembly of H_6-ELP-CCMV and H_6-ELP-CCMV in the disassembled state as a control. Note that in AF_4, the elution order is inverted compared to SEC

proteins in the dimer state in pH 7.5 dimer buffer can be injected (*see* Fig. 4).

4 Notes

1. The reader is referred to the original literature for the nature of the H_6-ELP-CCMV expression plasmid and details on the construction of the pET-15b-H_6-ELP[$V_4L_4G_1$–9]-CCMV(ΔN26) plasmid [8]. Gene ID wild-type CCMV sequence: 962152. In the plasmid used, the first 26 N-terminal amino acids were removed and replaced by an ELP consisting of nine pentapeptides. A hexahistidine tag was placed at the N-terminus of the resulting construct for purification purposes.
2. For the experiments mentioned in this chapter, a Wyatt Eclipse AF_4 was used, equipped with the short channel with a regenerated cellulose 10 kDa membrane (Millipore) and a spacer of 350 μm. This system was combined with a Shimadzu LC-20A Prominence system with degasser and Shimadzu CTO20A injector. The system was connected to a series of detectors: a Wyatt DAWN HELEOS II light scattering detector, using a

laser operated at 664.5 nm; a Wyatt Optilab Rex refractive index detector and a Shimadzu SPD20A UV-Vis detector.

3. To avoid having to do the transformation each time before the expression, a glycerol stock can be made using the overnight culture. To do so, a small volume of the overnight culture should be diluted to 15–30% glycerol and stored at −80 °C. The stock is now stable for years, as long as it is stored at −80 °C. During the next expression, the transformation can be skipped and the overnight culture can be inoculated using a sterile pipette tip, which can be used to scrape off some of the glycerol stock.
4. Cell pellets can be lysed directly or stored at −20 °C for at least a couple of days. If the cell pellets are not used directly, thaw the cell pellets on ice for 15 min before resuspending them in lysis buffer.
5. Because heat is being produced during the ultrasonic disruption, the cell suspension should be kept on ice during this process. In between the cycles, time should be taken for the suspension to cool down (at least 5–10 s).
6. The number and length of the ultrasonic disruption cycles depend on the output of the machine being used. Generally, the lysis should be continued until the suspension is homogeneous and nonviscous.
7. The protein concentration and purity (A260/A280 ratio) of the column fractions can be measured by, for instance, a NanoDrop spectrophotometer. A value of approximately 0.6 for the ratio of A260/A280 is expected for a solution of pure protein. A higher ratio could indicate nucleic acid contamination. In order to prevent interactions between nucleic acids and the capsid proteins, a relatively high NaCl concentration of 1.3 M is used in all buffers during protein purification.
8. After the last overnight dialysis step, some protein precipitation is often observed. This can be minimized by including some more diluted column fractions, thereby lowering the overall protein concentration during dialysis. A maximum protein concentration of approximately 5 mg/mL during dialysis is recommended.
9. For SEC analysis of CCMV capsid assembly, a Superose 6 10/300 GL column from GE Healthcare is recommended, with pH 5.0 capsid buffer as the eluent. For TEM analysis of the capsids, formvar/carbon-coated grids (FCF200-Cu) from EMS are normally used.
10. Always store H_6-ELP-CCMV in the assembled state in pH 5.0 capsid buffer. In this buffer, the protein can be stored for months at 4 °C. The protein should be not stored in pH 7.5 buffer, as it is not stable for prolonged periods of time in the

dimer state. The protein is not stable upon freezing; thus, protein aliquots cannot be stored at −20 °C or −80 °C.

11. An AF_4 method involves three stages: injection, focus, and elution. The durations and flow rates of these stages can be varied to obtain an optimal separation of the sample.

Acknowledgments

The Ministry of Education, Culture, and Science (Gravitation program 024.001.035) is acknowledged for financial support.

References

1. Vriezema DM, Comellas Aragonès M, Elemans JAAW et al (2005) Self-assembled nanoreactors. Chem Rev 105:1445–1489
2. Doll TAPF, Raman S, Dey R, Burkhard P (2013) Nanoscale assemblies and their biomedical applications. J R Soc Interface 10:20120740
3. Schoonen L, van Hest JCM (2016) Compartmentalization approaches in soft matter science: from nanoreactor development to organelle mimics. Adv Mater 28:1109–1128
4. Douglas T, Young M (2006) Viruses: making friends with old foes. Science 312:873–875
5. Lavelle L, Gingery M, Phillips M et al (2009) Phase diagram of self-assembled viral capsid protein polymorphs. J Phys Chem B 113:3813–3819
6. Caspar DLD, Klug A (1962) Physical principles in the construction of regular viruses. Cold Spring Harb Symp Quant Biol 27:1–24
7. Speir JA, Munshi S, Wang G et al (1995) Structures of the native and swollen forms of cowpea chlorotic mottle virus determined by X-ray crystallography and cryo-electron microscopy. Structure 3:63–78
8. van Eldijk MB, Wang JC-Y, Minten IJ et al (2012) Designing two self-assembly mechanisms into one viral capsid protein. J Am Chem Soc 134:18506–18509
9. van Eldijk MB, Schoonen L, Cornelissen JJLM et al (2016) Metal ion-induced self-assembly of a multi-responsive block copolypeptide into well-defined Nanocapsules. Small 12:2476–2483
10. Ueda EKM, Gout PW, Morganti L (2003) Current and prospective applications of metal ion-protein binding. J Chromatogr A 988:1–23
11. Gill SC, von Hippel PH (1989) Calculation of protein extinction coefficients from amino acid sequence data. Anal Biochem 182:319–326

Chapter 6

Modification of CCMV Nanocages for Enzyme Encapsulation

Lise Schoonen and Jan C. M. van Hest

Abstract

In cellular systems, compartmentalization plays an important role in the protection and regulation of enzymes. Controlled encapsulation of enzymes in nanocompartments is crucial in understanding biocatalytic processes in the cellular environment. We have recently described an enzymatic method to covalently attach enzymes, equipped with a small recognition peptide, to the interior of viral capsids. Viral capsids are especially interesting in this respect, as they form very well-defined nanoparticles with a uniform size and shape. Here, we describe the relevant experimental procedures to encapsulate a model enzyme into the interior of a viral capsid, purify the resulting viral capsids, and measure the catalytic activity of the encapsulated enzymes.

Key words *Candida antarctica* lipase B, Cowpea chlorotic mottle virus, Esterase activity, Self-assembly, Sortagging, Sortase A, Viral capsid

1 Introduction

Over the years, it has become more and more clear that cells are incredibly complex systems. Although researchers have managed to get an understanding of some isolated processes that take place in a cellular environment, the spatiotemporal arrangement and interaction of multistep biocatalytic pathways have only recently become a topic of intensive investigation in the fields of systems and chemical biology. For instance, it is still not fully understood how catalysis takes place inside the crowded interior of cells. What is known is that compartmentalization plays a crucial role in catalytic processes. It allows the protection of catalysts against inhibitors and regulates the flow of substrates and products to and from the catalysts. In order to fully understand how catalysis takes place in a cellular system, it is important to recreate the environment of the catalyst in the best possible way in a model system.

A common strategy which has been applied in this field is the encapsulation of catalysts in small nanocompartments. Virus-like

Andrew K. Udit (ed.), *Protein Scaffolds: Design, Synthesis, and Applications*, Methods in Molecular Biology, vol. 1798, https://doi.org/10.1007/978-1-4939-7893-9_6, © Springer Science+Business Media, LLC, part of Springer Nature 2018

particles (VLPs), i.e., viruses without their genetic material, are especially interesting in this respect, since these consist of proteins, making them biocompatible. Furthermore, their capsids always consist of a specific amount of protein, and they are therefore monodisperse and well-defined [1]. In particular, the cowpea chlorotic mottle virus (CCMV) capsid is an interesting VLP, since it can be reversibly assembled and disassembled upon a pH shift. At pH 7.5, the protein exists in the dimer form. When the pH is lowered to pH 5.0, 90 dimers assemble into a capsid of 28 nm in diameter with $T = 3$ icosahedral symmetry [2, 3]. The N-terminus of the capsid proteins is located on the interior of the final capsid structure, which therefore provides an interesting handle for the covalent attachment of catalysts, toward the formation of CCMV-based nanoreactors. Upon encapsulation, the cargo will automatically end up on the inside of the capsid structure.

Recently, we reported on a method to covalently attach any given cargo to the N-terminus [4]. In order to selectively modify the N-terminus over other free amine groups present in the protein in the form of lysines, the enzyme Sortase A (SrtA) was utilized. SrtA is a transpeptidase, which anchors surface proteins to the bacterial cell wall of Gram-positive bacteria [5]. This enzyme recognizes proteins containing the sorting signal LPXTG (where X is any natural amino acid), and cleaves the bonds between threonine and glycine, forming a thioester conjugate intermediate. Subsequently, a nucleophile containing an N-terminal glycine can attack the LPXT-SrtA intermediate, eliminating the SrtA enzyme and resulting in the coupled product.

SrtA was used to couple a model enzyme, *Candida antarctica* lipase B (CalB), to the N-terminus of CCMV, followed by encapsulation in the CCMV capsids (Fig. 1) [6]. These capsids could be used to investigate the effect of enzyme encapsulation on its catalytic performance. The use of the CCMV capsid in this case, however, restricts the conditions of these types of studies to nonphysiologically relevant conditions, since the capsid is most stable at pH 5.0, but not at neutral pH. To extend the assembly conditions of CCMV, its capsid protein was coupled to a short elastin-like polypeptide (ELP), to create H_6-ELP-CCMV [7]. ELPs are stimulus-responsive peptides, consisting of repeats of pentapeptide VPGXG, where X is any amino acid, except proline. These peptides can be reversibly switched from a water-soluble state to an insoluble, collapsed state, by changes in the environment, such as the temperature, pH or salt concentration of the surrounding solution. The attachment of the ELP block to the N-terminus of the CCMV capsid protein resulted in a second assembly pathway: ELP-induced assembly. Upon an increase of the salt concentration or temperature, the ELPs aggregate, resulting in subsequent capsid assembly, even at neutral pH. By using ELP-CCMV as a nanoreactor, the enzymatic behavior of its cargo could also be investigated at physiological pH [6].

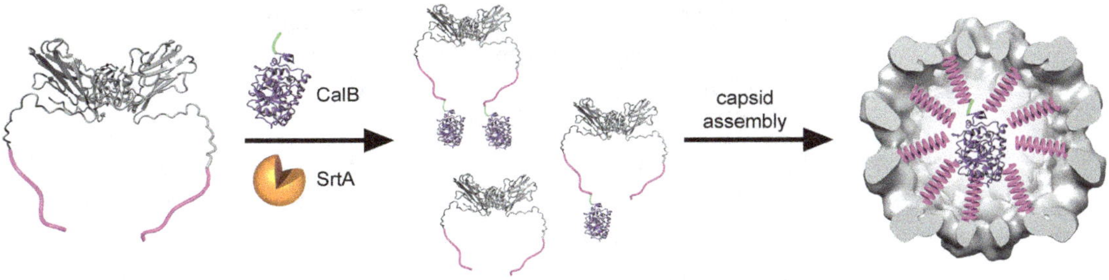

Fig. 1 Schematic overview of sortase A-mediated N-terminal modification of H_6-ELP-CCMV with the enzyme CalB-LPETG, and subsequent enzyme encapsulation via CCMV capsid formation. Reproduced from Schoonen et al. 2016 with permission from The Royal Society of Chemistry [6]

Here, we will discuss the relevant experimental procedures and techniques that are needed to prepare ELP-CCMV capsids with encapsulated CalB-LPETG and to investigate the effect of encapsulation on the catalytic properties of the CalB enzyme. We believe these procedures are readily extendable for the encapsulation of other enzymes, equipped with an LPXTG sorting signal, or even other types of catalysts.

2 Materials

2.1 Expression and Purification of H_6-ELP-CCMV Capsid Protein

1. *E. coli* BLR(DE3)pLysS competent cells.
2. Expression vector pET-15b-H_6-ELP[$V_4L_4G_1$–9]-CCMV(ΔN26) (*see* **Note 1**).
3. Sterile LB agar plates containing 100 mg/L ampicillin and 50 mg/L chloramphenicol.
4. Autoclaved 250 mL and 5 L Erlenmeyer flasks.
5. Sterile LB medium containing 100 mg/L ampicillin and 50 mg/L chloramphenicol.
6. Sterile 2× TY medium containing 100 mg/L ampicillin.
7. 1 M isopropyl β-D-1-thiogalactopyranoside (IPTG) stock solution in sterile Milli-Q water.
8. Lysis buffer: 50 mM NaH_2PO_4, 1300 mM NaCl, 10 mM imidazole, pH adjusted to 8.0 using NaOH (filtered over a 0.2 μm filter and autoclaved).
9. Wash buffer: 50 mM NaH_2PO_4, 1300 mM NaCl, 20 mM imidazole, pH adjusted to 8.0 using NaOH (filtered over a 0.2 μm filter and autoclaved).
10. Elution buffer: 50 mM NaH_2PO_4, 1300 mM NaCl, 250 mM imidazole, pH adjusted to 8.0 using NaOH (filtered over a 0.2 μm filter and autoclaved).
11. Ni^{2+}-NTA agarose beads.

12. Disposable plastic columns (approximately 15 × 120 mm).
13. pH 7.5 dimer buffer: 50 mM Tris–HCl, 500 mM NaCl, 10 mM $MgCl_2$, 1 mM EDTA, pH adjusted to 7.5 using HCl (filtered over a 0.2 μm filter).
14. pH 5.0 capsid buffer: 50 mM sodium acetate, 500 mM NaCl, 10 mM $MgCl_2$, 1 mM EDTA, pH adjusted to 5.0 using HCl (filtered over a 0.2 μm filter).
15. Dialysis tubing (12–14 kDa molecular weight cutoff) and clamps.

2.2 Expression and Purification of H_6-SrtA

1. *E. coli* BL21 AI competent cells.
2. Expression vector pQE30-H_6-SrtA (*see* **Note 2**).
3. Autoclaved 25 mL and 2.5 L Erlenmeyer flasks.
4. Sterile LB medium containing 100 mg/L ampicillin.
5. Sterile TB medium containing 100 mg/L ampicillin.
6. Lysis buffer: 50 mM NaH_2PO_4, 300 mM NaCl, 10 mM imidazole, 1 mM phenylmethanesulfonyl fluoride, pH adjusted to 8.0 using NaOH (filtered over a 0.2 μm filter and autoclaved).
7. Wash buffer: 50 mM NaH_2PO_4, 300 mM NaCl, 20 mM imidazole, pH adjusted to 8.0 using NaOH (filtered over a 0.2 μm filter and autoclaved).
8. Elution buffer: 50 mM NaH_2PO_4, 300 mM NaCl, 250 mM imidazole, pH adjusted to 8.0 using NaOH (filtered over a 0.2 μm filter and autoclaved).
9. Ni^{2+}-NTA agarose beads.
10. Disposable plastic columns (approximately 15 × 120 mm).
11. Sortase buffer: 50 mM HEPES, 150 mM NaCl, 5 mM $CaCl_2$, pH adjusted to 7.5 using NaOH (filtered over a 0.2 μm filter).
12. Amicon Ultra-15 mL centrifugal filters with a 10 kDa molecular weight cutoff.

2.3 Expression and Purification of CalB-LPETG-H_6

1. *E. coli* BLR(DE3)pLysS competent cells.
2. Expression vector pET-22b-CalB-LPETG-H_6 (*see* **Note 3**).
3. Sterile LB agar plates containing 100 mg/L ampicillin and 50 mg/L chloramphenicol.
4. Autoclaved 250 mL and 5 L Erlenmeyer flasks.
5. Sterile LB medium containing 100 mg/L ampicillin and 50 mg/L chloramphenicol.
6. Sterile 2× TY medium containing 100 mg/L ampicillin.
7. 1 M Isopropyl β-D-1-thiogalactopyranoside (IPTG) stock solution in sterile Milli-Q water.
8. Lysis buffer: 50 mM NaH_2PO_4, 300 mM NaCl, 10 mM imidazole, pH adjusted to 8.0 using NaOH (filtered over a 0.2 μm filter and autoclaved).

9. Wash buffer: 50 mM NaH_2PO_4, 300 mM NaCl, 20 mM imidazole, pH adjusted to 8.0 using NaOH (filtered over a 0.2 μm filter and autoclaved).
10. Elution buffer: 50 mM NaH_2PO_4, 300 mM NaCl, 250 mM imidazole, pH adjusted to 8.0 using NaOH (filtered over a 0.2 μm filter and autoclaved).
11. Ni^{2+}-NTA agarose beads.
12. Disposable plastic columns (approximately 15 × 120 mm).
13. CalB buffer: 50 mM NaH_2PO_4, 150 mM NaCl, pH adjusted to 7.0 using NaOH (filtered over a 0.2 μm filter).
14. Amicon Ultra-15 mL centrifugal filters with a 10 kDa molecular weight cutoff.
15. Superdex 75 PC 10/300 column from GE Healthcare.

2.4 SrtA-Mediated Coupling Experiment

1. Sortase buffer: 50 mM HEPES, 150 mM NaCl, 5 mM $CaCl_2$, pH adjusted to 7.5 using NaOH (filtered over a 0.2 μm filter).
2. Amicon Ultra-0.5 mL centrifugal filters with a 10 kDa molecular weight cutoff.
3. 5 mM stock solution of ethylenediaminetetraacetic acid (EDTA) in Milli-Q water.

2.5 pH-Induced Capsid Assembly and SEC Purification after SrtA Reaction

1. pH 5.0 capsid buffer: 50 mM sodium acetate, 500 mM NaCl, 10 mM $MgCl_2$, 1 mM EDTA, pH adjusted to 5.0 using HCl (filtered over a 0.2 μm filter).
2. pH 7.5 capsid buffer: 50 mM Tris–HCl, 2000 mM NaCl, 10 mM $MgCl_2$, 1 mM EDTA, pH adjusted to 7.5 using HCl (filtered over a 0.2 μm filter).
3. Amicon Ultra-0.5 mL centrifugal filters with a 10 kDa molecular weight cutoff.
4. Superose 6 10/300 column from GE Healthcare.

2.6 Activity Assay with Encapsulated CalB

1. 100 mM stock solution of *p*-nitrophenyl acetate (*p*-NPA) in DMSO.
2. 96-wells microplates.

3 Methods

3.1 Expression and Purification of H_6-ELP-CCMV Capsid Protein

1. Transform the pET-15b-H_6-ELP[$V_4L_4G_1$–9]-CCMV(ΔN26) vector into *E. coli* BLR(DE3)pLysS competent cells and plate onto an LB agar plate containing 100 mg/L ampicillin and 50 mg/L chloramphenicol. Incubate at 37 °C for 16–24 h.
2. Use a single colony to inoculate 50 mL of LB medium containing 100 mg/L ampicillin and 50 mg/L

chloramphenicol. Grow at 37 °C for 16–24 h while shaking vigorously (200–250 rpm) (*see* **Note 4**).

3. Dilute the overnight culture in 1 L of 2× TY medium containing 100 mg/L ampicillin and grow at 37 °C while shaking vigorously (200–250 rpm) until a culture density of OD_{600} of 0.4–0.6 is reached (approximately 1–2 h).
4. Induce protein expression by the addition of IPTG (final concentration 1 mM). Incubate the culture at 30 °C for 4–6 h while shaking vigorously (200–250 rpm).
5. Harvest the cells by centrifugation (4000 × *g*, 4 °C, 15 min) (*see* **Note 5**).
6. Resuspend the cells in 25 mL of cold lysis buffer.
7. Lyse the cells by ultrasonic disruption (3 times 30 s) (*see* **Notes 6** and **7**).
8. Centrifuge the lysate to pellet the cellular debris (15,000 × *g*, 4 °C, 15 min).
9. Add 1.5 mL of Ni^{2+}-NTA agarose beads resuspended in lysis buffer to the supernatant. Incubate the mixture at 4 °C for at least 1 h while mixing gently to allow binding of H_6-tagged ELP-CCMV to the beads.
10. Load the mixture into a column and flow the lysate through the column.
11. Wash the beads 2 times with 10 mL of cold wash buffer.
12. Elute H_6-ELP-CCMV with cold elution buffer (first time with 0.5 mL to remove residual wash buffer, then 7 times with 1.5 mL).
13. Combine the column fractions containing the desired protein (*see* **Notes 8** and **9**). Dialyze the protein to 100 mL of pH 7.5 dimer buffer at 4 °C to remove the excess imidazole, using a dialysis membrane with a 12–14 kDa molecular weight cutoff. Dialyze for 3 h and refresh the buffer approximately every hour.
14. Continue with dialysis to 100 mL of pH 5.0 capsid buffer at 4 °C. Refresh the buffer after the first and second hours of dialysis. Leave the protein solution at 4 °C for 16–72 h during the last dialysis step (*see* **Note 10**).
15. Verify the purity of the protein by SDS-PAGE analysis. Measure the protein mass by electrospray ionization time-of-flight (ESI-TOF) mass spectrometry to validate the expected molecular mass of H_6-ELP-CCMV. The expected SDS-PAGE and ESI-TOF mass spectrometry results are shown in Fig. 2.
16. The self-assembling properties of the protein can be verified by size exclusion chromatography (SEC) and transmission electron microscopy (TEM) (*see* **Note 11**).
17. Determine the protein yield using the molar extinction coefficient at 280 nm (23,470 M^{-1} cm^{-1}) [8]. A yield of

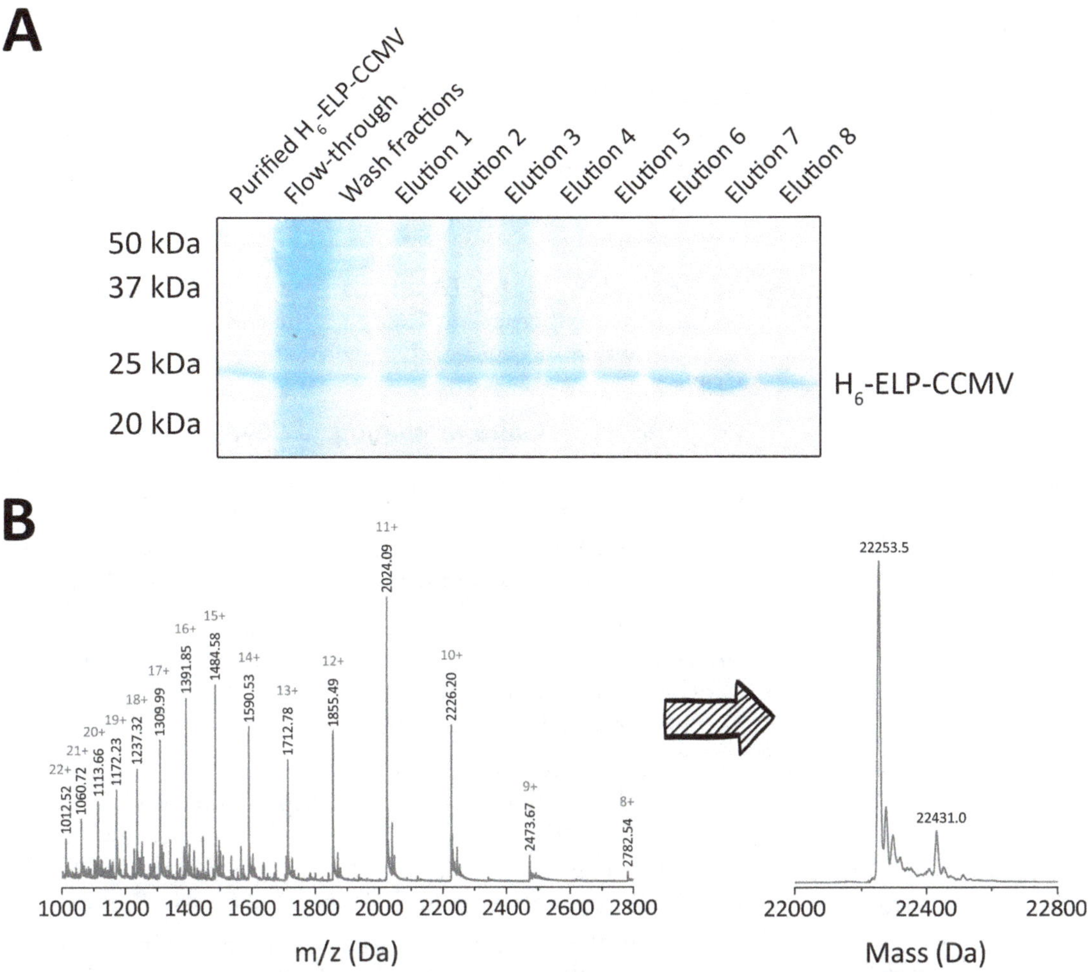

Fig. 2 (**a**) SDS-PAGE of the affinity purification of H_6-ELP-CCMV and purified H_6-ELP-CCMV. (**b**) ESI-TOF mass spectrometry of purified H_6-ELP-CCMV: multiply charged ion series (left) and deconvoluted total mass spectrum (right). The expected molecular weight is 22253.4 Da

50–100 mg/L of bacterial culture is expected for H_6-ELP-CCMV under these conditions.

18. Store the protein solution at 4 °C (*see* **Note 12**).

3.2 Expression and Purification of H_6-SrtA

1. Transform the pQE30-H_6-SrtA vector into *E. coli* BL21 AI competent cells using a standard heat-shock protocol and inoculate 1 mL LB medium containing 100 mg/L ampicillin. Grow at 37 °C for 1 h while shaking vigorously (200–250 rpm).
2. Transfer the cells into 4 mL of fresh LB medium containing 100 mg/L ampicillin and grow at 37 °C for 4 h while shaking vigorously (200–250 rpm).
3. Dilute the preculture into 500 mL of TB medium containing 100 mg/L ampicillin and grow at 37 °C for 24 h while shaking vigorously (200–250 rpm).

4. Harvest the cells by centrifugation (4000 × *g*, 4 °C, 15 min) (*see* **Note 5**).
5. Resuspend the cells in 15 mL of cold lysis buffer.
6. Lyse the cells by ultrasonic disruption (3 times 30 s) (*see* **Notes 6** and **7**).
7. Centrifuge the lysate to pellet the cellular debris (15,000 × *g*, 4 °C, 15 min).
8. Add 750 μL of Ni^{2+}-NTA agarose beads resuspended in lysis buffer to the supernatant. Incubate the mixture at 4 °C for at least 2 h while mixing gently to allow binding of His_6-tagged ELP-CCMV to the beads.
9. Load the mixture into a column and flow the lysate through the column.
10. Wash the beads 2 times with 10 mL of cold wash buffer.
11. Elute H_6-SrtA with cold elution buffer (first time with 0.5 mL to remove residual wash buffer, then 6 times with 1.5 mL).
12. Combine the column fractions containing the desired protein (*see* **Note 8**). Dialyze the protein to Sortase buffer at 4 °C to remove excess imidazole, using a centrifugal filter with a 10 kDa molecular weight cutoff.
13. Verify the purity of the protein by SDS-PAGE analysis. Measure the protein mass by electrospray ionization time-of-flight (ESI-TOF) mass spectrometry to validate the expected molecular mass of H_6-SrtA. The expected SDS-PAGE and ESI-TOF mass spectrometry results are shown in Fig. 3.
14. Determine the protein yield using the molar extinction coefficient at 280 nm (15,930 M^{-1} cm^{-1}) [8]. A yield of 10–12.5 mg/L of bacterial culture is expected for H_6-SrtA under these conditions.
15. Store the protein solution as aliquots at −20 °C or −80 °C in the presence of glycerol (12%).

3.3 Expression and Purification of CalB-LPETG-H_6

1. Transform the pET-22b-CalB-LPETG-H_6 vector into *E. coli* BLR(DE3)pLysS competent cells and plate onto an LB agar plate containing 100 mg/L ampicillin and 50 mg/L chloramphenicol. Incubate at 37 °C for 16–24 h.
2. Use a single colony to inoculate 50 mL of LB medium containing 100 mg/L ampicillin and 50 mg/L chloramphenicol. Grow at 37 °C for 16–24 h while shaking vigorously (200–250 rpm) (*see* **Note 4**).
3. Dilute the overnight culture into 1 L of 2× TY medium containing 100 mg/L ampicillin and grow at 37 °C while shaking vigorously (200–250 rpm) until a culture density of OD_{600} of 0.4–0.6 is reached (approximately 1–2 h).

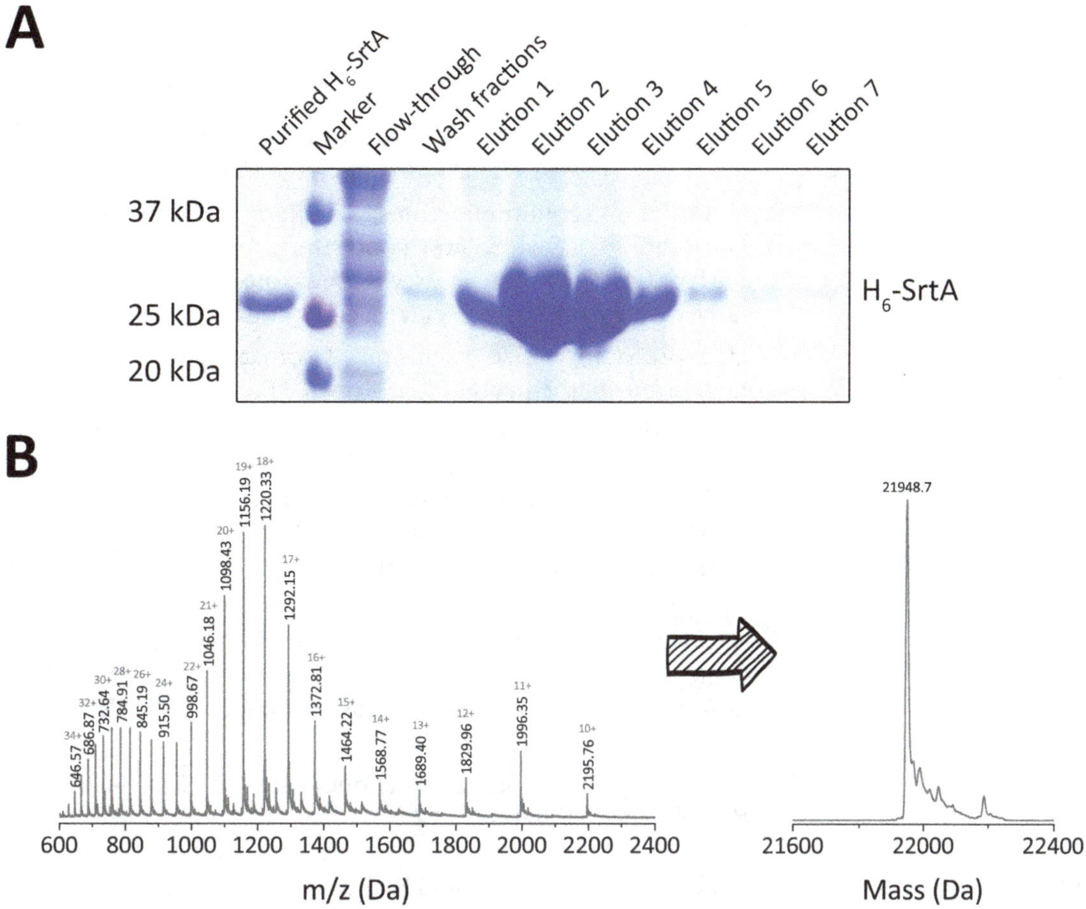

Fig. 3 (**a**) SDS-PAGE of the affinity purification of H_6-SrtA and purified H_6-SrtA. (**b**) ESI-TOF mass spectrometry of purified H_6-SrtA: multiply charged ion series (left) and deconvoluted total mass spectrum (right). The expected molecular weight is 21947.5 Da

4. Induce protein expression by the addition of IPTG (final concentration 1 mM). Incubate the culture at 30 °C for 4–6 h while shaking vigorously (200–250 rpm).
5. Harvest the cells by centrifugation (4000 × *g*, 4 °C, 15 min) (*see* **Note 5**).
6. Resuspend the cells in 25 mL of cold lysis buffer.
7. Lyse the cells by ultrasonic disruption (3 times 30 s) (*see* **Notes 6** and 7).
8. Centrifuge the lysate to pellet the cellular debris (15,000 × *g*, 4 °C, 15 min).
9. Add 1.5 mL of Ni^{2+}-NTA agarose beads resuspended in lysis buffer to the supernatant. Incubate the mixture at 4 °C for at least 1 h while mixing gently to allow binding of His_6-tagged CalB-LPETG to the beads.

10. Load the mixture into a column and flow the lysate through the column.
11. Wash the beads 2 times with 10 mL of cold wash buffer.
12. Elute CalB-LPETG-H_6 with cold elution buffer (first time with 0.5 mL to remove residual wash buffer, then 5 times with 1.5 mL).
13. Combine the column fractions containing the desired protein (*see* **Note 8**). Dialyze and concentrate the protein to CalB buffer at 4 °C to remove excess imidazole, using a centrifugal filter with a 10 kDa molecular weight cutoff.
14. Use preparative size exclusion chromatography (SEC) to further purify the protein (*see* **Note 13**).
15. Verify the purity of the protein by SDS-PAGE analysis. Measure the protein mass by electrospray ionization time-of-flight (ESI-TOF) mass spectrometry to validate the expected molecular mass of CalB-LPETG-H_6. The expected SDS-PAGE and ESI-TOF mass spectrometry results are shown in Fig. 4.
16. Determine the protein yield using the molar extinction coefficient at 280 nm (39,970 M^{-1} cm^{-1}) [8]. A yield of 2–4 mg/L of bacterial culture is expected for CalB-LPETG-H_6 under these conditions.
17. Store the protein solution as aliquots at −20 °C or −80 °C in the presence of glycerol (12%).

3.4 SrtA-Mediated Coupling Experiment

1. Dialyze H_6-*E*LP-CCMV, H_6-SrtA and CalB-LPETG-H_6 to Sortase buffer at 4 °C, using centrifugal filters with a 10 kDa molecular weight cutoff (*see* **Note 14**). Centrifuge three times for 10 min at 13,000 rpm (15,928 × *g*) in a microcentrifuge (*see* **Note 15**).
2. Determine the protein concentrations using the respective molar extinction coefficients [8].
3. Add all components together toward a final concentration of 50 μM of H_6-SrtA, 50 μM of H_6-ELP-CCMV, and 50 μM of CalB-LPETG-H_6 in Sortase buffer (*see* **Note 16**).
4. Incubate the solution at 21 °C for 3–24 h at 400 rpm in a thermoshaker (*see* **Note 17**).
5. Follow the reaction progress using SDS-PAGE analysis (*see* **Note 18**). An example of a SrtA-mediated coupling reaction of CalB-LPETG-H_6 to H_6-ELP-CCMV, followed by SDS-PAGE, is shown in Fig. 5.

3.5 pH-Induced Capsid Assembly and SEC Purification after SrtA Reaction

1. Dialyze a SrtA reaction mixture to the desired capsid buffer (pH 5.0 capsid buffer or pH 7.5 capsid buffer) at 4 °C to assemble CCMV capsids, using centrifugal filters with a

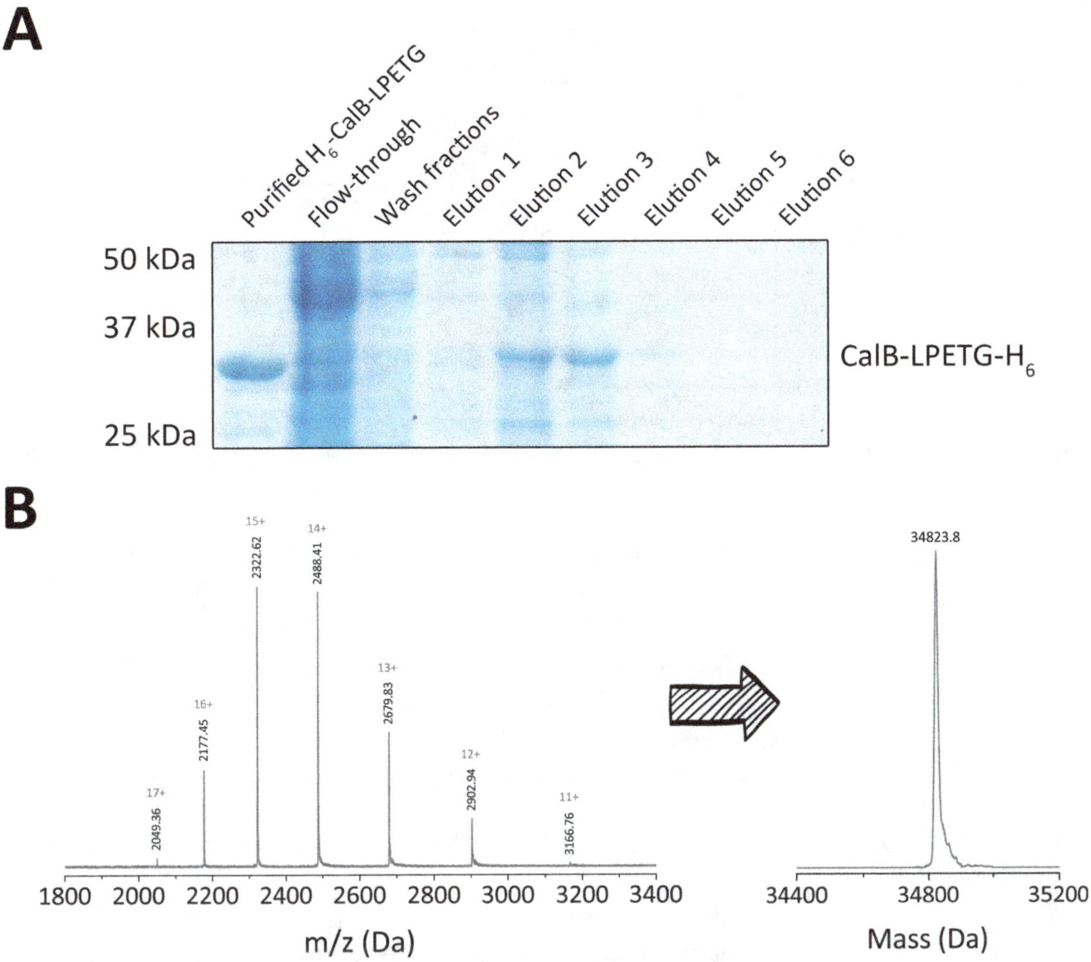

Fig. 4 (**a**) SDS-PAGE of the affinity purification of CalB-LPETG-H_6 and purified CalB-LPETG-H_6. (**b**) ESI-TOF mass spectrometry of purified CalB-LPETG-H_6: multiply charged ion series (left) and deconvoluted total mass spectrum (right). The expected molecular weight is 34824.3 Da

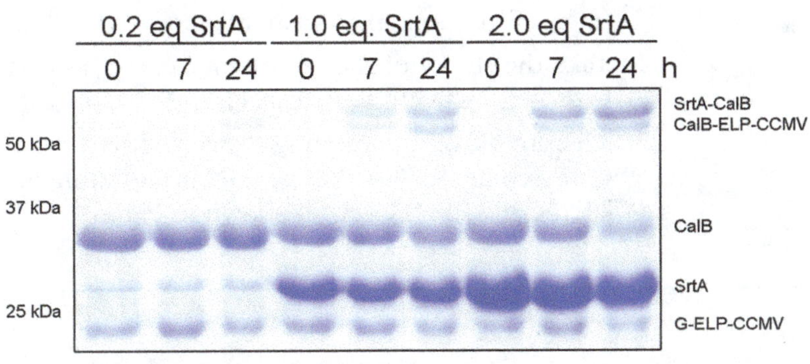

Fig. 5 SDS-PAGE analysis of the SrtA-mediated coupling of CalB-LPETG-H_6 to H_6-ELP-CCMV, using different equivalents of H_6-SrtA. The gel was stained by Coomassie blue staining. Reproduced from Schoonen et al. 2016 with permission from The Royal Society of Chemistry [6]

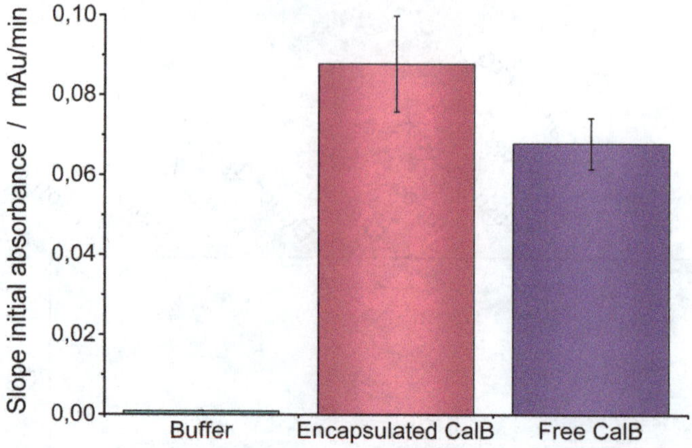

Fig. 6 Hydrolytic activity in the hydrolysis of *p*-NPA to *p*-nitrophenol of equal concentrations of encapsulated and free CalB-LPETG

10 kDa molecular weight cutoff (*see* **Note 19**). Centrifuge three times for 10 min at 13,000 rpm (15,928 × *g*) in a microcentrifuge.

2. Use preparative size exclusion chromatography (SEC) to isolate the CCMV capsids (*see* **Note 20**).
3. Use SDS-PAGE analysis to determine the CalB-loading in the capsids of the different column fractions (*see* **Note 21**).

3.6 Activity Assay with Encapsulated CalB

1. Pipette 0.5 μL of the *p*-NPA stock solution into the wells of a 96-wells plate.
2. Add a column fraction containing ELP-CCMV capsids with encapsulated CalB to the *p*-NPA substrate (in triplicate) (*see* **Note 22**). As a control, either buffer or free CalB-LPETG-H_6 (in the same concentration as in the capsids) can be added to the substrate.
3. Monitor the formation of *p*-nitrophenol by measuring the absorbance at 405 nm for 5 min at a constant temperature (e.g., 25 °C) in a microplate reader.
4. Take the slopes of the absorbance curves as a measure of the hydrolytic activity of the capsids.
5. An example of the resulting hydrolytic activity of encapsulated CalB, compared to the same concentration of nonencapsulated CalB-LPETG-H_6, is shown in Fig. 6.

4 Notes

1. The reader is referred to the original literature for the nature of the H_6-ELP-CCMV expression plasmid and details on the construction of the pET-15b-H_6-ELP[$V_4L_4G_1$–9]-CCMV(ΔN26) plasmid [7]. Gene ID wild-type CCMV

sequence: 962152. In the plasmid used, the first 26 N-terminal amino acids were removed and replaced by an ELP consisting of nine pentapeptides. A hexahistidine tag was placed at the N-terminus of the resulting construct for purification purposes.

2. The reader is referred to the original literature for the nature of the H_6-SrtA expression plasmid and details on the construction of the pQE30-H_6-SrtA plasmid [9]. Gene ID wild-type *Staphylococcus aureus* Sortase A: 3921273. In the plasmid used, a hexahistidine tag was placed at the N-terminus of the protein for purification purposes.
3. The reader is referred to the original literature for the nature of the CalB-LPETG-H_6 expression plasmid and details on the construction of the pET-22b-CalB-LPETG-H_6 plasmid [6]. PDB ID wild-type CalB: 4K6G. In the plasmid used, an LPETG recognition sequence was placed at the C-terminus of the protein, to allow coupling using SrtA. A hexahistidine tag was placed at the C-terminus of the resulting construct for purification purposes.
4. To avoid having to do the transformation each time before the expression, a glycerol stock can be made using the overnight culture. To do so, a small volume of the overnight culture should be diluted to 15–30% glycerol and stored at −80 °C. The stock is now stable for years, as long as it is stored at −80 °C. During the next expression, the transformation can be skipped and the overnight culture can be inoculated using a sterile pipette tip, which can be used to scrape off some of the glycerol stock.
5. Cell pellets can be lysed directly or stored at −20 °C for at least a couple of days. If the cell pellets are not used directly, thaw the cell pellets on ice for 15 min before resuspending them in lysis buffer.
6. Because heat is being produced during the ultrasonic disruption, the cell suspension should be kept on ice during this process. In between the cycles, time should be taken for the suspension to cool down (at least 5–10 s).
7. The number and length of the ultrasonic disruption cycles depend on the output of the machine being used. Generally, the lysis should be continued until the suspension is homogeneous and nonviscous.
8. The protein concentration and purity (A260/A280 ratio) of the column fractions can be measured by for instance a NanoDrop spectrophotometer. A value of approximately 0.6 for the ratio of A260/A280 is expected for a solution of pure protein. A higher ratio could indicate nucleic acid contamination.

9. In order to prevent interactions between nucleic acids and the CCMV capsid proteins, a relatively high NaCl concentration of 1.3 M is used in all buffers during protein purification.
10. After the last overnight dialysis step, some protein precipitation is often observed. This can be minimized by including some more diluted column fractions, thereby lowering the overall protein concentration during dialysis. A maximum protein concentration of approximately 5 mg/mL during dialysis is recommended.
11. For SEC analysis of CCMV capsid assembly, a Superose 6 10/300 GL column from GE Healthcare is recommended, with pH 5.0 capsid buffer as the eluent. For TEM analysis of the capsids, formvar/carbon-coated grids (FCF200-Cu) from EMS are normally used.
12. Always store H_6-ELP-CCMV in the assembled state in pH 5.0 capsid buffer. In this buffer, the protein can be stored for months at 4 °C. The protein should be not stored in pH 7.5 buffer, as it is not stable for prolonged periods of time in the dimer state. The protein is not stable upon freezing; thus protein aliquots cannot be stored at −20 °C or −80 °C.
13. For preparative SEC of CalB-LPETG, a Superdex 75 PC 10/300 column from GE Healthcare is recommended, with CalB buffer as the eluent.
14. Of course, other types of substrates, containing an LPXTG-peptide can also be used as a substrate in the SrtA reaction. It is, however, required that the bond C-terminally of the LPXTG-motif is an amide. So, either a C-terminal amide peptide should be used, i.e., R-LPXTG-NH_2, or one or more additional amino acids should be present C-terminally of the LPXTG motif.
15. Usually, some protein material is lost during the centrifugation steps. Therefore, it is recommended to start with approximately twice the amount of protein needed, so that enough protein material is left after centrifugation for the experiment.
16. As the sortase reaction is an equilibrium, a high catalyst loading is required to achieve a significant conversion. The H_6-SrtA concentration can be tuned, however, depending on the specific application. As a control, a sample can be prepared without any H_6-SrtA.
17. The reaction time can be optimized, depending on the desired conversion and amount of cargo encapsulation. The amount of H_6-SrtA added to the reaction can also be varied for this purpose.
18. Quench the SDS-PAGE samples with EDTA (1 μL of a 5 mM EDTA solution in Milli-Q water, per 5 μL reaction mixture), to remove the Ca^{2+} ions from the solution. This will inhibit the activity of H_6-SrtA.

19. pH-induced assembly will lead to $T = 3$ capsids and salt-induced assembly will lead to $T = 1$ capsids. In the latter case, capsid assembly will be more efficient due to the additional interactions of the ELPs, compared to pH-induced assembly. The assembly will be most efficient if a high protein concentration is achieved during spin filtration, as this will aid in the assembly process of the partially modified CCMV capsid proteins. If no capsids are observed after dialysis to pH 5.0 capsid buffer, it might help to start with more protein material (thus a higher reaction volume), as the concentration factor during spin filtration will then be higher.
20. For preparative SEC of CCMV capsids, a Superose 6 10/300 column from GE Healthcare is recommended, with either pH 5.0 capsid buffer or pH 7.5 capsid buffer as the eluent. Product elution can best be followed by the protein absorption at 280 nm.
21. In order to determine the CalB loading in a capsid fraction using SDS-PAGE, the capsid fractions can be loaded on an SDS-PAGE gel, together with a calibration curve of H_6-ELP-CCMV. This calibration curve can then be used to calculate the amount of H_6-ELP-CCMV in the capsid fraction, which in turn can be used to calculate the amount of CalB encapsulated in the capsids.
22. If the capsid concentration in the column fractions is very low, it is recommended to combine some column fractions and concentrate those by spin filtration.

Acknowledgments

The Ministry of Education, Culture, and Science (Gravitation program 024.001.035) is acknowledged for financial support.

References

1. Schoonen L, van Hest JCM (2014) Functionalization of protein-based nanocages for drug delivery applications. Nanoscale 6:7124–7141
2. Caspar DLD, Klug A (1962) Physical principles in the construction of regular viruses. Cold Spring Harb Symp Quant Biol 27:1–24
3. Speir JA, Munshi S, Wang G et al (1995) Structures of the native and swollen forms of cowpea chlorotic mottle virus determined by X-ray crystallography and cryo-electron microscopy. Structure 3:63–78
4. Schoonen L, Pille J, Borrmann A et al (2015) Sortase A-mediated N-terminal modification of cowpea chlorotic mottle virus for highly efficient cargo loading. Bioconjug Chem 26:2429–2434
5. Clancy KW, Melvin JA, McCafferty DG (2010) Sortase transpeptidases: insights into mechanism, substrate specificity, and inhibition. Biopolymers 94:385–396
6. Schoonen L, Nolte RJM, van Hest JCM (2016) Highly efficient enzyme encapsulation in a protein nanocage: towards enzyme catalysis in a cellular nanocompartment mimic. Nanoscale 8:14467–14472
7. van Eldijk MB, Wang JC-Y, Minten IJ et al (2012) Designing two self-assembly mechanisms into one viral capsid protein. J Am Chem Soc 134:18506–18509
8. Gill SC, von Hippel PH (1989) Calculation of protein extinction coefficients from amino acid sequence data. Anal Biochem 182:319–326
9. Ton-That H, Liu G, Mazmanian SK et al (1999) Purification and characterization of sortase, the transpeptidase that cleaves surface proteins of Staphylococcus aureus at the LPXTG motif. Proc Natl Acad Sci 96:12424–12429

Chapter 7

Synthesis of Cross-Linked 2-Aminoethyl Methacrylate in P22 Viral Capsid via Atom-Transfer Radical Polymerization

Shefah Qazi

Abstract

Virus-like particles (VLPs) resemble viruses, but are devoid their genetic material, rendering them as noninfectious, hollow protein shells. VLPs are ideal templates to synthesize nanoparticles because they have homogeneous size and their empty cavity can provide a confined environment for selectively directed synthesis. Atom-transfer radical polymerization (ATRP) is well suited for directed synthesis of polymers inside VLPs. In addition to being rapid, monomer-promiscuous, and resulting in products with relatively low polydispersity, the simplicity of the ATRP initiator allows it to be readily modified for amending to biomolecules. This chapter describes the polymerization of 2-aminoethyl methacrylate (AEMA) via ATRP in a viral capsid derived from the bacteriophage P22.

Key words P22 bacteriophage, Virus-like particles (VLPs), Atom transfer radical polymerization (ATRP), Protein scaffolds, Nanoparticles

1 Introduction

The use of protein–polymer composite materials for medical and materials applications is a growing field that aims to take advantage of the exquisite monodispersity and bioactivity of biomolecules, and also impart new material properties via polymer conjugation. A simple route to achieve an anchored, addressable polymer on a protein is via atom transfer radical polymerization (ATRP), which allows for continuous polymerization of simple monomers from an easily modified initiator.

The interior of bacteriophage P22 virus-like particles (P22) provides a scaffold upon which ATRP can be achieved, resulting in addressable cross-linked strands with high labeling capacity [1]. To effectively direct polymer growth to the virus-like particle (VLP) interior, an internal cysteine-containing viral coat protein, $P22_{S39C}$, is modified with a thiol reactive ATRP initiator [2, 3] to be used as a macro initiator for ATRP. The location of the initiator on the

Andrew K. Udit (ed.), *Protein Scaffolds: Design, Synthesis, and Applications*, Methods in Molecular Biology, vol. 1798, https://doi.org/10.1007/978-1-4939-7893-9_7,

protein cage is critical to achieve confined polymerization inside the cavity. According to the available P22 structural models, Ser^{39} is expected to be exposed to the interior cavity [4] and is therefore suitable for attachment of the ATRP initiator [1, 5]. ATRP is therefore initiated from a genetically introduced cysteine position 39 on the coat protein, $P22_{S39C}$. The protein shell acts as a barrier, limiting polymer size and confining growth to the interior of P22, leaving only the protein exposed to the bulk solution.

Selecting a monomer rich in primary amines results in a P22-confined polymer which acts as a scaffold for the attachment of primary amine-reactive molecules of interest, thus complementing the polyvalency of the scaffold (Fig. 1). One such example is labeling the scaffold with Gd-DTPA-NCS, a common small molecule contrast agent. The introduced polymer scaffold results in a significantly increased number of labeling sites per VLP compared to other VLP-based systems. Consequently, the contrast offered per particle, which is directly correlated to the number of GD-DTPA-NCS bound per particle of P22, is also significantly improved. The improvement in labeling is beneficial for the delivery of materials on a per-particle basis, allowing for high concentration delivery of molecules of interest. Due to the simplicity, modular nature, and loading level of the ATRP based approach taken to make these P22–polymer internal composites,

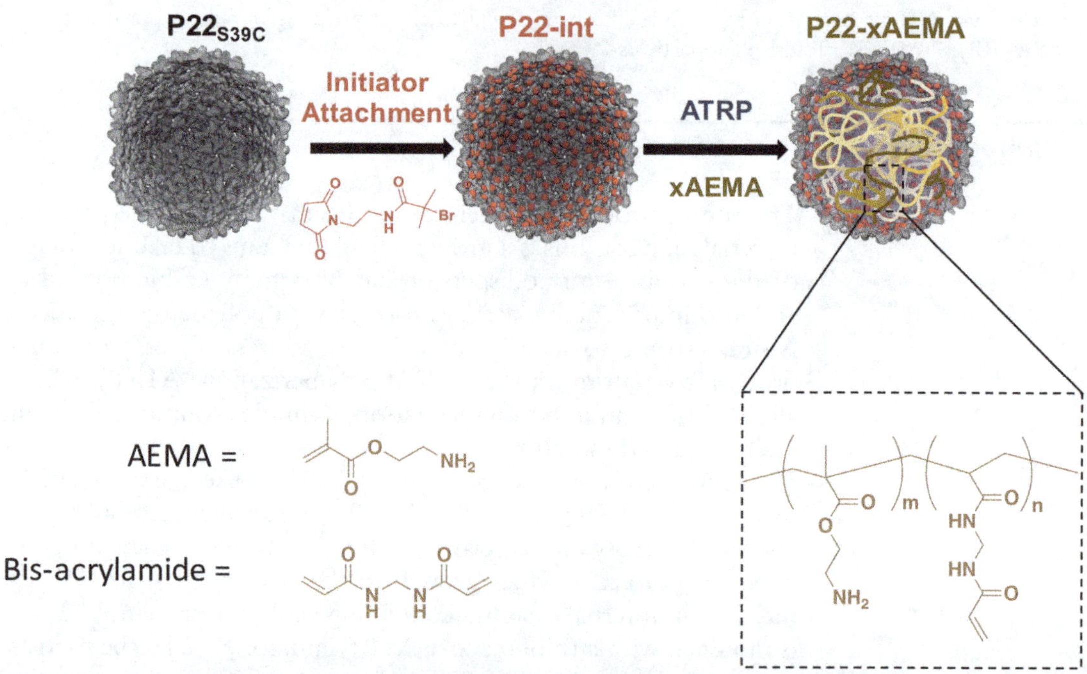

Fig. 1 Reaction scheme for the synthesis of P22-xAEMA. P22 is modified with a cysteine reactive ATRP initiator, which is subsequently used as a macro-initiator for polymer growth on the inside of the P22 virus-like particle (VLP) using a mixture of 2-aminoethyl methacrylate (AEMA) and bis-acrylamide monomers

this same method can be employed to make a range of novel VLP–polymer composites with biomedical and catalytic applications.

There are three forms of P22, procapsid (PC), expanded (EX), and wiffleball (WB). The procapsid form (PC) of $P22_{S39C}$ is heterologously expressed in and purified from *E. coli*. The PC of $P22_{S39C}$ is transformed to the expanded form (EX) in preparation for labeling. The protocol herein utilizes the expanded form which has more internal volume and therefore a higher loading capacity [1].

2 Materials

Prepare all solutions using ultrapure water (prepared by purifying deionized water to attain a sensitivity of 18 MΩ cm at 25 °C).

2.1 Cloning, Expression, Purification of $P22_{S39C}$ EX

1. Plasmid pET11a (Novagen) containing P22 gene 5 (coat protein) with S39C mutation and gene 8 (scaffolding protein) (*see* **Note 1**) [6]. Store at −20 °C.
2. BL21 (DE3) *E. coli.*
3. Ampicillin: 100 mg of the sodium salt of ampicillin dissolved in 1 mL of H_2O. Store at −20 °C.
4. LB Agar plates: Add 10 g of Bacto tryptone, 5 g of yeast extract, and 10 g of NaCl to 800 mL of H_2O. Adjust the pH to 7.5 with NaOH. Add 15 g of agar and melt the agar into solution in the microwave. Adjust the volume to 1 L with H_2O. Sterilize by autoclaving. Cool to 50 °C, add ampicillin to 50 μg/mL, and pour into plates.
5. LB Broth: Add 10 g of Bacto tryptone, 5 g of yeast extract, and 10 g of NaCl to 800 mL of H_2O. Adjust the pH to 7.5 with NaOH. Adjust the volume to 1 L with H_2O. Sterilize by autoclaving.
6. Isopropyl β-D-1-thiogalactopyranoside (IPTG).
7. Centrifuges and rotors: Thermo Scientific Sorvall Legend X1 centrifuge, TX-400 Swinging Bucket Rotor, Fiberlite F15-8x50cy Fixed Angle Rotor, Sorvall WX 80+ Ultracentrifuge, Fiberlite F50L-8x39 Fixed Angle Rotor, and Cole-Parmer Eppendorf Microcentrifuge 24-Tube Rotor (*see* **Note 2**).
8. Phosphate buffered saline (PBS): 100 mM sodium phosphate, 50 mM sodium chloride. Weigh 31.2 g of sodium phosphate monobasic dihydrate and 5.84 g of sodium chloride and add to 1900 mL of water. Make three such mixtures. Mix and adjust the pH with NaOH to target pH's (7.0, 7.6, and 8.0). Make each up to 2 L with water. Filter through a 0.22 μm membrane filter. Store at room temperature.
9. DNAse.

10. RNAse.
11. Lysozyme.
12. Sonicator.
13. 35% w/v sucrose solution.
14. 2 M sodium hydroxide.
15. 0.5 M guanidine–HCl: Weigh 47.8 g of guanidine–HCl and add to 900 mL of PBS, pH 7.0. Dissolve and make up to 1.0 L with PBS pH 7.0.
16. Dialysis tubing (100 kDa MWCO).

2.2 Synthesis of x-AEMA in P22 Viral Capsid

1. 2-bromoisobutyryl aminoethyl maleimide: The ATRP initiator, 2-bromoisobutyryl aminoethyl maleimide, is synthesized by a modification of the procedures reported previously [7, 8]. Mix N-2-aminoethylmaleimide (250 mg, 0.98 mmol) with dry triethylamine (300 μL, 2.2 mmol) in 5 mL of dry dichloromethane on ice. Add 2-bromo-2-methylpropionylbromide (200 μL, 1.6 mmol) to the reaction dropwise. Allow the reaction to warm to room temperature and then extract three times with dichloromethane and water, followed by drying over anhydrous sodium sulfate. Purify the crude product using column chromatography (silica gel, 10% ethyl acetate in dichloromethane) and further characterize by NMR and MS (*see* **Note 3**).
2. Dimethylsulfoxide (DMSO).
3. 40 mM dithiothreitol (DTT).
4. 2-aminoethyl methacrylate hydrochloride (AEMA).
5. Bis-acrylamide.
6. 10 mL crimp-top vials and caps with septa.
7. Metal catalyst solution for ATRP: Degas ultrapure water in advance by gently bubbling argon gas through the solution for 30 min. Weigh 19.2 mg of CuBr (0.134 mmol), 29.9 mg of $CuBr_2$ (0.134 mmol), and 83.5 mg of 2,2′-bipyridine (0.535 mmol) into a 10 mL crimp-top vial, seal the vial, and degas with argon gas flow. After 30 min, add 10 mL of degassed water to the vial and sonicate for 5 min. Repeat as necessary until solution turns dark-brown to black in color (*see* **Note 4**). The solution is prepared just prior to the polymerization experiment with P22.
8. High salt buffer solution: 250 mM sodium sulfate, 100 mM sodium phosphate, pH 8.0. Weigh 35.5 g of sodium sulfate and 15.6 g of sodium phosphate monobasic dihydrate and add to 900 mL of water. Dissolve well and adjust to pH 8.0 with 2 M NaOH. Adjust the volume to 1.0 L with water. Filter

through a 0.22 μm membrane filter. Store at room temperature.

9. Apparatus and reagents for standard SDS-PAGE.

2.2.1 Multiangle Light Scattering (MALS)

1. WTC-0100S (Wyatt Technologies) size-exclusion column.
2. HPLC with UV-Vis detector.
3. MALS buffer: 50 mM sodium phosphate, 500 mM sodium chloride, pH 8.0.
4. Wyatt HELEOS MALS detector.
5. Wyatt Optilab rEX differential refractometer.
6. Astra 5.3.14 software (Wyatt Technologies).

2.2.2 Analytical Ultracentrifugation

1. XL-A ultracentrifuge with AN-60Ti rotor (Beckman Coulter).
2. Epon double-sector centerpiece centrifuge tubes (Beckman).
3. SEDFIT software.
4. Sednterp software.

3 Methods

3.1 Cloning, Expression, and Purification of $P22_{S39C}$ EX

1. Transform the plasmid into *E. coli* strain BL21 (DE3) and select for ampicillin resistance.
2. Grow transformed plasmid overnight in 1 mL of LB medium at 37 °C in the presence of ampicillin (100 μg/mL).
3. Transfer the 1 mL overnight culture into 1 L of LB medium with ampicillin (100 μg/mL) and grow at 37 °C.
4. Induce expression of the genes by adding IPTG to a final concentration of 0.5 mM once the cells reach mid log phase (OD_{600} = 0.6). Shake vigorously overnight.
5. Centrifuge the culture at 3700 × *g* for 20 min (Thermo Scientific Sorvall Legend X1 centrifuge, TX-400 Swinging Bucket Rotor).
6. Resuspend the cell pellet (~4 g) into PBS pH 7.0 (100 mL) and incubate with DNAse (20 μg/mL), RNAse (30 μg/mL), and lysozyme (15 μg/mL) for 30 min on ice.
7. Lyse the cells further by sonication on ice (40% duty cycle, 5 min three times with a 2 min cool down in between).
8. Remove the cell debris via centrifugation at 12,000 × *g* for 45 min (Thermo Scientific Sorvall Legend X1 centrifuge, Fiberlite F15-8×50cy Fixed Angle Rotor).
9. Load the supernatant (~25 mL) on a 35% (w/v) sucrose cushion (5 mL) and centrifuge at 209,000 × *g* for 50 min in an ultracentrifuge (Sorvall WX 80+ Ultracentrifuge, Fiberlite F50L-8×39 Fixed Angle Rotor).

10. Resuspend the resulting virus pellet in 5 mL of PBS pH 7.0, spin at 17,000 × *g* for 20 min in a microcentrifuge and collect the supernatant.
11. Ultracentrifuge the sample at 209,000 × *g* for 50 min to pellet the capsid. Resuspend the pellet in 0.5 M guanidine–HCl by rocking gently for 1–2 h at 4 °C. Ultracentrifuge the protein again at 209,000 × *g* for 50 min [6]. Repeat this cycle of pelleting and resuspension 4 times (*see* **Note 5**).
12. Dialyze the sample against PBS, pH 7.0 overnight.
13. Heat-treat the sample for 20 min at 65 °C [1, 9, 10].
14. Purify $P22_{S39C}$ EX by pelleting using an ultracentrifuge at 209,000 × *g* for 50 min followed by resuspension into PBS pH 7.6 to bring the $P22_{S39C}$ EX concentration to 5 mg/mL in preparation for protein labeling (*see* **Notes 6** and 7) [9].

3.2 Synthesis of x-AEMA in P22 Viral Capsid

1. Prepare the initiator solution by dissolving 2-bromoisobutyryl aminoethyl maleimide (4.63 mg, 0.016 mmol) in 400 μL of DMSO to a concentration of 40 mM.
2. Add 322 μL of the initiator solution (prepared above) to 24 mL of $P22_{S39C}$ EX (5 mg/mL) in PBS, pH 7.6 (this corresponds to a fivefold excess per P22 subunit). Gently shake the mixture to react for 3 h at room temperature.
3. After 3 h, quench the reaction with 322 μL of 40 mM DTT.
4. Pellet the $P22_{S39C}$ EX by ultracentrifugation at 209,000 × *g* (Fiberlite F50L-8×39 Fixed Angle Rotor) for 50 min, to remove any excess soluble initiator and DTT, and then resuspend the pellet in 1 mL of PBS, pH 7.6.
5. Dissolve 160 mg of 2-aminoethyl methacrylate (AEMA) and 40 mg of bis-acrylamide in 4 mL of PBS pH 8.0. Adjust back to pH 8 with 2 M NaOH and adjust the volume to 5.0 mL. Spin down the mixture for 1 min at 4000 × *g* to pellet and remove any aggregate.
6. Transfer 4 mL of the AEMA–bis-acrylamide solution to a 10 mL crimp-top vial and add 2.2 mL of PBS, pH 8. Add 1.2 mL of 8.0 mg/mL $P22_{S39C}$ EX with the initiator molecule prepared in **step 2** to this mixture. Seal the vial, followed by a cycle of pumping with a vacuum and back-filling with argon gas 4 times through the septum to deaerate the mixture (*see* **Note 8**).
7. Add 0.6 mL of the metal catalyst solution to the monomer–$P22_{S39C}$ EX mixture anaerobically (under N_2 or Ar) and maintain the temperature at 23 °C for the duration of the polymerization reaction (180 min) (*see* **Note 9**).
8. Quench the reaction by exposure to air and centrifuge for 5 min at 17,000 × *g* to remove any aggregates formed during polymerization.

9. Purify the $P22_{S39C}$ EX–polymer composite away from any unreacted monomer and the metal catalyst by pelleting the protein at 209,000 × *g* for 50 min by ultracentrifugation and resuspend in 1 mL of PBS, pH 8.
10. Add 2 mL of the high salt buffer and agitate gently for 3 h at 4 °C to remove any AEMA and bis-acrylamide monomer which might be non-covalently associated with the $P22_{S39C}$ EX.
11. Pellet the protein again by ultracentrifugation followed by resuspension in 1.0 mL PBS, pH 8.0.
12. Analyze P22–polymer composites using SDS-PAGE (*see* **Note 10**) [1, 5].

3.2.1 Multi-Angle Light Scattering

1. Run P22 samples over a WTC-0100S size-exclusion column at a flow rate of 0.7 mL/min using an HPLC in equilibrated with MALS buffer.
2. Detect elution profiles for mass analysis using a UV-Vis detector, a Wyatt HELEOS MALS detector and a Wyatt Optilab rEX differential refractometer (*see* **Notes 11** and **12**).
3. Use the elution-profile data to calculate the number-averaged molecular mass with Astra 5.3.14 software using a dn/dc for protein of 0.185 and a dn/dc for the polymer component of 0.153.

3.2.2 Analytical Ultracentrifugation

1. Perform sedimentation velocity experiments in an XL-A ultracentrifuge with AN-60Ti rotor.
2. Fill Epon double-sector centerpiece centrifuge tubes with 400 μL of sample and the corresponding reference buffer, respectively.
3. Centrifuge samples at 20 °C and 3500 × *g*.
4. Acquire absorbance data at 280 nm. Analyze the raw data using SEDFIT software, implementing the ls-g*(s) model. Input parameters such as buffer density and viscosity are determined using Sednterp software.

4 Notes

1. We obtained this plasmid ("assembler" plasmid) as a gift from Peter E. Prevelige. Standard mutagenesis techniques can be used to make the mutant, and standard cloning can be used to generate the plasmid. PDB ID for P22 gene 5 is 2XYY [4], for gene 8 it is 1GP8 [11].
2. Alternate ultracentrifuges and rotors: Beckman Coulter Allegra X-30 benchtop centrifuge (SX4400 swinging bucket rotor, F0630 fixed angle rotor), Beckman Coulter Optima XPN

ultracentrifuge (70Ti 8×39mL fixed angle rotor), Beckman Coulter Microfuge20 (FA241.5, 24 × 1.5/2.0 mL, aluminum rotor)

3. ^{1}H NMR (500 MHz, $CDCl_3$) δ = 1.89 (s, 6H, CH_3); 3.46 (dd, J = 5.5, 5.5, 2H, NCH_2); 3.72 (dd, J = 5.0, 1.5, 2H, NCH_2); 6.71 (s, 2H, CH_{vinyl}), 6.97 (s, NH). ^{13}C NMR δ = 32.54 (CH_3), 37.34 (NCH_2), 39.84 (NCH_2), 62.51 (C), 134.59 (CH_{vinyl}), 170.96 (CO), 172.74 (CO). Electrospray ionization mass spectrometry: *m/z* calculated 289.0188, 291.0167 (MH^+), found *m/z* 289.0192, 291.0179.
4. The catalyst solution should be dark brown to black in color. A slight blue color suggests insufficient deaeration and could cause poor polymerization.
5. The scaffolding protein of $P22_{S39C}$ procapsid (PC) is removed using 0.5 M guanidine–HCl to give $P22_{S39C}$ empty shell (ES). The removal of scaffold can be verified using SDS-PAGE on 10–20% gradient Tris–glycine gels. Protein can be detected by staining with Coomassie stain.
6. $P22_{S39C}$ EX runs differently on a native agarose gel compared to $P22_{S39C}$ PC and $P22_{S39C}$ ES. Transformation to $P22_{S39C}$ EX is verified on a 0.8% native agarose gel using 40 mM Tris base, 5 mM sodium acetate, 1 mM EDTA, pH 8.2 running buffer, run for 3 h at 65 V. Protein can be detected by staining with Coomassie Blue.
7. Capsid integrity can be verified using SDS-PAGE and agarose gel as described in **Notes 5** and **6**. Additionally, intact particles can be verified using transmission electron microscope (Leo 912ab) by negatively staining the sample with 1% uranyl acetate on Formvar carbon-coated grids, or by dynamic light scattering (diameter = 64 nm).
8. Tap the vial during the vacuum pumping cycle to remove air bubbles.
9. Cover entire vial with foil to avoid any photochemical reactions.
10. The degree of polymerization of P22-polymer composites can be qualitatively analyzed using SDS-PAGE on 10–20% gradient Tris–glycine gels. We get different amounts of polymerization from each initiation point on the $P22_{S39C}$ EX monomer. The result is streaking/higher bands with P22 polymer composite compared to $P22_{S39C}$ EX, indicating polymerization is occurring. Fluorescence imaging of the gels can also be carried out, with excitation at 488 nm and detection at 526 nm.
11. The elution profile for $P22_{S39C}$ EX and P22 polymer composites should overlap, indicating that particle size is not compromised and polymerization occurred on the capsid interior.
12. Overall we see a mass increase with P22 polymer composites but no difference in the size of the particles.

Acknowledgments

This work was supported with grants from the US Department of Energy, Office of Basic Energy Sciences, Division of Materials Science and Engineering DE-FG02-08ER46537 and the National Institutes of Health NIAID R01AI104905.

References

1. Lucon J, Qazi S, Uchida M et al (2012) Use of the interior cavity of the P22 capsid for site-specific initiation of atom-transfer radical polymerization with high-density cargo loading. Nat Chem 4:781–788
2. Heredia KL, Bontempo D, Ly T et al (2005) In situ preparation of protein-"smart" polymer conjugates with retention of bioactivity. J Am Chem Soc 127:16955–16960
3. Mantovani G, Lecolley F, Tao L et al (2005) Design and synthesis of N-maleimido-functionalized hydrophilic polymers via copper-mediated living radical polymerization: a suitable alternative to PEGylation chemistry. J Am Chem Soc 127:2966–2973
4. Chen DH, Baker ML, Hryc CF et al (2011) Structural basis for scaffolding-mediated assembly and maturation of a dsDNA virus. Proc Natl Acad Sci U S A 108:1355–1360
5. Lucon J, Edwards E, Qazi S et al (2013) Atom transfer radical polymerization on the interior of the P22 capsid and incorporation of photocatalytic monomer crosslinks. Eur Polym J 49:2976–2985
6. Prevelige PE, Thomas D, King J (1988) Scaffolding protein regulates the polymerization of P22 coat subunits into icosahedral shells in vitro. J Mol Biol 202:743–757
7. Mantovani G, Lecolley F, Tao L et al (2005) Design and synthesis of N-maleimido-functionalized hydrophilic polymers via copper-mediated living radical polymerization: a suitable alternative to PEGylation chemistry. J Am Chem Soc 127:2966–2973
8. Heredia KL, Bontempo D, Ly T et al (2005) In situ preparation of protein-"smart" polymer conjugates with retention of bioactivity. J Am Chem Soc 127:16955–16960
9. Kang S, Uchida M, O'Neil A et al (2010) Implementation of P22 viral capsids as nanoplatforms. Biomacromolecules 11: 2804–2809
10. Teschke CM, McGough A, Thuman-Commike PA (2003) Penton release from P22 heat-expanded capsids suggests importance of stabilizing penton-hexon interactions during capsid maturation. Biophys J 84:2585–2592
11. Sun Y, Parker MH, Weigele P et al (2000) Structure of the coat protein-binding domain of the scaffolding protein from a double-stranded DNA virus. J Mol Biol 297: 1195–1202

Chapter 8

Synthesis of Metal–Organic Frameworks on Tobacco Mosaic Virus Templates

Shaobo Li and Jeremiah J. Gassensmith

Abstract

Tobacco mosaic virus (TMV) has long been exploited as a robust biological scaffold for organic/inorganic modification owing to its anisotropic structure and chemically addressable amino acid residues on both the exterior and interior. We present the fabrication of a crystalline microporous metal–organic framework (MOF) shell on the exterior of TMV, which retains its rod-like morphology, and produces uniformly formed core–shell structures with high accessible surface area and pore volume. We also describe an exfoliation method that can recover the intact viral particle from the core–shell composite.

Key words Metal–organic frameworks, MOFs, Virus-like particles, VLPs, SURMOFs, Self-assembly, Biomimetic mineralization, Templated crystal growth, Tobacco mosaic virus, Virus nanotechnology, Chemical virology

1 Introduction

The preparation of synthetic anisotropic polymeric structures with near exact size control in the hundreds of nanometer range is a challenge hitherto unrealized. This is unfortunate, as soft anisotropic templates would be of considerable use in optics, nanoscale electronics, and biomedicine. While controlled assembly via supramolecular templating [1] has been able to produce rigid anisotropic rods many nanometers long and with zero polydispersity, the practicality of expanding these synthetic methods to the hundreds of nanometer regimes seems unlikely. Access to these seemingly privileged structures, however, is of little effort to nature. Tobacco mosaic virus (TMV, illustrated in Fig. 1a) is a rod-shaped virus composed of 2130 coat proteins that self-assemble around a single strand of RNA [2]. The self-assembled product is a tubular structure of approximately 300 × 18 nm with an inner longitudinal channel 4 nm in diameter. The coat protein is chemically addressable by diazonium coupling to surface exposed tyrosine residues on the exterior whilst the inner surface can be

Andrew K. Udit (ed.), *Protein Scaffolds: Design, Synthesis, and Applications*, Methods in Molecular Biology, vol. 1798, https://doi.org/10.1007/978-1-4939-7893-9_8,

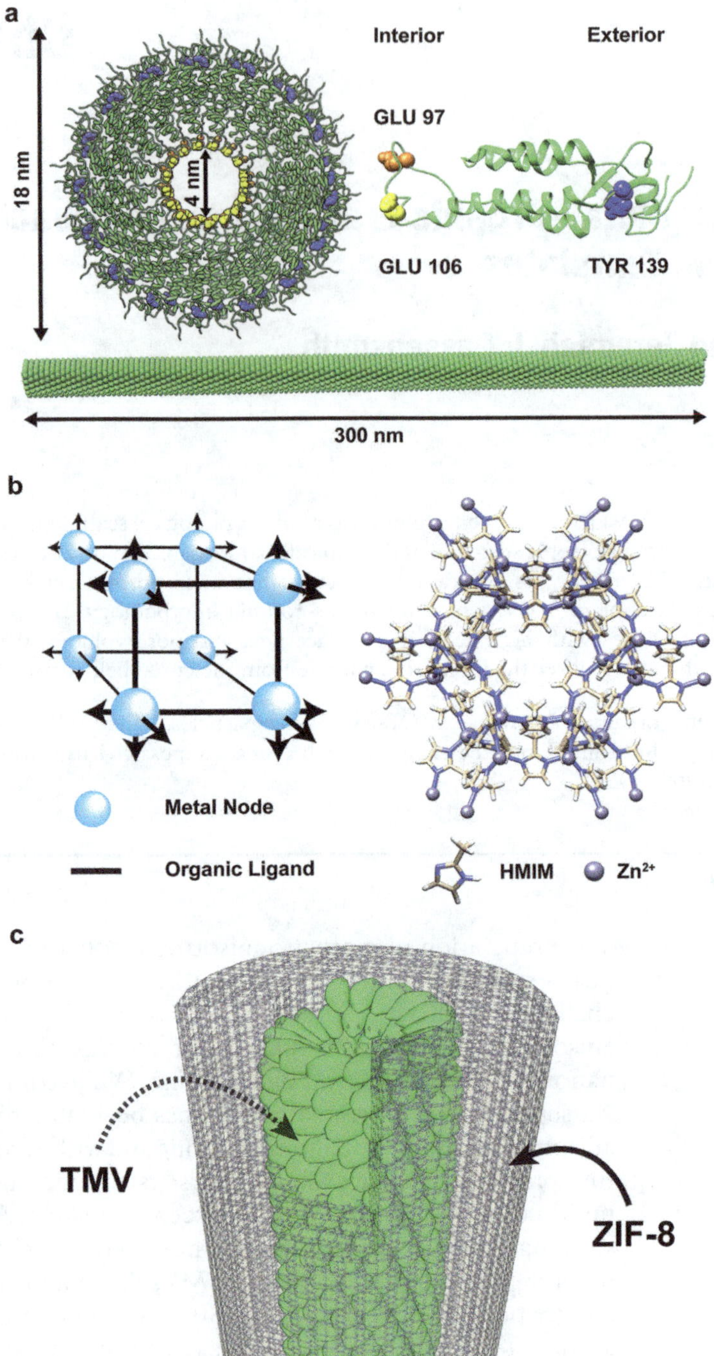

Fig. 1 (**a**) Simulated proteinaceous models that illustrate the morphology and structure of TMV: top-down view (left), single coat protein monomer (right), and integrated rod-like viral particle (bottom). (**b**) Illustration of MOFs. Left: a simplified representation of a typical MOF unit cell that repeats infinitely in all directions to form a porous matrix. Right: Crystal structure of the unit cell of zeolitic imidazolate framework-8 (ZIF-8), which is crystallized on the surface of TMV. (**c**) Schematic illustration of TMV@ZIF-8. Figure 1c is reproduced from Ref. [20] with permission from Wiley

modified using traditional amide coupling reactions to surface exposed glutamic acid residues [3]. This high aspect ratio tubular protein nanoparticle possesses extraordinary stability in a broad range of temperatures (up to 60 °C) and pH (2–10) [4]. Because these structures are templated by their RNA, it has been possible to precisely change the length of the viral rods via straightforward methods [5].

In the early 2000s, several pioneering studies demonstrated TMV's role as a versatile biological scaffold for directed synthesis of inorganic nanoparticles or hollow nanorods [4, 6–10]. Shenton et al. [4], for instance, investigated the capability of TMV to act as a template under broad mineralization conditions with various metal sources. They demonstrated that a uniform thin layer of SiO_2, PbS, CdS, and iron oxide nanoparticles could be coated on the outer surface of TMV at pH values ranging from 2.5 to 9. In another example, Dujardin et al. [7] used solutions of different pH to induce the formation of sub-10 nm metal nanoparticles selectively on either the exterior or interior surface of TMV by exploiting the differences in the pKa's of surface-exposed amino acids on either the outer or the inner surfaces. Inspired by these preliminary results, researchers have prepared metal/metal oxide-coated TMV nanorods and applied them in device fabrication and imaging applications [6, 9, 10].

Concurrent to and independent of the research focusing on metal/metal oxide deposition on viral nanoparticles, a new type of highly crystalline porous metal coordination polymer was being developed called "metal–organic frameworks" (Fig. 1b). Metal–organic frameworks (MOFs) possess three-dimensionally ordered channels composed of either micropores or mesopores and typically possess a high surface area [11]. Very recently, researchers have shown that it is possible to prepare MOFs under biocompatible conditions and have demonstrated that these advanced synthetic strategies allow MOF-coated biomaterials, such as enzymes, to retain their selectivity, catalytic efficiency, and stability even when the enzyme is buried within a single crystal of MOF [12–20]. Further, the MOF crystal affords incredible protection to the enzyme against proteolysis and denaturation as a result of exposure to harsh solvents or high temperatures [12].

From our initial perspective, therefore, the benefits for fabrication of MOFs on TMV were apparent: the porous materials have been widely employed in mass storage, mass transfer and catalysis applications [21–24]; the porous structure could protect the inlaid TMV from environmental attacks as well as leave space and pathways for surface exposed tyrosine residues for chemical modification under harsh synthetic organic conditions.

Here, we present the synthesis of MOF-encapsulated TMV bionanoparticles, which possesses a crystalline yet porous MOF shell that closely packs on the exterior of TMV [20] and is schematically shown in Fig. 1c. We demonstrated that zeolitic imidazolate framework-8 (ZIF-8) [25], a typical MOF with excellent thermal and chemical stability, can be homogeneously coated on the TMV and the thickness of the ZIF-8 layer can be tuned by changing the synthetic conditions. The as-obtained TMV@ZIF-8 bionanoparticles can be characterized by scanning electron microscopy (SEM), transmission electron microscopy (TEM), powder X-ray diffraction (PXRD), and gas sorption analysis. A key benefit of using a metal-based coordination polymer like a ZIF is that the coating is easily destroyed using a strong chelator like ethylenediamine tetraacetic acid (EDTA). This method returns the ZIF crystal to its small monomeric components neatly and rapidly. This allows us to reversibly capture and release the TMV from its protective shell. Herein we also describe the protocol for recovering the intact viral particles following exfoliation and dissolution of the ZIF coating with an aqueous solution of EDTA.

2 Materials

All chemicals are analytical grade and used without further purification. All aqueous solutions were prepared with ultrapure water (18.2 MΩ). All buffer solutions and TMV solutions are stored at 4 °C.

1. Tobacco seeds (*Nicotiana benthamiana*) from healthy plants.
2. Dedicated plant room at 25 °C, 65% humidity, and lighting period of 16 h.
3. 0.1 M Phosphate buffer: Dissolve 17.418 g of K_2HPO_4 in 100 mL ultrapure water to obtain a 1.0 M solution. Dissolve 13.609 g of KH_2PO_4 in 100 mL ultrapure water to obtain a 1.0 M solution. Mix 30.75 mL of the 1.0 M K_2HPO_4, 19.25 mL of the 1.0 M KH_2PO_4 mL and 450 mL ultrapure water to obtain a 0.1 M phosphate buffer pH 7.4. The pH of the buffer solution should be confirmed by pH meter. 0.01 M phosphate buffer can be prepared by diluting the 0.1 M solution.
4. Silicon carbide powder.
5. Extraction solution: Add 1 mL of β-mercaptoethanol to 499 mL of 0.1 M phosphate buffer to make a final volume of 500 mL. The concentration of β-mercaptoethanol is 0.2% (v/v).
6. Hamilton Beach Type B31 blender.
7. Cheesecloth.
8. Chloroform.

9. n-butanol.
10. Sodium chloride.
11. PEG-8000.
12. Triton X-100.
13. 40% (w/v) aqueous sucrose solution.
14. Thermo Scientific Sorvall Lynx 4000 Centrifuge or equivalent.
15. Thermo Scientific Sorvall wX+ Ultra Series Centrifuge or equivalent.
16. Fiberlite™ F10-4 × 1000 LEX Rotor.
17. Fiberlite™ F37 L-8 × 100 Fixed-Angle Rotor.
18. Beckman Coulter Microfuge 16 Centrifuge with FX241.5P Rotor.
19. G3000SW$_{XL}$ column on an Agilent 1100 series HPLC system.
20. Thermo Scientific NanoDrop 2000 Spectrophotometer.
21. GE Health Care PD midi Trap G-25 column.
22. ZIF-8 metal precursor solution: 20 mM zinc acetate in ultrapure water (*see* **Note 1**).
23. ZIF-8 ligand precursor solution: 400 mM 2-methylimidazole prepared in ultrapure water (*see* **Note 1**).
24. EDTA exfoliation solution (0.1 M, pH 7.0): Add 2.9224 g EDTA to 90 mL 0.1 M KOH aqueous solution, use a stir plate and magnetic stir bar to achieve best dissolution. Measure the pH of the as-prepared solution by pH meter and add 2.0 M HCl to adjust the pH to 7.0 if needed. Finally adjust the total volume to 100 mL with ultrapure water.
25. Negative staining solution: 2% (wt%) uranyl acetate in ultrapure water. The as-prepared solution should be stored at 4 °C (*see* **Note 2**).
26. Silicon wafer for scanning electron microscopy.
27. Acetone.
28. Methanol.
29. Double-sided copper tape for scanning electron microscopy.
30. Zeiss Supra 40 scanning electron microscope.
31. FCF200-Cu sample grid for transmission electron microscopy.
32. FEI Tecnai G2 Spirit Biotwin transmission electron microscope.
33. Bruker D8 Advance powder X-ray diffractometer.
34. Micromeritics ASAP 2020 surface area analyzer.

3 Methods

3.1 TMV Production and Purification

1. TMV is propagated in *N. benthamiana* in a dedicated plant room at 25 °C, 65% humidity and lighting period of 16 h. The tobacco plants are watered every other day (*see* **Note 3**).
2. When the size of the leaf is about 5 cm wide, conduct the infection on the day that the plants are supposed to be watered. Tap a small amount of silicon carbide with finger and sprinkle the powder onto the leaves. Add 30 μL of 0.17–0.27 mg/mL TMV in phosphate buffer to each leaf. Gently rub the leaf to spread the virus solution and let it enter the leaf (*see* **Note 4**).
3. Harvest the leaves after 7–10 days of infection. Weigh the leaves and store them in sealed plastic bag at −80 °C.
4. Pulverize the leaves by squeezing them in the frozen form. Homogenize the pulverized leaves in 250 mL of prechilled extraction buffer solution with a blender.
5. Collect the crude virus solution by filtering the homogenate through four layers of cheesecloth. Squeeze the cheesecloth to collect as much filtrate as possible.
6. Load the filtrate in centrifuge tubes. Balance the tubes with ultrapure water prior to centrifugation. Centrifuge at 10,500 rpm (11,800 × *g*) for 36 min at 4 °C with Fiberlite F10-4× 1000 LEX rotor.
7. Collect the supernatant via filtration with four layers of cheesecloth. Record the volume of the filtrate (*x* mL where *x* is the volume obtained).
8. To the filtrate, add ½(*x*) mL of chloroform and ½(*x*) mL of n-butanol to make a volume ratio ($V_{Aqueous}$: $V_{Organic}$) of 1:1. Stir the mixture for 30 min at 4 °C (adjust proper stirring speed to avoid foaming). Fill the centrifuge tubes again with the homogenized mixture and carefully balance the tubes before centrifugation.
9. Centrifuge at 6000 rpm (3871 × *g*) for 18 min at 4 °C with F10–4× 1000 LEX rotor.
10. Collect the crude TMV solution (top layer) via careful decantation. Keep the solution at 4 °C or on ice. Record the weight ($m_{virus\ solution}$) and volume ($V_{virus\ solution}$) of the crude virus solution.
11. To the crude virus solution, add NaCl to 0.2 M ($m_{NaCl} = 0.2$ mol/L × $V_{virus\ solution}$ × 55.44 g/mol), add PEG 8000 to 8% (w/w, $m_{PEG\ 8000} = m_{virus\ solution} \times 8\%$), add Triton X-100 surfactant to 1% (w/w, $m_{Triton\ x\text{-}100} = m_{virus\ solution} \times 1\%$). Stir at 4 °C for 30 min then store the solution for 1 h at the same temperature.
12. Load the solution into centrifuge tubes, carefully balancing the tubes with ultrapure water. Centrifuge at 10,500 rpm (11,800 × *g*) for 55 min at 4 °C with Fiberlite F10-4× 1000 LEX rotor.

13. Discard the supernatant and suspend the pellet in 20 mL phosphate buffer with the assistance of pipetting (*see* **Note 5**). Further homogenize the resuspension by gently stirring at 4 °C overnight.
14. To the centrifuge tube, add 60 mL of 40% sucrose solution. Carefully layer the suspended solution onto the top of the sucrose solution and fill the tube to the top. Two separate layers should be formed at this point.
15. Carefully balance the tubes with ultrapure water and centrifuge at 37,000 rpm (128,000 × g) for 160 min at 4 °C with Fiberlite F37 L-8× 100 rotor.
16. Discard the supernatant and suspend the pellet in 5–10 mL (*see* **Note 6**) of 0.01 M phosphate buffer overnight as described in **step 13**.
17. Split the suspended solution into 1.5 mL Eppendorf Microcentrifuge tubes. Centrifuge the solutions at 14,500 rpm (or 11,300 ×*g*) for 15 min in Beckman Coulter Microfuge 16 centrifuge.
18. Collect the supernatant as the purified TMV solution. Store the TMV solution at 4 °C.
19. The concentration of TMV is determined by UV-Vis measurement on NanoDrop spectrophotometer. The absorbance at 260 nm is used to calculate the concentration with Beer's law ($A = \varepsilon lc$) and $\varepsilon = 3.0$ mL mg^{-1} cm^{-1}. The ratio of absorbance at 260 nm and 280 nm (A260/A280) should be around 1.2.
20. The purity of TMV solution is evaluated by investigating the absorbance at 260 nm when employing size exclusion chromatography with G3000SW$_{XL}$ column on an Agilent 1100 series HPLC system.

3.2 Desalination of TMV Solution

1. Flush the GE Health Care PD midi Trap G-25 column with 4 mL of ultrapure water four times to prepare it for use.
2. Add 1 mL of virus solution into the column and wait for the solution to load onto the column. The flow-through can be discarded.
3. Add 1.5 mL of ultrapure water to the column and collect the eluate as desalted virus solution (*see* **Note 7**).

3.3 Preparation of Thin ZIF-8 Layer on TMV (TZ-Thin)

1. In a 1.5-mL microcentrifuge tube, add desalted TMV solution containing 0.0625 mg virus particles (*see* **Note 8**), followed by addition of 500 μL of the 400 mM HMIM aqueous solution and 500 μL of the 20 mM $Zn(OAc)_2$ aqueous solution (*see* **Note 9**). The reaction mixture will become turbid immediately.

2. Shake the tube by hand (horizontally back and forth) for 20 s. At this point a whitish flocculate can be seen clearly in the reaction solution.
3. Set the tube on the bench under ambient conditions overnight (approx. 16 h). The flocculates will sediment but the liquid will remain turbid over this reaction time.
4. Centrifuge the reaction mixture at 2656 ×*g* for 10 min to pellet the white sediment to the bottom of the tube.
5. Discard the clear supernatant gently with a pipette. Add 1000 μL of ultrapure water and suspend the flocculates via vortex.
6. Repeat **steps 4** and **5** to wash the product again. Finally, a suspension of TZ-thin in ultrapure water is obtained.

3.4 Preparation of Thick ZIF-8 Layer on TMV (TZ-Thick)

1. In a 20-mL scintillation vial, add desalted TMV solution that contains 0.111 mg virus particles (*see* **Note 8**), followed by three separate aliquots (3×) of 1000 μL of the 400 mM HMIM aqueous solution and three separate aliquots (3×) of 500 μL of the 20 mM $Zn(OAc)_2$ aqueous solution (*see* **Note 9**). The reaction mixture will turn turbid immediately.
2. Shake the vial (horizontally back and forth) for 20 s by hand and whitish flocculates should be visible at the bottom of the vial (*see* **Note 10**).
3. Set the vial on the bench under ambient conditions overnight (approx. 16 h). The flocculates sediment to the bottom but the liquid will stay turbid over the reaction time.
4. Transfer the reaction mixture into three 1.5-mL microcentrifuge tubes then centrifuge at 2656 × *g* for 10 min to pellet the white sediment to the bottom of the tube.
5. Discard the clear supernatant gently with a pipette. Add 1000 μL of ultrapure water to suspend the flocculates via vortex.
6. Repeat **steps 4** and **5** to wash the product again. Finally, a suspension of TZ-thick in ultrapure water is obtained.

3.5 Recover Virus Particles Via Exfoliation of ZIF-8 Shell

For best results, use at least 1 mg of TMV for TZ-thin/TZ-thick preparation and exfoliation. The exfoliation procedure described below was based on TZ-thin that was prepared using 1 mg of TMV.

1. Centrifuge and wash the as-prepared TZ-thin/TZ-thick as previously described in Subheading 3.3 (**steps 4–6**).
2. Add 1.0 mL of 1.0 M EDTA aqueous solution to the precipitate after the second wash. Vortex the mixture for 30 s to obtain a good suspension. Leave it on the benchtop.
3. After about 2 h the ZIF-8 should fully dissolve in the EDTA solution.

4. Add the as-obtained clear solution (1000 μL) to a preflushed GE Health Care PD midi Trap G-25 column.
5. Add 1.5 mL ultrapure water to the column and collect the eluate.
6. The concentration of recovered TMV is determined by UV-Vis measurement on NanoDrop. The absorbance at 260 nm is used to calculate the concentration with Beer's law ($A = \varepsilon lc$) and $\varepsilon = 3.0\ \text{mL mg}^{-1}\ \text{cm}^{-1}$. The ratio of absorbance at 260 nm and 280 nm (A260/A280) should be around 1.2.

3.6 Characterization

3.6.1 Scanning Electron Microcopy

1. Cut a silicon wafer into 3 mm × 5 mm chips. Clean the silicon chips in acetone with the assistance of sonication.
2. Place double-sided copper tape onto the SEM sample mount, then put the clean silicon chip on the copper tape.
3. Add 5 μL of TZ-thin/TZ-thick suspension onto silicon chip and wait for 20 s.
4. Remove the excess suspension with a piece of filter paper. Wait for 30 min to let sample air-dry.
5. Conduct SEM with a beam current of 2.5 kV and working distance of approximately 8 mm (Fig. 2a, c).

3.6.2 Transmission Electron Microscopy for TZ-Thin/TZ-Thick

1. Conduct glow-discharge for the FCF200-Cu TEM grids to prepare a clean and hydrophilic surface.
2. Add 5 μL of TZ-thin/TZ-thick suspension onto FCF200-Cu sample grid and wait for 30 s. Remove excess suspension by placing a piece of filter paper at the edge of the droplet.
3. Place the loaded TEM grid into the sample holder and conduct TEM characterization at 120 kV (Fig. 2b, d).

3.6.3 Transmission Electron Microscopy for Exfoliated TZ-Thin/TZ-Thick

1. Conduct glow-discharge for the TEM grids.
2. Add 5 μL of exfoliated TZ-thin/TZ-thick suspension onto FCF200-Cu sample grid and wait for 30 s. Remove excess suspension by placing a piece of filter paper at the edge of the droplet.
3. Add 5 μL of 2% uranyl acetate aqueous solution and wait for 30 s. Remove excess suspension by placing a piece of filter paper to the edge of the droplet. Discard filter paper in waste designated for heavy metals.
4. Place the loaded TEM grid into the sample holder and conduct TEM characterization at 120 kV (Fig. 3).

3.6.4 Powder X-Ray Diffraction (PXRD)

1. Wash freshly prepared TZ-thin/TZ-thick with methanol two times (the volume of methanol equals to the total volume of precursor solutions). Dry the product under high vacuum

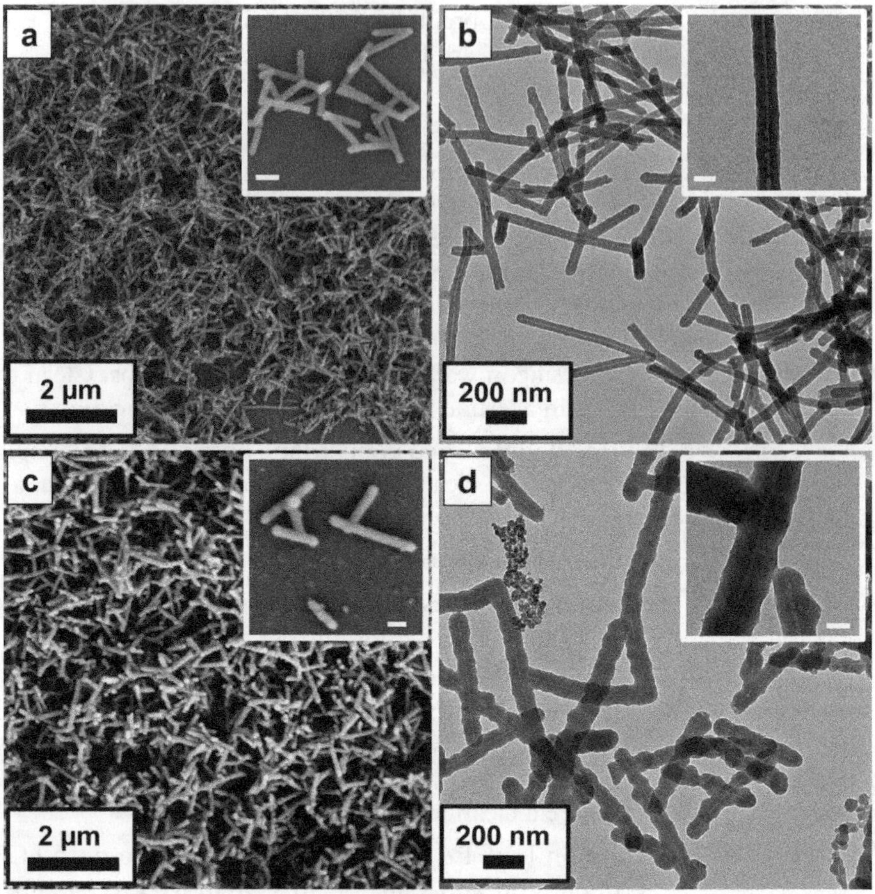

Fig. 2 Characterization of as-prepared TZ-thin in (**a**) SEM and (**b**) TEM; as-prepared TZ-thick in (**c**) SEM and (**d**) TEM. The inserted scale bar is 200 nm in (**a**) and (**c**); 50 nm in (**b**) and (**d**). Figure is reproduced from Ref. [20] with permission from Wiley

overnight (approx. 16 h). An amount of 20 mg of dried material should be used for PXRD characterization.

2. Grind the dry particulates prior to the PXRD measurement to obtain a fine powder without large granulated particles.
3. Conduct 2θ measurement from 5° to 80° with 0.02° step size (Fig. 4).

3.6.5 N_2 Sorption Analysis

1. Measure and record the mass of an empty sample container.
2. Load the TZ-thin/TZ-thick dry powder and measure the total mass. The loaded amount of sample should be at least 20 mg.
3. Heat the sample at 120 °C under vacuum overnight (approx. 16 h) to activate ZIF-8.

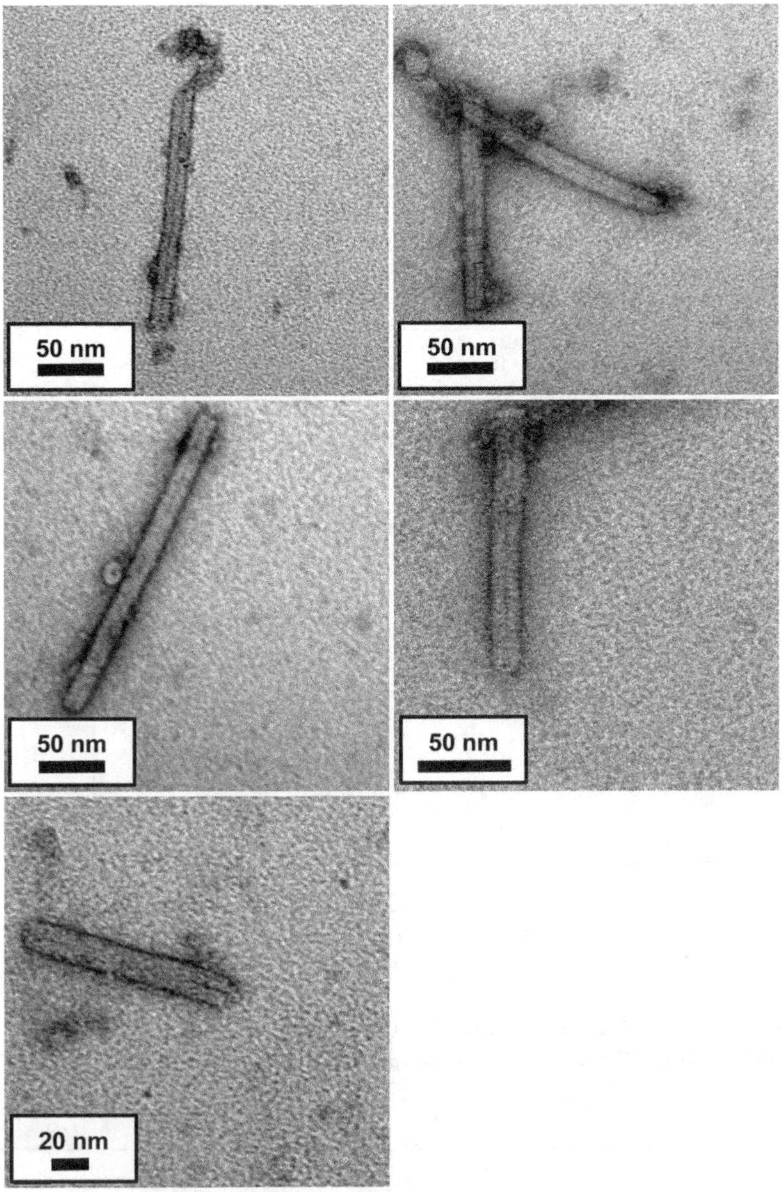

Fig. 3 TEM characterization of exfoliated TZ-thick. The sample is stained by 2% uranyl acetate solution. Figure is reproduced from Ref. [20] with permission from Wiley

4. Measure the total mass again after heating and use that value to calculate sample mass.
5. Conduct N_2 sorption analysis at 77 K in a liquid nitrogen bath.
6. Calculate Brunauer–Emmett–Teller (BET) surface area with P/P_0 range of 0.01–0.1 (Fig. 5).

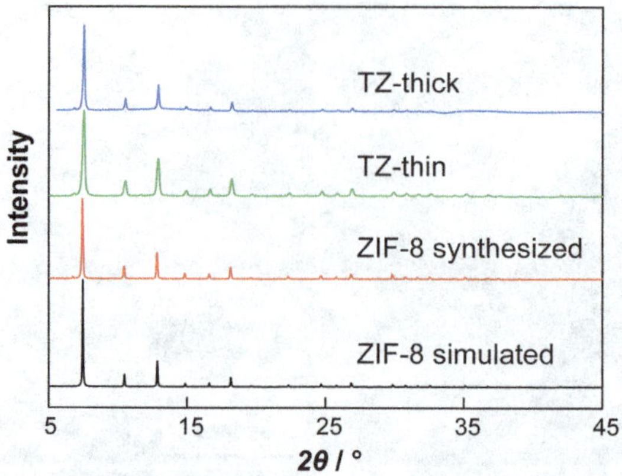

Fig. 4 PXRD pattern of as-synthesized ZIF-8, TZ-thin and TZ-thick. The simulated PXRD pattern of ZIF-8 is shown at the bottom. Figure is reproduced from Ref. [20] with permission from Wiley

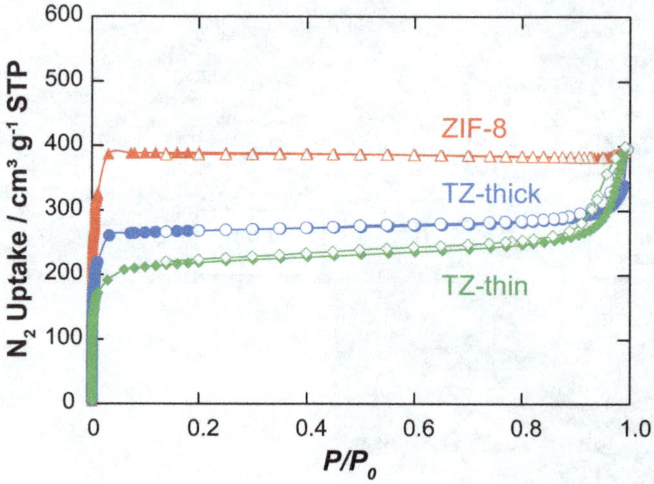

Fig. 5 N_2 sorption isotherm of ZIF-8, TZ-thin, and TZ-thick. The BET surface area of ZIF-8, TZ-thick, and TZ-thin is calculated as 1537, 1053, and 847 $m^2\ g^{-1}$, respectively. Figure is reproduced from Ref. [20] with permission from Wiley

4 Notes

1. To obtain optimal results, all precursor solutions should be freshly prepared.
2. The container should be covered with aluminum foil to avoid precipitation caused by light exposure. Disposal and handling should always be done in accordance with law.
3. It usually takes 8–10 weeks from seeding to being ready for infection.
4. The silicon carbide acts as an abrasive to break the leaf tissue. The particle size of silicon carbide powder is very small. Appropriate personal protective equipment should be used at

all times when handling it. Special care needs to be taken to avoid inhalation and contact with eyes. Gloves should be worn at all times. If too much silicon carbide is on the leaf, slightly tap the leaf to get rid of the excess powder; do not blow as this will aerosolize the powder. In general, the hand used for rubbing leaves should not touch the silicon carbide or the pipette. It is recommended that two people work together so that one can sprinkle the silicon carbide and add infection solution while the other rubs the leaves.

5. To resuspend the pellet, gently agitate the solution via suction and injection with a pipette. It is helpful if the stream of solution is pointed directly at the pellet to facilitate the resuspension.
6. We prefer to add the least amount of phosphate buffer as possible to make a concentrated virus solution, since it will take extra steps and produce more waste to a make a more concentrated virus solution later.
7. The eluate is collected from the first drop. UV-Vis can be used to check fractions to make sure that recovery is complete.
8. We recommend using TMV solutions with concentrations between 2 and 9 mg/mL and thus the volume of solution should fall within that range. The goal is to maintain the concentration of ZIF-8 precursors as the added volume of TMV solution is negligible and typically on the order of 10–30 μL in total when using TMV solutions between 2 and 9 mg/mL. If the TMV solution is too dilute then a large volume of TMV solution is needed, which could cause a significant change in the concentration of precursor solution. The concentration of precursor solution is crucial to the success of making TZ-thin/TZ-thick.
9. It is important to follow the order of addition as written. If the $Zn(OAc)_2$ solution is added first, discrete rod-like nanoparticles will not be obtained. Moreover, the $Zn(OAc)_2$ solution should be added as quickly as possible after addition of the HMIM solution. Delaying the addition of metal precursor will produce rods with an incorrect morphology.
10. The reaction mixture should be much thicker than the resulting suspension of TZ-thin.

Acknowledgments

We are grateful to the generous assistance we received from Professor Gerald Stubbs and Professor Nicole F. Steinmetz when starting our group for providing *N. benthamiana* seeds and TMV stock solutions. We would have been unable to conduct this research without the assistance they provided when others declined.

References

1. Belowich ME, Valente C, Stoddart JF (2010) Template-directed syntheses of rigid oligorotaxanes under thermodynamic control. Angew Chem Int Ed 49:7208–7212
2. Fan XZ, Pomerantseva E, Gnerlich M et al (2013) Tobacco mosaic virus: a biological building block for micro/nano/bio systems. J Vac Sci Technol A 31:050815
3. Schlick TL, Ding Z, Kovacs EW et al (2005) Dual-surface modification of the tobacco mosaic virus. J Am Chem Soc 127:3718–3723
4. Shenton W, Douglas T, Young M et al (1999) Inorganic–organic nanotube composites from template mineralization of tobacco mosaic virus. Adv Mater 11:253–256
5. Shukla S, Eber FJ, Nagarajan AS et al (2015) The impact of aspect ratio on the biodistribution and tumor homing of rigid soft-matter nanorods. Adv Healthc Mater 4:874–882
6. Bruckman MA, Randolph LN, Gulati NM et al (2015) Silica-coated Gd(DOTA)-loaded protein nanoparticles enable magnetic resonance imaging of macrophages. J Mater Chem B 3:7503–7510
7. Dujardin E, Peet C, Stubbs G et al (2003) Organization of metallic nanoparticles using tobacco mosaic virus templates. Nano Lett 3:413–417
8. Knez M, Sumser M, Bittner AM et al (2004) Spatially selective nucleation of metal clusters on the tobacco mosaic virus. Adv Funct Mater 14:116–124
9. Pomerantseva E, Gerasopoulos K, Chen X et al (2012) Electrochemical performance of the nanostructured biotemplated V_2O_5 cathode for lithium-ion batteries. J Power Sources 206:282–287
10. Royston E, Ghosh A, Kofinas P et al (2008) Self-assembly of virus-structured high surface area nanomaterials and their application as battery electrodes. Langmuir 24:906–912
11. Yaghi OM, O'Keeffe M, Ockwig NW et al (2003) Reticular synthesis and the design of new materials. Nature 423:705–714
12. Liang K, Ricco R, Doherty CM et al (2015) Biomimetic mineralization of metal-organic frameworks as protective coatings for biomacromolecules. Nat Commun 6:7240
13. Chulkaivalsucharit P, Wu X, Ge J (2015) Synthesis of enzyme-embedded metal-organic framework nanocrystals in reverse micelles. RSC Adv 5:101293–101296
14. Wu X, Ge J, Yang C et al (2015) Facile synthesis of multiple enzyme-containing metal-organic frameworks in a biomolecule-friendly environment. Chem Commun 51:13408–13411
15. Wu X, Yang C, Ge J et al (2015) Polydopamine tethered enzyme/metal-organic framework composites with high stability and reusability. Nanoscale 7:18883–18886
16. Liang K, Coghlan CJ, Bell SG et al (2016) Enzyme encapsulation in zeolitic imidazolate frameworks: a comparison between controlled co-precipitation and biomimetic mineralisation. Chem Commun 52:473–476
17. Lyu F, Zhang Y, Zare RN et al (2014) One-pot synthesis of protein-embedded metal–organic frameworks with enhanced biological activities. Nano Lett 14:5761–5765
18. Shieh F-K, Wang S-C, Yen C-I et al (2015) Imparting functionality to biocatalysts via embedding enzymes into nanoporous materials by a de novo approach: size-selective sheltering of catalase in metal–organic framework microcrystals. J Am Chem Soc 137:4276–4279
19. Liang K, Richardson JJ, Cui J et al (2016) Metal–organic framework coatings as cytoprotective exoskeletons for living cells. Adv Mater 28:7910–7914
20. Li S, Dharmarwardana M, Welch RP et al (2016) Template-directed synthesis of porous and protective core–shell bionanoparticles. Angew Chem Int Ed 55:10691–10696
21. Peng Y, Krungleviciute V, Eryazici I et al (2013) Methane storage in metal–organic frameworks: current records, surprise findings, and challenges. J Am Chem Soc 135:11887–11894
22. Gándara F, Furukawa H, Lee S et al (2014) High methane storage capacity in aluminum metal–organic frameworks. J Am Chem Soc 136:5271–5274
23. Karagiaridi O, Lalonde MB, Bury W et al (2012) Opening ZIF-8: a catalytically active zeolitic imidazolate framework of sodalite topology with unsubstituted linkers. J Am Chem Soc 134:18790–18796
24. He C, Lu K, Liu D et al (2014) Nanoscale metal–organic frameworks for the co-delivery of cisplatin and pooled siRNAs to enhance therapeutic efficacy in drug-resistant ovarian cancer cells. J Am Chem Soc 136:5181–5184
25. Park KS, Ni Z, Côté AP et al (2006) Exceptional chemical and thermal stability of zeolitic imidazolate frameworks. Proc Natl Acad Sci U S A 103:10186–10191

Chapter 9

TMV Disk Scaffolds for Making sub-30 nm Silver Nanorings

Serene Bayram, Omar Zahr, Julia Del Re, and Amy Szuchmacher Blum

Abstract

Nanosized bioscaffolds can be utilized to tackle the challenge of size reduction of metallic rings owing to their miniature features as well as their well-known biomineralization capacity. The tobacco mosaic virus coat protein is used as a command surface to grow and assemble silver nanoparticles into sub-30 nm rings. The versatility of TMV allows the formation of both solid silver rings and rings consisting of discrete silver nanoparticles. The pH-dependent coulombic surface map along with the annular geometry of the protein aggregate allow the generation of rings with or without a central nanoparticle. Our silver rings are believed to be the smallest to date, and they can offer a test material for existing theories on metallic nanorings of this heretofore unreached size scale.

Key words Silver, Nanorings, TMV, Photoreduction, Metamaterials, Bionanomaterials

1 Introduction

The smart design of metallic nanostructures continues to interest scientists owing to their ability to enhance light manipulation and confinement in small volumes. In particular, ring structures have stimulated rapid growth in the field of plasmonics due to their great impact on sensing, lasing, and nonlinear optics technologies [1–5]. Appropriately shaping light-scattering materials enables independent control of the effective permittivity and permeability, which when made negative concurrently, results in a negative index of refraction [6].

Numerous approaches have been undertaken to fabricate such metallic nanoring structures, including electron beam lithography [7], colloidal lithography [8], chemical synthesis, and nanoimprint lithography [9]. Bottom-up self-assembly of metallic nanoparticles remains the most cost-effective for mass production. The choice of template or command surface for bottom-up nanofabrication is crucial to define the size and geometry. Biotemplates have been successfully utilized to guide self-assembly owing to their

Andrew K. Udit (ed.), *Protein Scaffolds: Design, Synthesis, and Applications*, Methods in Molecular Biology, vol. 1798,
https://doi.org/10.1007/978-1-4939-7893-9_9,

monodispersity, abundance, robustness, ease in bioconjugation to introduce new functional groups, as well as their versatility in chemical functionality and shape induced by mutations [10–14].

Tobacco mosaic virus coat protein (TMVcp) has proven to be a useful template for organizing nanoparticles [13, 15, 16]. In its complete virion form, it assembles around RNA into 300 nm rods with a 4 nm diameter inner channel. The assembly dynamics of the isolated coat protein have also been well established and understood [17]. The RNA-free viral coat protein is capable of self-assembly into 18 nm stacked disks in a narrow pH region (6.5–7.5) and into helices below pH 6.5 under the appropriate ionic strength and temperature conditions. At higher pH values, it disassembles into monomers or small oligomers. The phase diagram presented in Fig. 1 depicts the predominant TMVcp aggregate states under different pH and ionic strength combinations [18].

Nearly all the methods developed to utilize TMV as a bioscaffold have employed the complete virion or the helical rod form of the capsid protein as the templating platform. The inner channel and outer surfaces of the virus are hydrophilic displaying glutamic and aspartic acid, lysine and arginine to the solvent. The differential pKa values of the outer surface residues, mainly lysines and N-termini, from those in the inner channel like glutamic and aspartic acid were used to selectively complex and mineralize

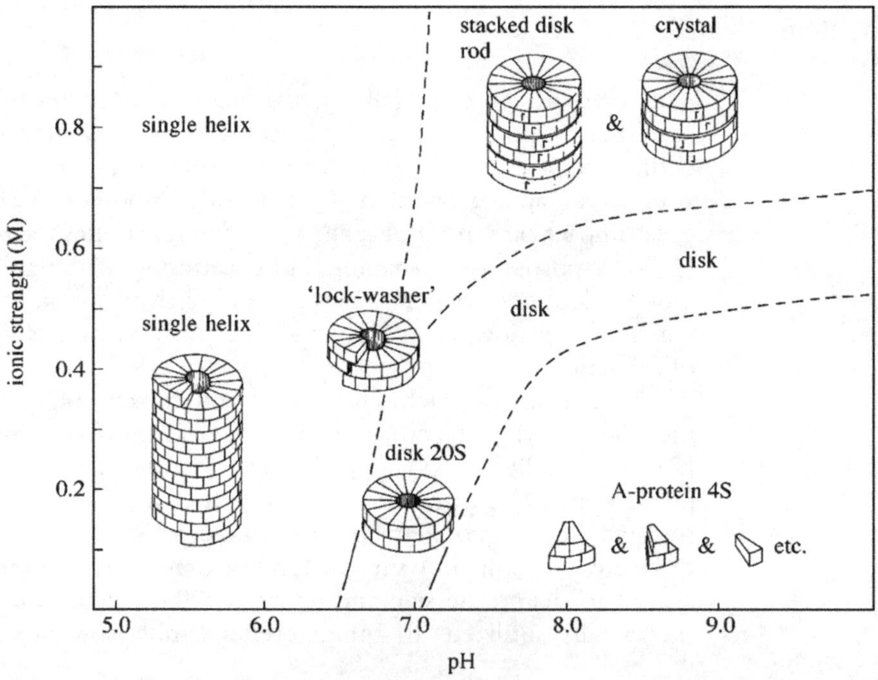

Fig. 1 Phase diagram of TMVcp predominant aggregate states under different pH and ionic strength conditions. Reprinted from Kegel et al. (2006) with permission from Elsevier [18]

several mineral salts [19]. The relative positive and negative charge of the TMV outer and inner surfaces can be altered by tuning pH to allow for surface-specific metallization [20–22]. The isolated coat protein has also been utilized to create functional assemblies by virtue of its inherent self-assembly properties. A Ser^{123} to Cys TMV mutant was used to bind thiol-reactive dyes to the mutant at the introduced Cys. Donor and acceptor dyes were then bound to TMV which assembled into a helical geometry to behave as a light-harvesting entity [23]. Tyrosine residues on the TMV surface have also been modified into diazonium salts by electrophilic substitutions to incur new functionalities such as alkynes [24]. Alkynes can then be reacted with azide-containing molecules by the highly specific copper catalyzed azide-alkyne cycloaddition. Tyrosines were targeted to create TMV-based enhanced MRI contrast agents [25], multifunctional viral nanoparticles [26, 27], thermally stable virus-polymer conjugates in hydrophobic solvents [28], and thin film sensors for volatile organic compounds [29]. The TMV rods themselves act as self-assembling units when they are assembled into hyperstructures by the in situ polymerization of aniline [30]. TMV coat protein mutated with a hexahistidine tag at the C-terminus has been assembled side-by-side through pi-stacking into long fibers or hexagonal disk arrays depending on the conditions of assembly [31]. Tunable, thin TMV film assemblies were also fabricated via layer-by-layer assembly with polyelectrolytes [32]. TMV-based composite materials have found a broad spectrum of applications including drug delivery [33–36], energy and catalysis [37–42], electronics [43–46], and optical resonators [47].

The disk phase is of particular interest, as it gives access to a size scale unreachable by other techniques. It was successfully employed to electrostatically assemble negatively charged gold nanoparticles into sub-30 nm rings with and without a central nanoparticle [13]. In this chapter, we detail a protocol for using the TMVcp disks as a platform to fabricate silver rings by the in situ photoreduction of silver under UV light (365 nm). We present two types of TMV-based silver nanorings: disk-like rings formed from plasmonically coupled discrete nanoparticles using the wild-type coat protein (TMVcp-WT), and solid donut-like silver rings using a cysteine mutant form of TMVcp (TMVcp-S123C). The latter displays an increased density of nucleation sites, the cysteine residues, which irreversibly bind silver ions providing enough proximity between the photoreduced metallic silver to realize their soldering into fused rings. The 365 nm illumination used excites the silver plasmons to create local "hot spots" capable of inducing the soldering event. Spectroscopically, we confirm the rings' optical plasmonic properties via UV–Vis spectroscopy.

2 Materials

Prepare all solutions using deionized water (18 MΩ at 25 °C). Store all solutions at 4 °C until ready for use to slow bacterial contamination (*see* **Note 1**).

1. UVGL-55 lamp (6 W) with 365 nm wavelength.
2. TEA buffer: 20 mM triethanolamine, adjusted to pH 7 using HCl.
3. Tobacco Mosaic Virus capsid protein (TMVcp) stock solution, either wild-type (TMVcp-WT) or cysteine mutant (TMVcp-S123C) in TEA buffer at a minimum concentration of 0.1–0.15 mg/mL (*see* **Note 2**).
4. Silver nitrate solution: 10 mM silver nitrate ($AgNO_3$) in deionized water (*see* **Note 3**).
5. Snakeskin dialysis membrane with 3.5 kDa molecular weight cutoff (*see* **Note 4**).
6. UV-Vis Spectrophotometer with cuvettes that have a useable range between 200 and 800 nm.
7. Transmission Electron Microscope (recommended High Tension 120 kV).
8. Centrifuge with free bucket rotor that can obtain a speed of 3000 × *g*.
9. Orbital mixer.
10. Centrifuge Spin-X UF concentrator tubes, 3500 MWCO.
11. Zeba Spin size exclusion desalting columns, 7000 MWCO.
12. TEM grids (35–50 nm-thick carbon film).

3 Methods

1. Dialyze the TMVcp stored at 4 °C in TEA buffer against at least 4 L of deionized water ($R > 18.2$ MΩ) for at least 24 h (*see* **Note 5**).
2. Determine the protein concentration by measuring its absorbance at 282 nm ($\varepsilon = 1.27$ mL mg^{-1} cm^{-1}) [48] using a spectrophotometer and blanking against water (*see* **Note 6**).
3. If the protein concentration is less than 0.1 mg/mL, use the spin-X concentrators to concentrate it to 0.1–0.15 mg/mL by centrifugation at 3000 × *g* for 30 min.
4. Leave the protein solution to sit at room temperature for 24 h for its assemblies to reach equilibrium (*see* **Note 7**).

5. To bind the silver to the disks, incubate 20–50 μL of 10 mM $AgNO_3$ with 1 mL of TMVcp-WT (0.1 mg/mL), and at least 50 μL of 10 mM $AgNO_3$ with 1 mL of TMVcp-S123C (0.1 mg mL^{-1}) for 30 min at room temperature.
6. Eliminate the excess unbound silver nitrate by centrifuging the mixture using Zeba spin size-exclusion desalting columns for 20 min at 3000 × *g* (*see* **Note 8**).
7. Illuminate the incubated mixture using a UVGL-55 lamp (6 W) to photoreduce the TMV-bound silver ions into silver nanoparticles. The color changes to yellow–orange within 1 min of top illumination.
8. Prepare TEM grids by exposing them to UV-light for 1 h to remove surface contaminants.
9. Plate 20 μL of the silver ring solution onto clean TEM grids and keep it for 1 min before wicking it with filter paper.
10. Visualize the rings using TEM. Sample rings formed using TMV-WT and TMV-S123C are depicted in Figs. 2 and 3, respectively (*see* **Note 9**).
11. Measure the extinction spectrum of the colloids to observe the plasmon shift and broadening (Fig. 4).

4 Notes

1. Protein solutions can be stored in 0.05% sodium azide to further protect from bacteria. The sodium azide will be dialyzed out before the TMVcp is put to self-assemble.
2. The PDB ID of TMV is 1EI7. The S123C mutant construct can be made by site-specific mutagenesis [23]. The protein can be expressed in *E. coli* cells and purified according to the protocol detailed by Zahr et al. [13].
3. The silver solution must be freshly prepared in deionized water to have a clear solution. Photoreduction fails if a 1-day old silver solution is used. Minimal chloride concentrations in the water lead to the formation of insoluble silver chloride complexes, which can be noticed if the solution turns turbid.
4. The TMV capsid protein exists as rods, disks, or multimers of individual subunits in equilibrium. The population of each structure depends on the ionic strength, pH, and temperature of the environment. When choosing a molecular weight cutoff, it is important to use a cutoff that is smaller than the smallest protein subunit in the system utilized. Use a molecular weight cutoff that is at least three times smaller than the biological template. If one fails to do this, protein subunits will be lost during dialysis and the system will begin to disassemble larger assemblies to maintain the population equilibrium.

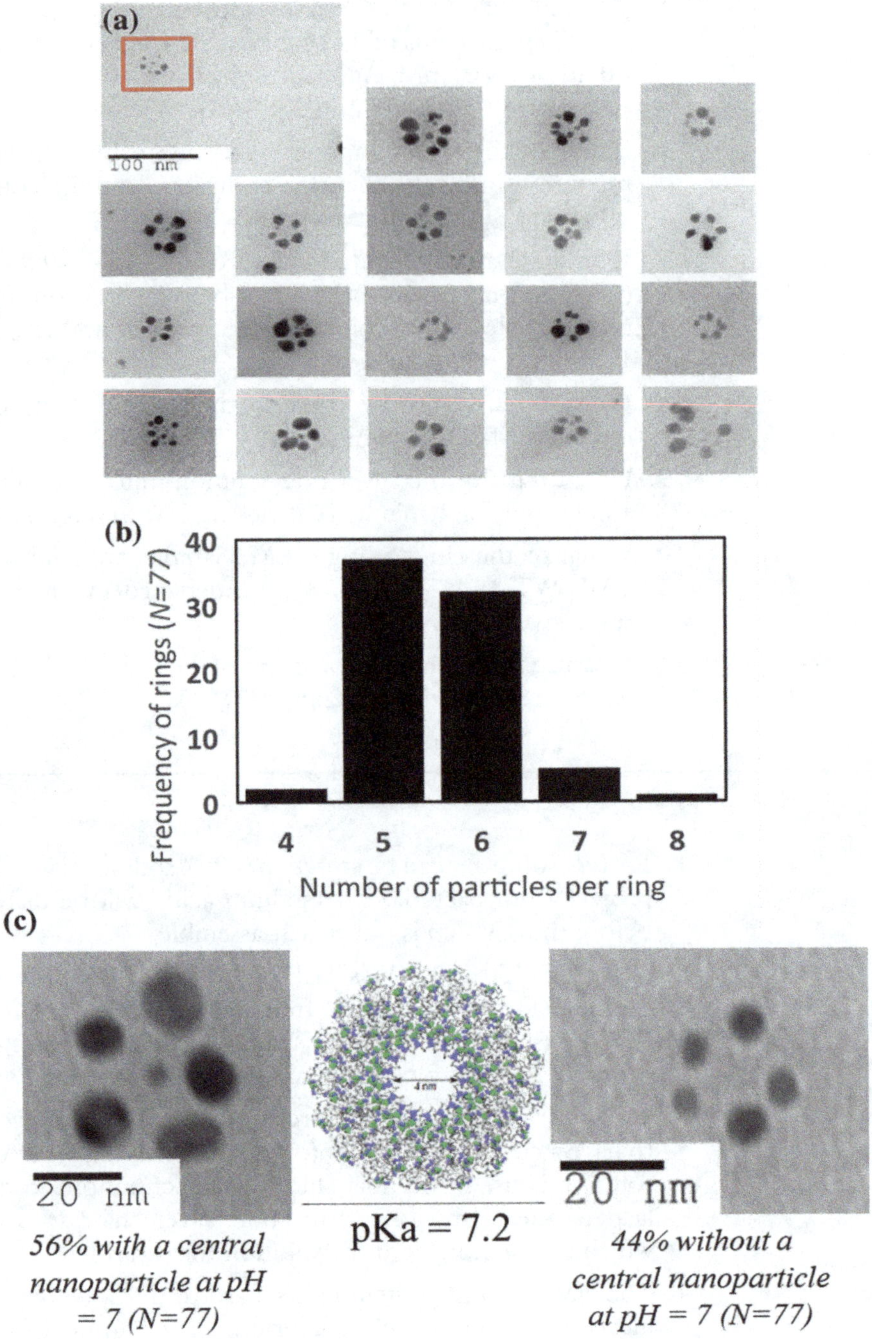

Fig. 2 (**a**) Sample silver nanorings formed on TMV-WT. The number of particles per ring varies between 5 and 6. (**b**) Histogram showing the predominance of rings possessing 5–6 nanoparticles per ring (**c**) Close-up on nanorings with and without central nanoparticle. These rings were created at pH 7, which approaches the pKa of the carboxylates lining the internal pore of the TMV disks that have an anomalous pKa value of 7.2 [49]. Accordingly, around 56% of the nanoring population exhibit a central nanoparticle, whilst 44% do not. Reprinted from Bayram et al. (2016) with permission from IOP Publishing [50]

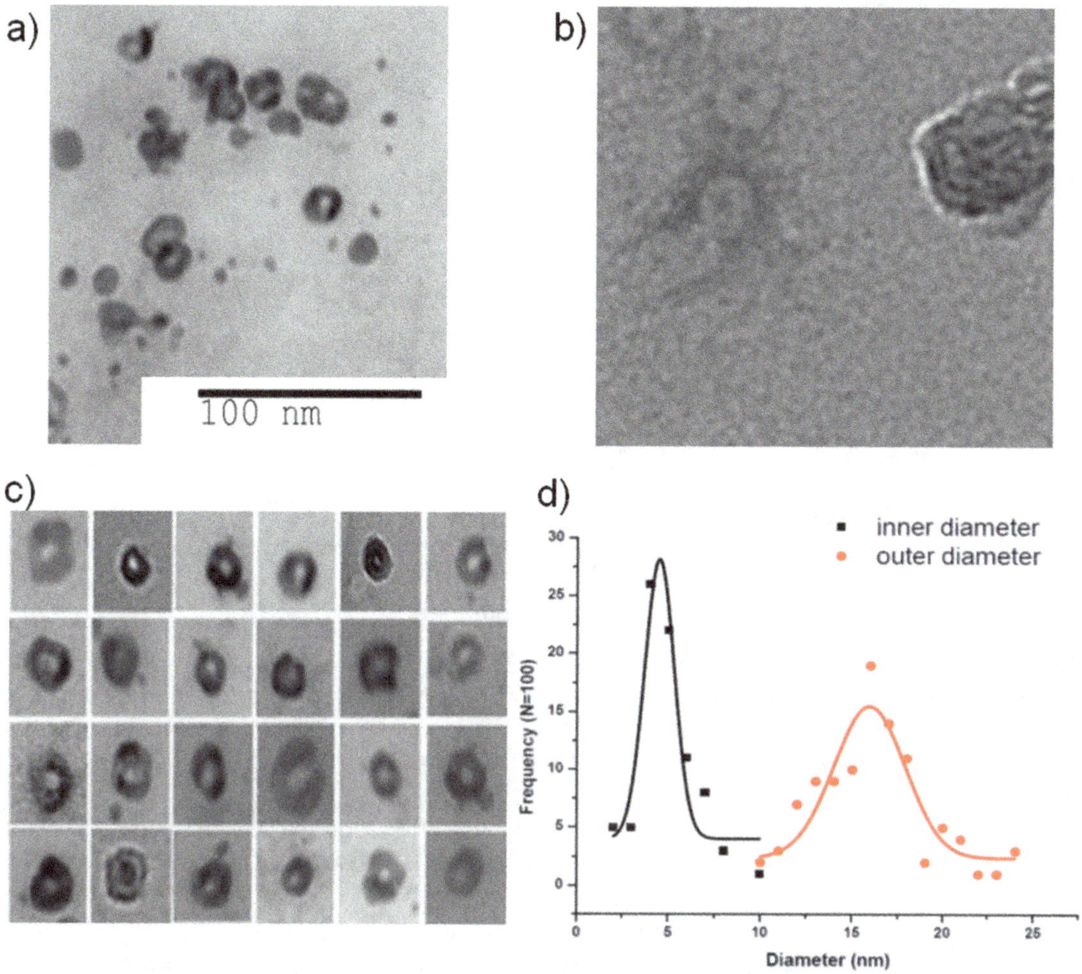

Fig. 3 (**a**) TEM images of a population of uncrowded (left) and (**b**) close-packed (right) solid silver nanorings. (**c**) Zoomed in TEM images of solid silver nanorings. (**d**) Size distribution of their inner and outer diameters for $N = 100$. The solid rings have a 4.5 ± 0.1 nm and 16 ± 0.2 nm average inner and outer diameters respectively. Reprinted from Bayram et al. (2016) with permission from IOP Publishing [50]

5. The TMVcp self-assembles in water to a lesser extent than in buffer. However, the concentration used in this protocol is enough to trigger the assembly of TMV-WT and TMV-S123C in water at room temperature. The cysteine mutation in the latter incurs more disk resilience against disassembly. Higher ionic strength conditions would screen the charge repulsion between TMV disks and induce their aggregation, which is not recommended if one wants to realize well-dispersed rings. Also, silver ions form insoluble complexes with the most common buffers used to provide the neutral condition required to have disks.
6. The background absorbance (i.e., above 400 nm) should not be above 0.01 a.u. if one wants to proceed. A high background signifies aggregation of the protein; one potential reason could

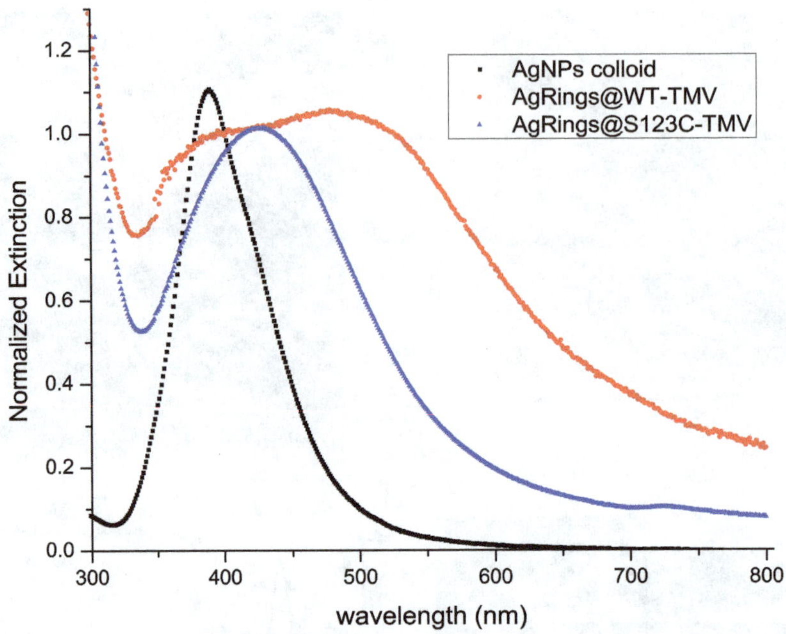

Fig. 4 Extinction spectra of the TMV-WT (red) and TMV-S123C (blue) based silver nanorings colloids compared to that of un-assembled colloidal AgNPs in water (black). The AgNPs colloid was prepared by the common sodium borohydride chemical reduction method for silver ions in water. It exhibits the characteristic plasmon band at 394 nm. Reprinted Bayram et al. (2016) with permission from IOP Publishing [50]

be improper folding of the protein. In the case of high background absorbance, the TMVcp solution must be filtered through 0.2 micron syringe filters before its concentration is quantified and adjusted again.

7. Temperature plays an important role in the assembly of TMV into disks. TMV coat protein assembly is an endothermic process and is primarily driven by hydrophobic interactions. At room temperature, it is a spontaneous process that occurs due to the net entropic gain upon the liberation of solvent molecules. Therefore, lowering the temperature would disfavor TMV assembly and results in a diminished population of disks and quenched yield of rings.
8. The desalting columns should be equilibrated with water before they are used to desalt the mixture to avoid mismatched conditions. This is done by loading the columns with water and centrifuging for 15 min at $3000 \times g$.
9. Due to the equilibrium state of the TMVcp disks with monomers, imaging reveals the presence of nonassembled isolated nanoparticles. Most of the rings observed will have a sub-30 nm diameter dimension consistent with their formation on the TMVcp disks.

References

1. Nordlander P (2009) The ring: a leitmotif in plasmonics. ACS Nano 3(3):488–492
2. Ye J, Shioi M, Lodewijks K et al (2010) Tuning plasmonic interaction between gold nanorings and a gold film for surface enhanced Raman scattering. Appl Phys Lett 97(16):163106
3. Jahani S, Jacob Z (2016) All-dielectric metamaterials. Nat Nanotechnol 11(1):23–36
4. Song M, Wang C, Zhao Z et al (2016) Nanofocusing beyond the near-field diffraction limit via plasmonic Fano resonance. Nanoscale 8(3):1635–1641
5. Kuchmizhak A, Gurbatov S, Vitrik O et al (2016) Ion-beam assisted laser fabrication of sensing plasmonic nanostructures. Sci Rep 6:19410
6. Shalaev VM (2007) Optical negative-index metamaterials. Nat Photonics 1(1):41–48
7. Near R, Tabor C, Duan J et al (2012) Pronounced effects of anisotropy on plasmonic properties of nanorings fabricated by electron beam lithography. Nano Lett 12(4):2158–2164
8. Vogel N, Fischer J, Mohammadi R et al (2011) Plasmon hybridization in stacked double crescents arrays fabricated by colloidal lithography. Nano Lett 11(2):446–454
9. Cui B, Veres T (2007) Fabrication of metal nanoring array by nanoimprint lithography (NIL) and reactive ion etching. Microelectron Eng 84(5):1544–1547
10. Steinmetz NF, Lomonossoff GP, Evans D (2006) Decoration of cowpea mosaic virus with multiple, redox-active, organometallic complexes. Small 2(4):530–533
11. Dedeo MT, Duderstadt KE, Berger JM et al (2009) Nanoscale protein assemblies from a circular permutant of the tobacco mosaic virus. Nano Lett 10(1):181–186
12. Paavola C, Chan S, Li Y, Mazzarella K, McMillan R, Trent J (2006) A versatile platform for nanotechnology based on circular permutation of a chaperonin protein. Nanotechnology 17(5):1171
13. Zahr OK, Blum AS (2012) Solution phase gold nanorings on a viral protein template. Nano Lett 12(2):629–633
14. Blum AS, Soto CM, Wilson CD et al (2004) Cowpea mosaic virus as a scaffold for 3-D patterning of gold nanoparticles. Nano Lett 4(5):867–870
15. Khan AA, Fox EK, Górzny MŁ et al (2013) pH control of the electrostatic binding of gold and iron oxide nanoparticles to tobacco mosaic virus. Langmuir 29(7):2094–2098
16. Dujardin E, Peet C, Stubbs G et al (2003) Organization of metallic nanoparticles using tobacco mosaic virus templates. Nano Lett 3(3):413–417
17. Klug A (1999) The tobacco mosaic virus particle: structure and assembly. Philos Trans R Soc Lond B Biol Sci 354(1383):531–535
18. Kegel WK, Van der Schoot P (2006) Physical regulation of the self-assembly of tobacco mosaic virus coat protein. Biophys J 91(4):1501–1512
19. Lim JS, Kim SM, Lee SY et al (2010) Quantitative study of au (III) and Pd (II) ion biosorption on genetically engineered tobacco mosaic virus. J Colloid Interface Sci 342(2):455–461
20. Górzny MŁ, Walton AS, Evans SD (2010) Synthesis of high-surface-area platinum nanotubes using a viral template. Adv Funct Mater 20(8):1295–1300
21. Balci S, Bittner A, Hahn K et al (2006) Copper nanowires within the central channel of tobacco mosaic virus particles. Electrochim Acta 51(28):6251–6257
22. Tsukamoto R, Muraoka M, Seki M et al (2007) Synthesis of CoPt and FePt3 nanowires using the central channel of tobacco mosaic virus as a biotemplate. Chem Mater 19(10):2389–2391
23. Miller RA, Presley AD, Francis MB (2007) Self-assembling light-harvesting systems from synthetically modified tobacco mosaic virus coat proteins. J Am Chem Soc 129(11):3104–3109
24. Bruckman MA, Kaur G, Lee LA et al (2008) Surface modification of tobacco mosaic virus with "click" chemistry. Chembiochem 9(4):519–523
25. Bruckman MA, Hern S, Jiang K et al (2013) Tobacco mosaic virus rods and spheres as supramolecular high-relaxivity MRI contrast agents. J Mater Chem B 1(10):1482–1490
26. Hu J, Wang P, Zhao X et al (2014) Charge-transfer interactions for the fabrication of multifunctional viral nanoparticles. Chem Commun 50(91):14125–14128
27. Chen L, Zhao X, Lin Y et al (2013) A supramolecular strategy to assemble multifunctional viral nanoparticles. Chem Commun 49(83):9678–9680
28. Holder PG, Finley DT, Stephanopoulos N et al (2010) Dramatic thermal stability of virus–polymer conjugates in hydrophobic solvents. Langmuir 26(22):17383–17388
29. Bruckman MA, Liu J, Koley G et al (2010) Tobacco mosaic virus based thin film sensor for detection of volatile organic compounds. J Mater Chem 20(27):5715–5719

30. Niu Z, Bruckman M, Kotakadi VS et al (2006) Study and characterization of tobacco mosaic virus head-to-tail assembly assisted by aniline polymerization. Chem Commun (28):3019–3021
31. Bruckman MA, Soto CM, McDowell H et al (2011) Role of hexahistidine in directed nano-assemblies of tobacco mosaic virus coat protein. ACS Nano 5(3):1606–1616
32. Steinmetz NF, Findlay KC, Noel TR et al (2008) Layer-by-layer assembly of viral nanoparticles and polyelectrolytes: the film architecture is different for spheres versus rods. Chembiochem 9(10):1662–1670
33. Czapar AE, Zheng YR, Riddell IA et al (2016) Tobacco mosaic virus delivery of phenanthriplatin for cancer therapy. ACS Nano 10(4):4119
34. Finbloom JA, Han K, Aanei IL et al (2016) Stable disk assemblies of a tobacco mosaic virus mutant as nanoscale scaffolds for applications in drug delivery. Bioconjug Chem 27(10):2480–2485
35. Tian Y, Gao S, Wu M et al (2016) Tobacco mosaic virus-based 1D nanorod-drug carrier via the integrin-mediated endocytosis pathway. ACS Appl Mater Interfaces 8(17):10800–10807
36. Shukla S (2016) Serum albumin 'camouflage'of plant virus based nanoparticles prevents their antibody recognition and enhances pharmacokinetics. Biomaterials 89:89–97
37. Bhanushali S, Ghosh P, Ganesh A et al (2015) 1D copper nanostructures: progress, challenges and opportunities. Small 11(11):1232–1252
38. Royston E, Ghosh A, Kofinas P et al (2008) Self-assembly of virus-structured high surface area nanomaterials and their application as battery electrodes. Langmuir 24(3):906–912
39. Lee SY, Choi J, Royston E et al (2006) Deposition of platinum clusters on surface-modified tobacco mosaic virus. J Nanosci Nanotechnol 6(4):974–981
40. Yang C, Meldon JH, Lee B et al (2014) Investigation on the catalytic reduction kinetics of hexavalent chromium by viral-templated palladium nanocatalysts. Catal Today 233:108–116
41. Kim YH, Jung H, Yoon TS et al (2016) Electrical charging characteristics of palladium nanoparticles synthesized on tobacco mosaic virus Nanotemplate for organic memory device. ECS J Solid State Sci Technol 5(9):Q226–Q230
42. Yang C, Manocchi AK, Lee B et al (2010) Viral templated palladium nanocatalysts for dichromate reduction. Appl Catal Environ 93(3):282–291
43. Chu S, Gerasopoulos K, Ghodssi R (2016) Tobacco mosaic virus-templated hierarchical Ni/NiO with high electrochemical charge storage performances. Electrochim Acta 220:184–192
44. Chen X, Guo J, Gerasopoulos K et al (2012) 3D tin anodes prepared by electrodeposition on a virus scaffold. J Power Sources 211:129–132
45. Chen X, Gerasopoulos K, Guo J et al (2011) A patterned 3D silicon anode fabricated by electrodeposition on a virus-structured current collector. Adv Funct Mater 21(2):380–387
46. Atanasova P, Rothenstein D, Schneider JJ et al (2011) Virus-templated synthesis of ZnO nanostructures and formation of field-effect transistors. Adv Mater 23(42):4918–4922
47. Fan X, Naves L, Siwak N et al (2015) Integration of genetically modified virus-like-particles with an optical resonator for selective bio-detection. Nanotechnology 26(20):205501
48. Fraenkel-Conrat H (1957) Degradation of tobacco mosaic virus with acetic acid. Virology 4(1):1–4
49. Lu B, Stubbs G, Culver JN (1996) Carboxylate interactions involved in the disassembly of tobacco mosaic tobamovirus. Virology 225(1):11–20
50. Bayram SS, Zahr OK, Del Re J et al (2016) Nanoring formation via in situ photoreduction of silver on a virus scaffold. Nanotechnology 27(48):485603

Chapter 10

Multivalent Display Using Hybrid Virus Nanoparticles

Steven D. Brown

Abstract

Many important biological interactions are multivalent and often sensitive to spatial organization. Nonenveloped viruses are a natural source of scaffolds for building multivalent ligands to probe these types of interactions which avoid complex synthetic schemes required for other types of scaffolds. The coat protein (CP) of bacteriophage Qβ can be fused to protein domains and coexpressed with the unfused CP to produce hybrid nanoparticles with high exterior loading of xenogenic protein domains. These hybrid nanoparticles are simple to produce in large quantity. Starting from cDNAs for the virus CP and a codon-optimized ligand domain of interest, bulk purification can be completed in as little as 3 weeks. Major phases of the work involve the cloning of cDNAs into plasmid vectors, test expressions for hybrid nanoparticle formation, and purification by selective precipitation and ultracentrifugation. For uncomplicated protein domains, laboratory culture yields as high as 50 mg/L and 30 protein domains per particle have been routinely achieved.

Key words Nanoparticle, Protein expression, Protein purification, Multivalency, Gene engineering

1 Introduction

Multivalent ligands are ubiquitous in biological systems, wherein multiple low single-molecule affinities are covalently combined to form a high-avidity interaction [1]. In immunology, IgM antibodies form pentamers which increases the functional affinity between the antibody and new or recently introduced antigens [2]. In virology, nonenveloped viruses have hundreds of copies of coat proteins which can bind to multiple receptors on cell surfaces, decreasing their risk of disassociating before entering the cell to initiate an infection [3, 4]. Similarly, it is easier to elicit desirable immune responses with multivalent displays of the target than with the target alone [5–8]. Cells themselves are multivalent scaffolds in their presentation of numerous copies of membrane proteins or lipids. In bacteria, carbohydrate receptors brought together by multivalent displays of sugars can induce stronger motility responses than individual sugars binding to isolated receptors [9]. Multivalent displays of amino acids can convert a repellant response to the

Andrew K. Udit (ed.), *Protein Scaffolds: Design, Synthesis, and Applications*, Methods in Molecular Biology, vol. 1798, https://doi.org/10.1007/978-1-4939-7893-9_10,

monomeric amino acid into an attractant response [10]. The bacteriophage P22 binds with high avidity to the lipopolysaccharide (LPS) which comprises the outer membrane of *S. enterica* Typhimurium. Avidity through multivalency provides the discriminatory power for the phage to remain infectious despite shed monomeric LPS in the growth media [11].

Hybrid virus nanoparticles bridge the gap between classical but poorly characterized multivalent displays (e.g., keyhole limpet hemocyanin) and well-defined monomeric antigens (e.g., peptides). Hybrid virus nanoparticles are composed of two versions of a single virus coat or structural protein which are co-assembled in the bacterial cytoplasm after induction of protein expression. The 14 kDa coat protein (CP) of bacteriophage Qβ spontaneously assembles 180 copies of itself to form nanoparticles approximately 28 nm in diameter [12–14]. The native gene structure of the CP includes a C-terminal extension which is expressed in low abundance (1–5 copies/virion) through tryptophan missense of the TGA stop codon. The method presented in this paper takes advantage of this natural attachment point for CP extensions by removing the stop codon entirely and replacing it with a linker sequence and a protein domain to be displayed. This coat protein-display domain fusion (CP-DD) coassembles with the unfused coat protein to produce intact nanoparticles with a variable number of extensions. In the icosahedral structure of the final nanoparticles, the C-termini of CPs are organized in clusters of three, suggesting a maximum of 60 displayed proteins per hybrid nanoparticle. In practice, hybrid nanoparticles displaying a complement protein C5 epitope contained 50 copies per nanoparticle, close to this theoretical maximum [5].

Removing barriers to translation is key to achieving high yields from microbial expression cultures. Eukaryotic cDNAs often include codons for low-abundance tRNAs in *E. coli*. These codons can induce ribosome stalling and subsequent degradation of the nascent polypeptide, severely reducing their quantity relative to the unfused CP. The Graphical Codon Usage Analyzer (GCUA, http://gcua.schoedl.de/sequential_v2.html [15]) has worked well in identifying problematic codons in our experience. Employing plasmids which overexpress rare tRNAs is a fast and economical approach to relieving the rarity of these codons [16]. However, this method does not directly increase the quantities of aminoacyl tRNA synthetases or upstream enzymatic machinery. A more robust approach to relieving rare-use codon limitations is to convert rare-use codons to codons preferred by the recombinant expression host. Commercial suppliers (e.g., Integrated DNA Technologies, Inc.) can synthesize sequences for increasingly reasonable costs, and the resulting DNA can be used directly in downstream cloning procedures. Designing codon-optimized sequences for gene synthesis also affords an opportunity to reduce or eliminate restriction endonuclease sites in the nucleotide sequence. Numerous software tools exist for this purpose; DNAWorks (https://hpcwebapps.cit.nih.gov/dnaworks/ [17]) is one such application which has worked well for our prior experiments. DNAWorks also produces short oligonucleotide sequences which are synthesis ready if whole-gene assembly by PCR is a more desirable approach. The codon usage table for "*E. coli* class II" derives codon abundance from genes which are highly and continuously expressed during exponential growth and has performed well for us in the past [18].

Folding is critical for functional display of protein domains on hybrid nanoparticles. One or more changes to the heterologous host environment may improve yields and activity for particularly difficult fusion partners. In these cases, there are numerous chemical changes which can be made to the growth media to support proper folding. In cases where the expressed protein appears to be soluble and folded but aggregates and precipitates during extraction, including a reducing agent such as dithiothreitol (DTT) or β-mercaptoethanol (BME) in the initial extraction procedure (*see* Subheading 3.5, **step 2**) may improve protein stability [19]. If the expressed protein is insoluble before extraction, decreasing the temperature of the culture during expression is a common adjustment to improve protein yields. Conversely, a brief heat shock at the start of expression induction can serve to induce host chaperones and aid in successful folding [20]. Inclusion of osmolytes ("chemical chaperones") such as sucrose, sorbitol, sodium chloride, and the osmoprotectant glycine betaine in the growth media may improve protein folding [21–26]. Sucrose and sorbitol in particular have performed very well in our prior experience. Other stress agents, such as ethanol and chloramphenicol, may also improve production of properly folded proteins [27, 28]. If the resulting protein is still nonfunctional, testing function in nanoparticles produced by a DE3 strain of *E. coli* specifically engineered to improve folding of highly oxidized proteins may be fruitful [29]. Finally, continuous overexpression of chaperones and foldases may ultimately be necessary to achieve adequate yields for still-troublesome targets [30]. For downstream applications, it may be helpful to investigate these changes in the context of *E. coli* expression strains lacking endotoxin [31].

The density of displayed protein domains on the produced nanoparticles can be indirectly altered by the plasmid backbone used as an expression vector and the selection markers and antibiotic drugs used to select for the growth of cells retaining the plasmid vector [32]. The pET-11b and pET-28b expression vectors (Novagen/EMD Millipore) are very similar, with the most salient differences in their expression cassette multiple cloning sites and their antibiotic markers. The pET-11b vector contains a β-lactamase selection marker whereas the pET-28b vector contains an aminoglycoside phosphotransferase marker. The pET-28b vector is more robust to varying growth conditions and may be better suited to circumstances where the displayed domain requires exotic growth conditions for proper folding. The pET-11b vector when used with carbenicillin sensitizes the nanoparticle assembly process to specific media conditions, which may be helpful in developing models for the relationship between valency and downstream biological activity. By creating both pET-28- and pET-11-based expression vectors, the CP-DD fusion protein encoded in pCDF-1b can be cloned into a single vector and used alternately with either CP-only vector depending on the circumstances. In either case, some fusion domains may have a native propensity for oligomerization, and formation of these oligomers in vivo may diminish their capacity for incorporation into nanoparticles [14, 33].

The specific growth media used during hybrid nanoparticle expression and assembly can have dramatic effects on the resulting nanoparticle content. The use of buffered media, particularly the MEM media described herein, readily increases the saturation density of cultures to OD600 greater than 5 and as high as 11. While the pET-28-CP expression vector leaves the nanoparticle assembly process relatively immune to changes in the media composition, the pET-11-CP vector used with carbenicillin can be used to tune the nanoparticle composition through media additives. However, other changes to the expression conditions such as the cell culture density at the time of induction may also cause unexpected changes in nanoparticle composition. In particular, adjustments made to the media to improve fusion protein solubility may also have effects on nanoparticle composition. The interplay of these effects is largely unexplored.

Purification of hybrid Qβ nanoparticles depends principally on the fusion protein or peptide being created. For peptides or robust protein domains, a workup similar to nonhybrid Qβ nanoparticles is likely to be successful. Typical workup includes extraction of lipids with organic solvents, precipitation of nanoparticles with poly(ethylene glycol) and/or ammonium sulfate, and sucrose gradient ultracentrifugation. Further purification may be achieved by size-exclusion chromatography. In addition to these methods, it may also be possible to take advantage of unique binding properties of the fused polypeptide. Hexahistidine tags at the C-terminus of the fused protein have been successfully used for nanoparticle purification by immobilized metal affinity chromatography, which may reduce or eliminate the need for ultracentrifugation [34, 35].

2 Materials

1. Water: all water should be deionized or distilled in origin (ddH_2O) and sterilized prior to use.
2. Orbital shaking air incubator (37 °C, 0.75–2 in. diameter orbit) with mounts for 16 or 17 mm culture tubes, 25 mL culture flasks, and 2 L culture flasks.
3. 1.7–2 mL microcentrifuge tubes (sterile, DNA/pyrogen-free).
4. 50 mL conical tubes (sterile, DNA/pyrogen-free).
5. 10 cm plastic petri dishes with lids (sterile).
6. Microdroplet spectrophotometer (e.g., NanoDrop, Take3 microplate).
7. 50 °C incubator for microcentrifuge tubes (dry block or water bath).
8. Refrigerated centrifuges and rotors for microcentrifuge tubes, 50 mL conical tubes or convenient larger preparatory tubes, and larger volumes for bulk expression.
9. Probe sonicator and polished probe tips without pitting damage.
10. Ultracentrifuge with swinging bucket and fixed-angle rotors (e.g., SW41).
11. Phusion High-Fidelity DNA Polymerase (New England Biolabs; includes buffers).
12. 10 mM dNTP mix.
13. pET cloning vectors: pET-11d, pET-28b(+), pCDF-1b (EMD Millipore/Novagen).
14. cDNA for Qβ bacteriophage coat protein (*see* **Note 1**).
15. Oligonucleotide primers for CP fusion generation (restriction sites in bold): CP-F (5′- GATC**CCATGG**CAAAATTAGAGACTGTTACTTTAG -3′), CP-Fuse-R (5′- ATACGCTGGGTTCAGCTG -3′), CP-R (5′- ATCGAT**GGATCCGCTCAGC**AATATACGCTGGGTTCAGCT -3′).
16. cDNA and/or protein sequence data for the domain to be displayed on the nanoparticle (available via the UniProt database, http://www.uniprot.org/).
17. Agarose for molecular biology.
18. SYBR Safe (Life Tech, 10,000× as supplied by the manufacturer) or ethidium bromide solution (5 mg/mL stock, 0.5 μg/mL in agarose).
19. 5× DNA Gel Loading Dye (*see* **Note 2**).
20. dsDNA MW ladder spanning 250–2000 bp.

21. 50× TAE buffer: combine 242 g Tris base, 57.1 mL of glacial acetic acid, 100 mL of 0.5 M EDTA, and add water to 0.75 L. Stir to dissolve, and add water to 1 L.
22. Clean, unused razor blades.
23. Gel DNA Extraction and PCR Purification Reagents (*see* **Note 3**).
24. Distilled H_2O, pH adjusted to 8.0 no more than 1 week prior to use (*see* **Note 4**).
25. Glycerol stocks of *E. coli* strains for cloning (*see* **Note 5**).
26. Glycerol stocks of *E. coli* BL21(DE3) cell for protein expression.
27. Restriction endonucleases (*see* **Note 6**).
28. T4 DNA ligase (10× reaction buffer usually included by manufacturer with the enzyme).
29. Bacterial cell electroporator capable of delivering 1.7–2.0 kV/mm cuvette gap at 200 ohms and 25 uF to a 1 mm-gap cuvette.
30. Electroporation cuvettes, 1 mm gap length.
31. Electroporation Buffer (EPB): 0.1 mM MOPS, 10% (w/v) glucose, pH 7.5, sterilized by filtration and stored at 4 °C (*see* **Note 4**).
32. Lennox LB: 10 g/L tryptone, 5 g/L yeast extract, 5 g/L NaCl, sterilized by autoclave.
33. Lennox LB Agar 10 cm Plates: Lennox LB, 15 g/L bacteriological agar, both without any antibiotics and with any of 50 μg/mL kanamycin, 100 μg/mL carbenicillin, or 100 μg/mL spectinomycin as needed.
34. Carbenicillin: 100 mg/mL in water, sterilized by filtration.
35. Kanamycin: 50 mg/mL in water, sterilized by filtration.
36. Spectinomycin: 50 mg/mL in water, sterilized by filtration.
37. Defined Media Components (*see* **Note** 7):
    (a) 25× M salts: 167.5 g of Na_2HPO_4-$7H_2O$, 85 g of KH_2PO_4, 67 g of NH_4Cl, 17.75 g of Na_2SO_4 added sequentially to 500 mL of hot ddH_2O with vigorous stirring, bring to 1 L with ddH_2O, sterilize by autoclaving.
    (b) 40% glucose: 40 g of glucose with ddH_2O to 100 mL total volume, sterilize by autoclaving.
    (c) 80% glycerol: 80 g of glycerol with ddH_2O to 100 mL total volume, sterilize by autoclaving.
    (d) 25% aspartate: 25 g of aspartic acid plus 50 mL of ddH_2O, stir vigorously with slow addition of approximately 8 g of NaOH to dissolve the aspartic acid, bring to a final volume of 100 mL with ddH_2O, sterilize by autoclaving.

(e) 17aa solution: to 80 mL ddH_2O add 1 g each of E, D, K, R, H, A, P, G, T, S, Q, N, V, L, I, F, W, add ddH_2O to 100 mL, sterilize by filtration.

(f) 25 mg/mL methionine, sterilize by filtration.

(g) 18aa solution: 10 mL of 17aa solution, 4 mL of 25 mg/mL methionine.

(h) 1 M $MgSO_4$, sterilize by autoclaving.

(i) 0.1 M $FeCl_3$ in 0.12 M HCl.

(j) 1000× metals: 36 mL of ddH_2O, 50 mL of 0.1 M $FeCl_3$ in 0.12 M HCl, 2 mL of 1 M $CaCl_2$, 1 mL of 1 M $MnCl_2$-$4H_2O$, 1 mL of 1 M $ZnSO_4$-$7H_2O$, 1 mL of 0.2 M $CoCl_2$-$6H_2O$, 2 mL of 0.1 M $CuCl_2$-$2H_2O$, 1 mL of 0.2 M $NiCl_2$-$6H_2O$, 2 mL of 0.1 M Na_2MoO_4-$2H_2O$, 2 mL of 0.1 M Na_2SeO_3-$5H_2O$, 2 mL of 0.1 M H_3BO_3.

38. Minimal expression media (MEM; recipe is for 100 mL and can be linearly scaled as needed): 4 mL of 25× M salts, 200 μL of 1 M $MgSO_4$, 20 μL of 1000× metals, 400 μL of 25% aspartic acid, 2.8 mL of 18aa mix, 1 mL of 40% glucose, 91.6 mL of ddH_2O, antibiotics as appropriate.
39. Plasmid DNA extraction kit or reagents (e.g., Qiagen, Invitrogen kits or similar kits with silica gel spin columns).
40. 10× TE buffer: 100 mM Tris, 10 mM EDTA, pH 8.0 (*see* **Note 4**).
41. 1× TE buffer: 1 mL 10× TE buffer added to 9 mL ddH_2O.
42. Sterile culture tubes for microbiology with caps: 16 mm or 17 mm diameter tubes made from polystyrene, polypropylene, or Pyrex glass.
43. 2 L baffled culture flasks with polypropylene caps.
44. 1 M IPTG solution, sterile filtered.
45. 10× phosphate-buffered saline (10× PBS): 80 g of NaCl, 2 g of KCl, 14.4 g of Na_2HPO_4, 2.4 g of KH_2PO_4 per liter of final 10× solution), sterilize by autoclaving.
46. 1× phosphate-buffered saline (1× PBS): 100 mL of 10× PBS, 900 mL of ddH_2O, sterilized by autoclaving.
47. 0.1 M potassium phosphate buffer, pH 7.
48. Ammonium sulfate (dry solid).
49. Poly(ethylene glycol) 8000 MW (PEG8000; dry solid).
50. Sucrose (dry solid).
51. Agilent BioAnalyzer 2100 and Protein 80 Analysis Kit.
52. FPLC system (e.g., Bio-Rad BioLogic Workstation, GE Healthcare Life Sciences AKTAExplorer).
53. Superose 6 10/300 resin.

3 Methods

In the following methods, the use of several specific software programs is described. These programs have worked well for this author in dozens of experiments and are suggested as an established, successful workflow. Numerous other software programs are available and may work as well as or better than those detailed here.

3.1 Design of Codon-Optimized Display Domains

1. Identify a cDNA clone of the domain of interest.
2. Load the GCUA website and go to the "each triplet position vs. usage table" option.
3. Enter a convenient name for the sequence for "sequence-name." This is for your reference only, but is required.
4. For "originating organism," select "Escherichia_coli_K12".
5. Copy the nucleotide sequence data into the "sequence" box.
6. For "codon usage rules to apply" again select "Escherichia_coli_K12".
7. Leave all other options as their defaults.
8. Click "submit."
9. Examine the output for red or missing bars.
10. Select a strategy for correcting rare-use codons based on the number of such codons (*see* **Note 8**).
11. To design a synthetic open reading frame, load the DNAWorks website.
12. Under "Job Name," enter a convenient name for the sequence (*see* **Note 9**).
13. Under "Codon Frequency Table," select Option 1 and "E. coli class II".
14. If the gene will be synthesized by PCR from oligonucleotides, verify that the "Parameters" box contains accurate information for your reagents and oligo provider.
15. Under "Restriction Site Screen," check the boxes for common restriction endonucleases and any that may be desirable for further cloning work. Also check the boxes for NcoI, XhoI, and BamHI in "Non-degenerate sites" and BlpI in "Degenerate sites."
16. Under "Sequences," select "protein."
17. In the text box, enter one line containing a FASTA record identifier (e.g., ">fusion_protein").
18. On a new line after the FASTA identifier, enter a linker sequence (e.g., GGASESGG; *see* **Note 10**) followed by the target protein sequence.

19. Click "Design Oligos."
20. The resulting logfile and email message will contain an optimized codon sequence and a list of oligonucleotides to order for PCR-based gene synthesis.
21. To the 5′ end of the resulting oligonucleotide sequence, append nucleotides corresponding to the last seven amino acids of the Qβ coat protein (5'-CAGCTGAACCCAGC GTAT-3′) to the of the program's output.
22. To the 3′ end of the sequence, append a TAA stop codon and a nucleotide sequence corresponding to an appropriate restriction endonuclease site (*see* **Note 11**).
23. For self-synthesis of the gene by PCR, append the same sequences to the oligonucleotides produced by DNAWorks before ordering.
24. If there are no problematic codons in the display domain cDNA sequence, design primers for amplifying the cDNA in preparation for creating the CP-display fusion protein (*see* **Note 12**).

3.2 Production of Coat Protein Gene

If a complete synthetic CP-DD gene was procured, proceed directly to Subheading 3.2.4 to generate a clean dsDNA product for ligation. If a suitable cDNA of the domain to be displayed is available, amplify it from that source using PCR to ensure a fully double-stranded starting material starting at Subheading 3.2.3 below. Assembly of oligonucleotides into a synthetic gene has been described but typically requires multiple cycles of PCR to synthesize a 1 kilobase gene product [36–38]. Subheading 3.2.1 is an example of gene assembly by PCR which we used to assemble an 850-bp gene for a fluorescent protein.

3.2.1 Assembly of Fused Domain from Individual Oligonucleotides

1. Upon receipt of the oligonucleotides, dissolve each to a final concentration of 100 μM in 0.1× TE buffer.
2. Create a master mix of the primers designed from DNAWorks by mixing 5 μL of each 100 μM primer stock in a 1.7–2.0 mL microcentrifuge tube.
3. Set up a PCR using the Phusion polymerase reagents: 37.5 μL of water, 1 μL of the primer mix, 10 μL of 5× Phusion HF buffer, 1 μL of 10 mM dNTP mix, and 0.5 μL of the Phusion DNA polymerase.
4. Perform 55 cycles of 98 °C for 15 s, 50 °C for 15 s, and 72 °C for 15 s per kilobase of final DNA length, rounded up to the nearest integer number of kilobases.
5. Purify the long dsDNA product from any remaining primers using a commercial PCR purification kit.

3.2.2 Amplification of Full-Length Assembled Fusion Domain

1. To generate the final DNA sequence, set up a second-round PCR with 32.5 μL of water, 10 μL of 5× Phusion HF buffer, 2.5 μL of 10 μM 5′-most oligonucleotide, 2.5 μL of 10 μM 3′-most oligonucleotide, 1 μL of the purified eluent from Subheading 3.2.1, **step 5**, 1 μL of 10 mM dNTPs, and 0.5 μL of Phusion DNA polymerase.
2. Perform 35 cycles of 98 °C for 15 s, 60 °C for 15 s, 72 °C for the same time as used in Subheading 3.2.1, **step 4**. Add a final extension step of 72 °C for 1 min followed by a hold at 4 °C.
3. Make a 0.8% agarose gel in 0.5× TAE with SYBR Safe (to 1×) or ethidium bromide (to 0.5 μg/mL).
4. Combine half of the post-cycling PCR mixture with loading dye (e.g., 25 μL of reaction, 5 μL of 5× loading dye) and run it out on the gel alongside a molecular weight ladder (per manufacturer's instructions).
5. Excise the band corresponding to the anticipated final product length using a clean razor blade.
6. Extract the DNA from the gel slice using a commercial gel extraction kit and quantify the resulting DNA with the microdroplet spectrophotometer.

3.2.3 Generate a dsDNA Fragment for the CP Portion of the CP-Fusion

To generate the CP-only gene, substitute primer CP-R for CP-Fuse-R in this procedure.

1. Set up a PCR: 10 μL of Phusion HF buffer, 2.5 μL of 10 μM CP-F, 2.5 μL of 10 μM CP-Fuse-R, 1 μL of 10 mM dNTP mix, 1 μL of template DNA (10–100 ng), 0.5 μL of Phusion DNA polymerase.
2. Perform 35 cycles of 98 °C for 15 s, 58 °C for 15 s, and 72 °C for 15 s. Add a final extension step of 72 °C for 1 min followed by a hold at 4 °C.
3. Make a 0.8% agarose gel in 0.5× TAE with SYBR Safe (1×) or ethidium bromide (0.5 μg/mL).
4. Combine half of the post-cycling PCR mixture with loading dye (e.g., 25 μL of reaction, 5 μL of 5× loading dye) and run it out on the gel alongside a molecular weight ladder.
5. Excise the band corresponding to the anticipated final product length (~400 bp) using a clean razor blade.
6. Extract DNA from the gel slice using a commercial gel extraction kit and quantify the resulting DNA.

3.2.4 Generate a CP-Fusion

This procedure will use overlap-extension PCR to combine the products from Subheadings 3.2.2 and 3.2.3.

1. Set up a PCR: 10 μL of Phusion HF buffer, 2.5 μL of 10 μM CP-F, 2.5 μL of the 10 μM 3′ terminal gene synthesis primer, 1 μL of 10 mM dNTP mix, 1 μL of CP DNA (~10 ng) (product

from Subheading 3.2.3), 1 μL of display domain DNA (~10 ng) (product from Subheading 3.2.2), 0.5 μL Phusion DNA Polymerase.

2. Perform 35 cycles of 98 °C for 15 s, 58 °C for 15 s, and 72 °C for 15 s per kilobase of the final product length, rounded up to the nearest integer. Add a final extension step of 72 °C for 1 min followed by a hold at 4 °C.
3. Make a 0.8% agarose gel in 0.5× TAE with SYBR Safe (1×) or ethidium bromide (0.5 μg/mL).
4. Combine half of the post-cycling PCR mixture with loading dye (e.g., 25 μL of reaction, 5 μL of 5× loading dye) and run it out on the gel alongside a molecular weight ladder.
5. Excise the band corresponding to the anticipated final product length using a clean razor blade.
6. Extract DNA from the gel slice using a commercial gel extraction kit and quantify the resulting DNA.

3.3 Incorporation of CP and CP-DD Genes into Expression Vectors

The following methods apply to creating either CP-only cloning or CP-display fusion expression vectors. "Vector" thus indicates the target vector and "insert" thus indicates the CP or CP-fusion gene created by PCR.

3.3.1 Perform Ligation Between Insert and Vector

1. Set up restriction digests of the vector and insert using NcoI and the appropriate C-terminal enzyme (e.g., BlpI) according to the enzyme manufacturer's instructions.
2. Purify desired fragments (*see* **Note 13**).
3. Measure the concentration of the purified DNA fragments using the microdroplet spectrophotometer.
4. Calculate an approximately 3:1 insert–vector molar ratio by dividing the fragment DNA concentration by (654 times the number of DNA base pairs present in the fragment), giving an approximate concentration in nmol/μL. Use the vector concentration as the base value, aim for approximately 100 ng of input vector DNA. Scale the resulting volumes such that their sum is less than 17 μL (*see* **Note 14**).
5. Combine the insert and vector DNAs as calculated above and add water to bring the mixture to 17 μL. Add 2 μL of 10× T4 DNA ligase reaction buffer and 1 μL of T4 DNA ligase. Set up control ligations, where one control replaces the T4 DNA ligase with water and one control where the insert is replaced with water. Leave the three reactions at room temperature overnight.
6. Remove salts from the ligation reactions by purifying with a commercial DNA purification kit. Elute the DNA from the column with 30 μL of pH 8.0 water.

3.3.2 Transformation of the Ligation Products and Isolation of Clonal Plasmids

1. Streak out the cloning strain glycerol stock onto a Lennox LB plate lacking antibiotics and incubate overnight at 37 °C.
2. Place EPB solution and centrifuge rotors in a cold room or refrigerator to chill overnight.
3. After overnight ligation (*see* Subheading 3.3.1) and plate growth, inoculate a 25 mL baffled flask containing 5 mL of Lennox LB, 5 mM $MgSO_4$, and 0.2% (w/v) glucose with an isolate clone from the agar plate. Shake at 37 °C and approximately 250 rpm.
4. Place 4 Lennox LB Agar plates containing antibiotics appropriate for the ligated vector at 37 °C to warm up.
5. Monitor the optical density of the 5 mL culture of the cloning strain until it reaches an absorbance at 600 nm of approximately 0.4 at a 1 cm optical path length.
6. Transfer the culture to an ice bucket containing an ice–water slurry. Swirl intermittently over the course of approximately 15 min to chill well.
7. Label four electroporation cuvettes and four microcentrifuge tubes with "no DNA," "no T4," "no insert," and "exp." Transfer 1 μL from each desalted ligation to the bottom of the appropriately labeled microcentrifuge tubes. For the "no DNA" tube, place 1 μL of the water used for silica gel column elution at the bottom of the tube. Place the EPB buffer, 4 electroporation cuvettes, and 4 microcentrifuge tubes into wet ice, submerging the cuvettes only as far as the top of the conductive plates.
8. Transfer the chilled culture to a 50 mL conical tube and collect the cells by centrifugation in a chilled rotor (e.g., ≥7,000 × *g* for 10 min) (*see* **Note 15**).
9. Carefully decant the supernatant and invert the tube on a dry paper towel for approximately 30 s (*see* **Note 16**).
10. Resuspend the cell pellet in 1 mL of cold EPB, transfer to a microcentrifuge tube, and chill the microcentrifuge tube in the ice–water slurry for 1 min.
11. Collect the cells by centrifugation in a chilled microcentrifuge (e.g., 20,000 × *g* for 1 min).
12. Remove the cleared supernatant by aspiration with a micropipette and resuspend the cells in 1 mL of fresh, cold EPB.
13. Repeat the centrifugation and resuspension, **steps 11** and **12**, 2 more times.
14. Collect the cells by centrifugation as before (**step 11**), aspirate the cleared supernatant, and resuspend the cells in 200 μL of fresh, cold EPB.

15. For each labeled microcentrifuge tube prepared in **step 7**, perform **steps 17–21** in order, completing the steps for one sample before moving to the next sample.
16. Transfer 40 μL of the 200 μL cell slurry to the tube containing 1 μL of sample. Mix by pipetting up and down once.
17. Transfer the cells-sample mixture to the appropriate chilled electroporation cuvette. Confirm visually that the mixture contacts both electrodes of the cuvette.
18. Place the cuvette in the electroporator and press the electroporation start button.
19. Record the decay time constant displayed by the electroporator, and immediately flood the cuvette at the top of the electrode gap with 1 mL of Lennox LB containing 5 mM $MgSO_4$ (5 μL 1 M $MgSO_4$) and 0.2% w/v glucose (5 μL 40% glucose).
20. Mix the media with the cells by pipetting up and down several times or until the bulk of the media is cloudy with cells.
21. Transfer by micropipette as much of the media and cells which can be retrieved from the cuvette. This is typically greater than 900 μL.
22. Shake the cell cultures at 37 °C for 1.5 h.
23. Transfer 400 μL of each culture into labeled microcentrifuge tubes and collect the cells by centrifugation (~20,000 × *g* for 1 min). A chilled centrifuge is not required here.
24. Aspirate and dispose of the supernatant from each tube. Resuspend the cells in 100 μL of room-temperature EPB, and spread all cells onto a prewarmed Lennox LB Agar plate from **step 4**.
25. Incubate the plates overnight at 37 °C. Save the remaining recovery cultures from **step 23** at room temperature.
26. The following day, check the plates for colonies.
27. If there are no colonies on any plate, collect the remaining volume from the recovery cultures and plate them on four fresh, selective LB plates and incubate again overnight (*see* **Note 17**).

3.3.3 Analyze Resulting Colonies for Successfully Ligated Plasmids

1. The ratio of colonies on the experimental plate to the control plate with the highest number of colonies should be at least five. Streak six clones from the experimental plate onto new, selective agar plates and incubate overnight at 37 °C (*see* **Note 17**).
2. Inoculate six culture tubes containing 1 mL of Lennox LB plus 5 mM $MgSO_4$, 0.2% w/v glucose, and the appropriate antibiotic drugs with one subclone each from each restreaked clone. Shake the cultures at 37 °C and 250 rpm overnight.

3. The following day, make glycerol stocks of each culture by mixing 100 μL of the culture with 50 μL of 80% w/v glycerol. Freeze at −80 °C.
4. Extract plasmid DNA from the remaining volumes of cultures as specified by the protocol or kit manufacturer, eluting with 50 μL of elution buffer (either 1× TE or the manufacturer's buffer). Determine the DNA concentration using the microdroplet spectrophotometer and store the eluent at 4 °C (*see* **Note 18**).
5. Set up restriction digests with 20 μL of plasmid DNA using the same restriction endonucleases as used in Subheading 3.3.1 **step 1**.
6. After digestion of the DNA, check the fragment sizes with a 0.8% agarose gel as in Subheading 3.2.3, **steps 3–5**.
7. For clones whose fragment sizes appear to be correct, submit 10–20 μL of the remaining DNA for capillary sequencing using the T7F and T7R primers which most sequencing providers provide as a courtesy.
8. Confirm that the incorporated sequence and the expected sequence are the same (*see* **Note 19**).

3.3.4 Perform Transformation of *E. coli* BL21(DE3) for Expression

1. Prepare expression strains of *E. coli* BL21(DE3) by following Subheading 3.3.2, **steps 1–27**, substituting the BL21(DE3) strain for the cloning strain and extracted plasmid for ligation products. For transformation with one plasmid, use 1 μL of plasmid stock from Subheading 3.3.3, **step 4**. For transformation with two plasmids, both plasmids can be simultaneously co-transformed using 1 μL of each plasmid and plating the resulting recovery cultures on appropriately double-selective Lennox LB Agar. Prepare transformed strains with the following combinations of plasmid DNA:
2. pET-11-CP alone
3. pET-28-CP alone
4. pCDF-CP-DD alone
5. pET-11-CP + pCDF-CP-DD
6. pET-28-CP + pCDF-CP-DD
7. Confirm that the resulting transformants contain the expected plasmids by plasmid extraction, restriction digestion, and gel electrophoresis. For those which appear to be successful, create glycerol stocks as in Subheading 3.3.3, **step 3**.

3.4 Expression of CP/CP-DD Nanoparticles

The following steps cover protein expression for either test expressions or for large scale bulk expression. Test expressions should be performed for all five strains created in Subheading 3.3.4. After confirmation of nanoparticle integrity, scale-up to bulk expression is a simple matter of scaling up the volumes of media, culture flasks, and centrifuge rotors.

1. Streak out the glycerol stock of the desired strain onto Lennox LB Agar with the appropriate antibiotics. Incubate overnight at 37 °C.
2. The following day, prepare 100 mL of MEM.
3. Inoculate 1 mL of MEM in a 16/17 mm culture tube with a clone from the streaked cells. Shake at 37 °C overnight. This culture can serve as an inoculum culture for about 1 week (*see* **Note 20**).
4. The following morning, inoculate 5 mL of MEM in a 25 mL culture flask with 10 μL of the previous overnight culture. Shake the culture at 37 °C and 250 rpm.
5. Monitor the growth of the culture by OD_{600} every 20–30 min until it is no longer increasing.
6. Plot the $\log_2(OD_{600})$ data versus time in a spreadsheet. Note the doubling time of the culture by the time it takes the culture to move 1.0 in the *y*-axis.
7. Find the point at which the log2-transformed OD_{600} data is no longer linear. Subtract 1.5 culture density doubling times from that time and note the OD_{600}. This is an appropriate OD_{600} at which to induce protein expression (*see* **Note 21**).
8. The following day, inoculate a 5 mL culture as in **step 4** using the same 1 mL overnight culture as was used in **step 4**.
9. Monitor the growth of the culture until the OD_{600} reaches the level identified in **step** 7.
10. Collect 0.25 mL of cells from the culture in a microcentrifuge tube and then add 4.5 μL of 1 M IPTG solution to the remaining culture in the shaking incubator. Note the time. Record the OD_{600} of the 0.25 mL aliquot.
11. Pellet the cells from the 0.25 mL of culture media in a microcentrifuge at ~20,000 × *g* for 1 min, aspirate the supernatant, and freeze the cell pellet at −80 °C for later analysis.
12. Every 1 h after the time noted in **step 10** ending at approximately 6 h, collect another 0.25 mL of culture media. Check the OD_{600} and collect the cells by centrifugation. Aspirate the supernatant and store the cell pellet at −80 °C.
13. After 6 h, collect the remaining culture by centrifugation (e.g., in a 50 mL conical tube at ≥7,000 × *g* for 10 min) and store the resulting cell pellet at −80 °C.
14. Thaw the cell pellets from the 0.25 mL samples and resuspend the cell pellets in a volume of 1× PBS appropriate to bring each sample to roughly the same OD_{600}, using 100 μL for the first

(preinduction) sample. These volumes can be calculated from the OD_{600} values noted in **steps 10** and **12**.

15. Analyze the protein content of the cells by SDS-PAGE or BioAnalyzer Protein 80 chip. Confirm that the CP and CP-DD fusions are produced at the expected molecular weights.
16. Prepare recombinant nanoparticles from the remaining large terminal cell pellet by following steps in the next section to confirm intact nanoparticles.

3.5 Bulk Purification of CP/CP-DD Nanoparticles

1. Thaw a large pellet of post-induction cells to room temperature.
2. Resuspend the cells in approximately 200 mL of 0.1 M pH 7 potassium phosphate buffer per liter of final culture volume. Transfer the cells to a glass beaker if the volume is ≥50 mL, or a plastic conical tube for smaller volumes such that the depth of the sample is about twice the air-exposed surface diameter.
3. Chill the resuspended cells in an ice–water slurry and mount the container in the probe sonicator (*see* **Note 22**).
4. Sonicate the samples (*see* **Note 23**).
5. Remove cell debris by centrifugation (*see* **Note 24**).
6. Add 0.265 grams of ammonium sulfate per milliliter of cleared lysate. Stir or shake to dissolve, then incubate at 4 °C with inversion (e.g., on a rotisserie) for 15 min.
7. Collect precipitated material by centrifugation as in **step 5** for half the necessary time (e.g., 5 min in the example configuration of **step 5**).
8. Decant the supernatant, being careful not to disturb the packed precipitate.
9. Redissolve the precipitated material in 1 mL of 0.1 M pH 7 potassium phosphate buffer for each 25 mL of final culture volume (*see* **Note 25**).
10. Prepare a 10–40% sucrose gradient in tubes appropriate for the swinging bucket ultracentrifuge rotor.
11. Add up to 1 mL of redissolved particles to the top of each SW41-sized sucrose gradient.
12. For an SW41 rotor, run the gradients at 38,000 rpm (~287,000 × *g*) for 3 h. Check for a side-scattering band by illumination from below.
    (a) If the band has migrated approximately halfway through the gradient, carefully withdraw the band using a serological pipette inserted from the top. Alternatively, a syringe can be inserted through the side of the gradient tube to more carefully avoid smaller contaminants from the top of the gradient.

(b) If the band has not migrated sufficiently through the gradient, continue the centrifugation run until separation has been achieved.

13. Precipitate nanoparticles by the addition of PEG8000 to the recovered sucrose-nanoparticle solution to achieve a 10% w/v concentration.
14. Incubate the PEG-containing solution at 4 °C with mixing by inversion for 30 min.
15. Collect the precipitated nanoparticles by centrifugation as in **step 5**.
16. Aspirate or decant the supernatant and redissolve the particles in the desired storage buffer (e.g., 0.1 M pH 7 potassium phosphate).
17. Assess the protein content of the nanoparticles using a BioAnalyzer Protein 80 microchip (*see* **Note 26**).
18. Determine the ratio of displayed domains per particle by integrating the resulting monomer peaks (e.g., Protein 80 chip). Multiply the molar ratio of the CP-DD:CP by 180 to determine the number of domains displayed on the particles.
19. Further analysis by light scattering to determine particle size distribution or cryo-electron microscopy may be used to confirm homogeneity of the resulting nanoparticle samples.

4 Notes

1. Bacteriophage Qβ coat protein cDNA can be synthesized by commercial services (e.g., IDT, Inc.), amplified from existing vectors, or amplified after reverse transcription of the Qβ genomic RNA (NCBI Reference Sequence NC_001890.1).
2. Avoid loading dyes containing bromophenol blue, which will obscure the band corresponding to 300–500 bp fragments.
3. Any of a variety of kits may be used. Self-produced reagents for gel extraction (e.g., sodium iodide, 3 M sodium acetate pH 5, and 1× TE) have been used by the author with commercial silica gel spin columns without loss of yield.
4. Details of adjusting the pH of a given solution depends largely on the chemical reagents available in a given laboratory. For some buffers, such as Tris, it is equally common in our experience to find either the free amine base or hydrochloride salt as a stocked chemical. Thus, how the pH of a given solution is adjusted will depend on the nature of the stocked chemical. Continuing the example of Tris buffer, laboratories which stock the free amine will also need a solution of hydrochloric acid for titrating the pH of the solution resulting from the

dissolution of Tris. Laboratories which stock the hydrochloride salt of Tris will require a solution of a hydroxide base (e.g., NaOH) to titrate the solution resulting from the dissolution of Tris–HCl.

5. DH5α, DH10B, and JM109 are all excellent cloning strains. Any strain which is routinely used for molecular cloning is acceptable here, provided that they have an *endA* mutation and no intrinsic antibiotic resistances which conflict with the markers in the pET and pCDF plasmids (carbenicillin, kanamycin, spectinomycin).
6. Using only the provided sequence primers and the suggested pET vectors (e.g., only cloning the unfused CP), NcoI and BlpI are required. For fusion proteins, at least one other restriction endonuclease will be required based on the final nucleotide-level design.
7. The recipes and protocols found in F. W. Studier's "Stable Expression Clones and Auto-Induction for Protein Production in *E. coli*" [39] should be used for the preparation of these reagents. The recipe for 25× M salts is identical to the recipe for 50× M salts, but the final volume of the solution is doubled. This prevents the occasional precipitation of salts from the stock solution over time.
8. In the case where only one or two rare codons are present, it may be easiest and most economical to change these to high-use codons (class II [18]) by site-directed mutagenesis [40]. The PrimerX software (http://bioinformatics.org/primerx/cgi-bin/DNA_1.cgi) is an excellent option for designing the necessary primers. If there are three or more rare codons in the sequence, whole-gene synthesis is often a faster option.
9. Entering an email address is recommended to create a record of the project and results.
10. Experiments with a variety of linker sequence lengths of 4, 6, and 12 amino acids of similar AA composition revealed no substantial increases in the number of fusions incorporated per-particle and in some cases a slight decrease in the number of fusions per-particle. However, a given domain to be displayed may have particular requirements for linkers which have been previously established. Given previous results with linker sequence length, other predetermined linker sequences may reasonably be expected to be well tolerated.
11. Any unique site downstream from and excluding NcoI (e.g., KpnI, BamHI, SacI) plus four extra nucleotides to facilitate endonuclease activity.
12. The reverse primer is unremarkable except that it should contain the stop codon (preferably a TAA) and nucleotides for restriction digestion and subsequent ligation into the expression

vector. The forward primer is more specialized. Identify an appropriate-length sequence which balances the number of nucleotides for priming with the melting temperature of the sequence, ideally ending with a C or G at the 3′ end. Inclusion of the initiating methionine residue is optional. To the 5′ end of the priming sequence, append a Qβ overlap sequence (5′-CAGCTGAACCCAGCGTAT -3′) and a linker sequence such as GGASESGG (5′- GGTGGCGCGTCAGAATCGGGCGGT -3′; also *see* **Note 10**).

13. For digests of insert DNA created by PCR, the short fragments liberated from the termini of the PCR product can be removed by performing silica gel spin-column purification without an intermediary gel and band excision. For digests of vector DNA, gel purification is desirable. The liberated multiple cloning site fragments are well-separated from the significantly longer backbone DNA fragment and result in lower background of recircularized plasmid DNA after ligation. Alternatively, the digest can be treated with a phosphatase enzyme to reduce background recircularization, but this also reduces the rate of successful ligation.
14. In practice, the vector concentration is most often the limiting reagent.
15. Cells are stable at 4 °C in the growth medium for several hours. After the cells are first pelleted and resuspended in EPB for the first wash, competency decreases over the course of several hours. It is important that the cells stay ice-cold during the washing and electroporation process; a rise in temperature when the cells are not in growth media will significantly reduce the competency and viability of the cells.
16. The cell pellet is relatively loose and may be lost if excessive force is used to tap out remaining cleared media.
17. The experimental, noncontrol plate should contain at least 5 times more colonies than any of the control plates. The molecular factors behind unsuccessful ligation reactions are not as difficult to diagnose as they are time-consuming to diagnose. First, it is prudent to repeat the transformation twice before concluding that the enzymatic steps are problematic. There seems to be a "warm-up" period for individual researchers working independently such that the third attempt to generate a ligation clone is successful. The decay time constant from the electroporator provides a key metric for diagnosing problems with bench technique. A good time constant will vary between devices but generally should exceed 4.5 ms. To determine the upper limit for a particular electroporator, run the electroporation program using a cuvette containing only ddH_2O. When the transformation has failed 3 times, a lack of colonies or a poor ratio of colony counts may be the result of one or more

of the following potential problems: low input DNA concentration, poor electrocompetent cell preparation (e.g., cells warmed before post-shock recovery), poor time constant during electroporation, low resistivity of the electroporation buffer, incomplete digestion of insert or vector prior to ligation.

18. If DNA concentrations are less than 20 ng/μL, regrow the subclone, perhaps over several 1 mL cultures, and serially collect eluted DNA from each silica gel column using the same elution buffer volume.
19. The online BLAST software provided by the NCBI is useful in this regard (https://blast.ncbi.nlm.nih.gov/Blast.cgi).
20. *E. coli* which have exhausted nutrients of their growth medium enter the stationary phase of growth. After a long period of stationary-phase growth, the cells begin to slowly die completely by lysis. This process is hastened by the often low pH of complex mediums such as Lennox LB. MEM medium is formulated such that the pH of the exhausted medium is close to the ideal growth pH. As a result, the rate of cell death in the stationary phase is much lower in MEM than it is in complex mediums and the stationary culture can be used to inoculate further subcultures for many days.
21. In some cases, improved yield of nanoparticles can result from varying the cell density at which expression is induced. Prior experiments have successfully induced at an OD_{600} of 2.0 for cells growing in MEM, but if the accumulation of nanoparticles is not toxic to the cells (as revealed by a lack of change in growth rate after addition of IPTG), expression can be induced earlier if convenient.
22. A tip diameter of 2–3 mm is appropriate for small samples (e.g., <25 mL) and a larger tip diameter up to 1 cm may be used for larger samples. It is critical for good cell lysis that the probe tip is polished at the start of a sonication run.
23. The necessary power and time will vary depending on the model of sonicator, tip size, and sample volume. For a tip diameter of 2.5 mm, 30 W of power applied in 5 s bursts followed by 5 s of rest for a total of 5 min can sufficiently disrupt a sample.
24. Centrifugation is generally longer than required for collecting cells. For example, using a Beckman JA-17 rotor, 10 min of centrifugation at 15,000 rpm (30,913 × *g*) is sufficient to collect debris. If the volume of debris is roughly the same as the original cell pellet, sufficient lysis was not achieved. Repeat sonication with adjusted power and time to effect lysis.
25. If the displayed domain is not sensitive to organic solvents, lipids may be extracted at this point using a volume of 1:1

n-butanol–chloroform mixture. After vigorous mixing and centrifugation to separate the organic and aqueous layers, the upper aqueous layer contains the nanoparticles and may be carried forward to sucrose gradients.

26. If a BioAnalyzer instrument is not available, a careful concentration gradient of nanoparticle samples run on a denaturing SDS-PAGE gel may permit the quantitation of displayed domains. However, this method is technically challenging and usually less accurate than the BioAnalyzer integrated signals.

References

1. Mammen M, Choi S-K, Whitesides GM (1998) Polyvalent interactions in biological systems: implications for design and use of multivalent ligands and inhibitors. Angew Chem Int Ed 37(20):2754–2794
2. Crothers DM, Metzger H (1972) The influence of polyvalency on the binding properties of antibodies. Immunochemistry 9(3):341–357
3. Matrosovich MN, Mochalova LV, Marinina VP et al (1990) Synthetic polymeric sialoside inhibitors of influenza virus receptor-binding activity. FEBS J 1(2):209–212
4. Johansson SM, Arnberg N, Elofsson M et al (2005) Multivalent HSA conjugates of 3′-sialyllactose are potent inhibitors of adenoviral cell attachment and infection. Chembiochem 6(2):358–364. https://doi.org/10.1002/cbic.200400227
5. Zhang L, Qiu W, Crooke S et al (2017) Development of autologous C5 vaccine nanoparticles to reduce intravascular hemolysis in vivo. ACS Chem Biol 12(2):539–547. https://doi.org/10.1021/acschembio.6b00994
6. Astronomo RD, Kaltgrad E, Udit AK et al (2010) Defining criteria for oligomannose immunogens for HIV using icosahedral virus capsid scaffolds. Chem Biol 17(4):357–370. https://doi.org/10.1016/j.chembiol.2010.03.012
7. Toy R, Roy K (2016) Engineering nanoparticles to overcome barriers to immunotherapy. Bioeng Transl Med 1(1):47–62. https://doi.org/10.1002/btm2.10005
8. Kaltgrad E, Sen Gupta S, Punna S et al (2007) Anti-carbohydrate antibodies elicited by polyvalent display on a viral scaffold. Chembiochem 8(12):1455–1462. https://doi.org/10.1002/cbic.200700225
9. Lamanna AC, Gestwicki JE, Strong LE et al (2002) Conserved amplification of chemotactic responses through chemoreceptor interactions. J Bacteriol 184(18):4981–4987. https://doi.org/10.1128/jb.184.18.4981-4987.2002
10. Borrok MJ, Kolonko EM, Kiessling LL (2008) Chemical probes of bacterial signal transduction reveal that repellents stabilize and attractants destabilize the chemoreceptor array. ACS Chem Biol 3(2):101–109
11. Handa H, Gurczynski S, Jackson MP et al (2010) Immobilization and molecular interactions between bacteriophage and lipopolysaccharide bilayers. Langmuir 26(14):12095–12103. https://doi.org/10.1021/la1013413
12. Karnik S, Billeter M (1983) The lysis function of RNA bacteriophage Qβ is mediated by the maturation (A2) protein. EMBO J 2(9):1521–1526
13. Golmohammadi R, Fridborg K, Bundule M et al (1996) The crystal structure of bacteriophage Qβ at 3.5 angstrom resolution. Structure 4(5):543–554
14. Fiedler JD, Higginson C, Hovlid ML et al (2012) Engineered mutations change the structure and stability of a virus-like particle. Biomacromolecules 13(8):2339–2348. https://doi.org/10.1021/bm300590x
15. Fuhrmann M, Hausherr A, Ferbitz L et al (2004) Monitoring dynamic expression of nuclear genes in *Chlamydomonas reinhardtii* by using a synthetic luciferase reporter gene. Plant Mol Biol 55(6):869–881
16. Rosano GL, Ceccarelli EA (2009) Rare codon content affects the solubility of recombinant proteins in a codon bias-adjusted Escherichia coli strain. Microb Cell Factories 8:41. https://doi.org/10.1186/1475-2859-8-41
17. Hoover DM, Lubkowski J (2002) DNAWorks: an automated method for designing oligonucleotides for PCR-based gene synthesis. Nucleic Acids Res 30(10):e43
18. Hénaut A, Danchin A (1996) Analysis and predictions from *Escherichia coli* sequences, or *E. coli* in silico. In: Neidhardt FC (ed) *Escherichia coli* and *Salmonella*, vol 2
19. Structural Genomics C, China Structural Genomics C, Northeast Structural Genomics C

et al (2008) Protein production and purification. Nat Methods 5(2):135–146. https://doi.org/10.1038/nmeth.f.202

20. Oganesyan N, Ankoudinova I, Kim SH et al (2007) Effect of osmotic stress and heat shock in recombinant protein overexpression and crystallization. Protein Expr Purif 52(2):280–285. https://doi.org/10.1016/j.pep.2006.09.015
21. Blackwell JR, Horgan R (1991) A novel strategy for production of a highly expressed recombinant protein in an active form. FEBS 295(1–3):10–12
22. Sandee D, Tungpradabkul S, Kurokawa Y et al (2005) Combination of Dsb coexpression and an addition of sorbitol markedly enhanced soluble expression of single-chain Fv in Escherichia coli. Biotechnol Bioeng 91(4):418–424. https://doi.org/10.1002/bit.20524
23. Barth S, Huhn M, Matthey B et al (2000) Compatible-solute-supported periplasmic expression of functional recombinant proteins under stress conditions. Appl Environ Microbiol 66(4):1572–1579
24. Picaud S, Olsson ME, Brodelius PE (2007) Improved conditions for production of recombinant plant sesquiterpene synthases in Escherichia coli. Protein Expr Purif 51(1):71–79. https://doi.org/10.1016/j.pep.2006.06.025
25. Prasad S, Khadatare PB, Roy I (2011) Effect of chemical chaperones in improving the solubility of recombinant proteins in Escherichia coli. Appl Environ Microbiol 77(13):4603–4609. https://doi.org/10.1128/AEM.05259-11
26. Schaffner J, Winter J, Rudolph R et al (2001) Cosecretion of chaperones and low-molecular-size medium additives increases the yield of recombinant disulfide-bridged proteins. Appl Environ Microbiol 67(9):3994–4000. https://doi.org/10.1128/aem.67.9.3994-4000.2001
27. Kusano K, Waterman MR, Sakaguchi M et al (1999) Protein synthesis inhibitors and ethanol selectively enhance heterologous expression of P450s and related proteins in *Escherichia coli*. Arch Biochem Biophys 367(1):129–136
28. Chhetri G, Kalita P, Tripathi T (2015) An efficient protocol to enhance recombinant protein expression using ethanol in Escherichia coli. MethodsX 2:385–391. https://doi.org/10.1016/j.mex.2015.09.005
29. Lobstein J, Emrich CA, Jeans C et al (2012) SHuffle, a novel Escherichia coli protein expression strain capable of correctly folding disulfide bonded proteins in its cytoplasm. Microb Cell Factories 11:56. https://doi.org/10.1186/1475-2859-11-56
30. Geogiou G, Valax P (1996) Expression of correctly folded proteins in *Escherichia coli*. Curr Opin Biotechnol 7(2):190–197
31. Planesse C, Nativel B, Iwema T et al (2015) Recombinant human HSP60 produced in ClearColi BL21(DE3) does not activate the NFkappaB pathway. Cytokine 73(1):190–195. https://doi.org/10.1016/j.cyto.2015.01.028
32. Brown SD, Fiedler JD, Finn MG (2009) Assembly of hybrid bacteriophage Qβ virus-like particles. Biochemistry 48(47):11155–11157. https://doi.org/10.1021/bi901306p
33. Brown SD (2010) Bacteriophage Qβ: a versatile platform for Nanoengineering. The Scripps Research Institute, La Jolla, CA
34. Udit AK, Brown S, Baksh MM et al (2008) Immobilization of bacteriophage Qβ on metal-derivatized surfaces via polyvalent display of hexahistidine tags. J Inorg Biochem 102(12): 2142–2146. https://doi.org/10.1016/j.jinorgbio.2008.08.003
35. Udit AK, Hollingsworth W, Choi K (2010) Metal- and metallocycle-binding sites engineered into polyvalent virus-like scaffolds. Bioconjug Chem 21(2):399–404
36. TerMaat JR, Pienaar E, Whitney SE et al (2009) Gene synthesis by integrated polymerase chain assembly and PCR amplification using a high-speed thermocycler. J Microbiol Methods 79(3):295–300. https://doi.org/10.1016/j.mimet.2009.09.015
37. Xiong AS, Yao QH, Peng RH et al (2006) PCR-based accurate synthesis of long DNA sequences. Nat Protoc 1(2):791–797. https://doi.org/10.1038/nprot.2006.103
38. Marsic D, Hughes RC, Byrne-Steele ML et al (2008) PCR-based gene synthesis to produce recombinant proteins for crystallization. BMC Biotechnol 8:44. https://doi.org/10.1186/1472-6750-8-44
39. Studier FW (2014) Stable expression clones and auto-induction for protein production in *E. coli*. Methods Mol Biol 1091:17–32. https://doi.org/10.1007/978-1-62703-691-7_2
40. Liu H, Naismith JH (2008) An efficient one-step site-directed deletion, insertion, single and multiple-site plasmid mutagenesis protocol. BMC Biotechnol 8:91. https://doi.org/10.1186/1472-6750-8-91

Chapter 11

Synthesis of Triangular Silver and Gold Nanoprisms Using Consensus Sequence Tetratricopeptide Repeat Proteins

Xi Geng and Tijana Z. Grove

Abstract

Anisotropic metallic nanoparticles, such as Au and Ag nanoprisms (NPSMs), have received tremendous attention for their application in catalysis, molecular sensing, signal amplification, bioimaging, and therapeutic applications due to their shape-dependent optical and physical properties. Herein, we present a protein-enabled synthetic strategy for the seeded growth of silver and gold NPSMs with low shape polydispersity, narrow size distribution, and tailored plasmonic absorbance. During the initial seed nucleation step, *c*onsensus sequence *t*etratrico*p*eptide *r*epeat (CTPR) proteins are utilized as potent stabilizers to facilitate the formation of planar-twinned Ag seeds. High yield production of well-defined Ag/Au NPSMs is achieved, respectively, by adding CTPR-stabilized Ag seeds into the growth solutions containing metal precursor, mild reducing agent, sodium halide, and additional CTPR.

Key words Au nanoprisms, Ag nanoprisms, Repeat protein, Seed-mediated growth, Biomineralization, Plasmonic nanoparticles, Localized surface plasmon resonance

1 Introduction

Over the past decade, much attention and research have been devoted to the synthesis of plasmonic metallic silver (Ag) and gold (Au) nanoprisms (NPSMs). Such NPSMs are exceptionally well suited for biomedical applications such as biosensing [1], diagnostics and therapeutics [2] due to their highly anisotropic structure and localized surface plasmon resonance (LSPR) properties. Hitherto, a wide variety of colloidal routes have been developed to prepare well-defined Ag and Au NPSMs [3, 4]. For instance, photo-induced approaches through selective plasmon excitation enabled the efficient conversion of Ag nanospheres into NPSMs with low polydispersity [5–8]. Thermal approaches using synthetic polymers with terminal hydroxyl groups and polar organic solvents provide simple and rapid routes toward high-yield production of Ag NPSMs [9, 10]. On the other hand, the synthesis of planar Au nanoparticles (NPs) is typically achieved via cetyltrimethylammonium bromide/chloride (CTAB/CTAC)-based

Andrew K. Udit (ed.), *Protein Scaffolds: Design, Synthesis, and Applications*, Methods in Molecular Biology, vol. 1798, https://doi.org/10.1007/978-1-4939-7893-9_11, © Springer Science+Business Media, LLC, part of Springer Nature 2018

protocols [11–18]. In addition, seed-mediated growth along with the addition of shape-directing halides has been invoked as the most potent tools for directing the anisotropic growth of noble metal NPs [17–22]. The exquisite shape control is mainly realized through kinetic control as well as the preferential binding to the low index facets. However, a number of challenges remain. Despite high-precision control over the morphology of NPs, harsh experimental conditions and energy-consuming processes including elevated temperatures and hazardous chemicals are involved. Furthermore, biomedical applications necessitate tedious and stringent purification processes for the complete removal of toxic cationic surfactants [23, 24].

A growing interest has thus been focused on biocompatible and green synthetic approaches that mimic the natural biomineralizaton process at ambient conditions [3, 25]. In previous studies, a wide array of biopolymers has demonstrated dual functions as stabilizers and reducing agents to produce anisotropic Au and Ag nanoplates [26–32]. To our knowledge, current advances in biomimetic synthesis of anisotropic NPs are still centered on one-pot biomimetic methods, which typically produce NPs with rather broader shape and size polydispersity [26–31].

We have demonstrated that the incorporation of seed-mediated techniques into conventional biomineralizaton yields unprecedented control over NP size and shape while maintaining biocompatibility requirements [33, 34]. A modified seed-mediated synthetic strategy was developed for the growth of silver NPSMs with low shape polydispersity, narrow size distribution and tailored plasmonic absorbance. During the seed nucleation step, CTPR proteins are utilized as potent stabilizers to facilitate the formation of planar-twinned Ag seeds. Ag nanoprisms can be produced in high yield in a growth solution containing ascorbic acid and CTPR-stabilized Ag seeds [34]. Notably, this robust synthetic technique can be extended to fabricate gold NPSMs with tunable edge length in an analogous manner. The well-defined morphologies, tunable optical properties, high colloidal stability and biocompatibility are attributed to the synergistic roles of CTPR proteins, halide ions, and CTPR-stabilized Ag seeds [33]. Here we described the formation of NPSMs using CTPR3 (Fig. 1).

2 Materials

All aqueous solutions are made in deionized (DI) water (>18 MΩ/cm).

2.1 Chemicals and Reagents

1. CTPR proteins with three tandem repeating units (CTPR3, Fig. 1) [35, 36] (*see* **Notes 1** and **2**).
2. MOPS buffer: 100 mM 3-(N-morpholino)propanesulfonic acid (MOPS), pH 7.3.
3. Ampicillin.

(a) AEAWYNLGNAYYKQGDYQKAIEYYQKALELDPNN

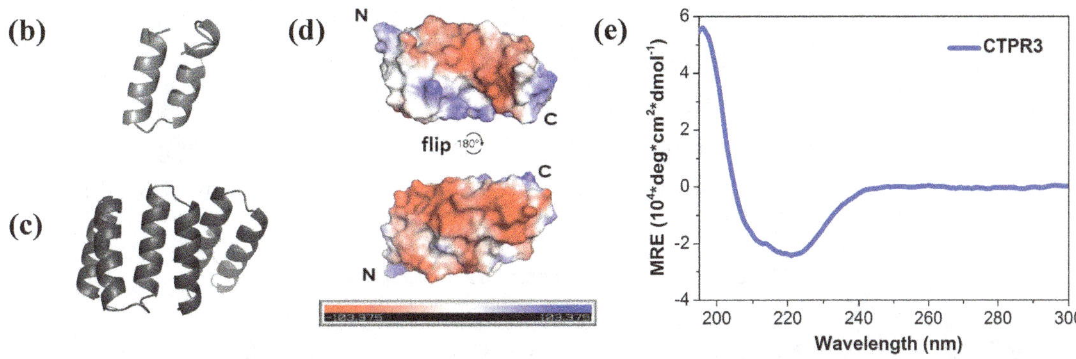

Fig. 1 (**a**) Single letter amino acid sequence of the CTPR repeat. (**b**) Cartoon representation of CTPR repeat unit with a helix-turn-helix structure (**c**) CTPR3 consisting of three tandem repeats. (**d**) An electrostatic potential map of CTPR3 with the surface charge shown; the color range from deep red to deep blue corresponds to an electrostatic potential range from −108 to +108 kT/e, where k is Boltzmann's constant, T is absolute temperature, and e is the charge of a proton. All figures were prepared from X-ray coordinates (PDB 1NA0) using Pymol. (N:N-terminus, C:C-terminus) (**e**) Circular dichroism (CD) spectrum of CTPR protein (MRE: mean residue ellipticity) (Reproduced from Geng et al. (2015) with permission from the Royal Society of Chemistry)

4. Ascorbic acid solution (0.15 mM).
5. Bovine serum albumin (BSA).
6. Dithiothreitol (DTT).
7. *E.Coli* BL21 (DE3).
8. Fetal bovine serum (FBS).
9. Hydrogen tetrachloroaurate(III) hydrate ($HAuCl_4$) solution (10 mM).
10. Isopropyl-β-D-thiogalactoside (IPTG) solution (1 M).
11. Luria–Bertani (LB) liquid media.
12. Lysozyme (activity >23,500 U/mg).
13. Silver nitrate ($AgNO_3$) solution (0.5 mM).
14. Sodium borohydride ($NaBH_4$) solution (10 mM).
15. Sodium bromide (NaBr) solution (1 M).
16. Citrate solution: 2.5 mM trisodium citrate.
17. Tobacco etch virus (TEV) protease (activity >10,000 U/mg protein, Sigma-Aldrich).
18. Ni-NTA Resin.
19. Lysis buffer: 50 mM Tris, 300 mM sodium chloride, 0.1 wt% Tween 20, pH 8.0.
20. Dialysis buffer: 10 mM sodium phosphate, pH 8.0.
21. Chromatography buffer: 50 mM sodium phosphate, 150 mM sodium chloride, pH 8.0.

2.2 Instrumentation

1. Incubator shaker.
2. Centrifuges.
3. Plastic centrifuge tubes with 1.5 mL and 15 mL capacity.
4. Ultrasonic homogenizer with microtip.
5. Fast protein liquid chromatography (FPLC) system.
6. Superdex 75 16/600 size-exclusion chromatography column.
7. Dialysis cassette with molecular weight cutoffs of 3.5 kDa.
8. UV-Vis spectrophotometer.
9. UV-Vis-NIR spectrophotometer.
10. NanoDrop spectrophotometer.
11. 1 cm path length PMMA or quartz cuvette.
12. Transmission electron microscope (TEM) (120 kV).
13. High resolution transmission electron microscope (HRTEM) (200 kV).
14. 300 mesh ultrathin carbon-coated Cu grids.
15. Atomic force microscope (AFM).
16. Field emission scanning electron microscopy (FESEM).
17. NIH Image J software.
18. Syringe pump.
19. 20 mL glass scintillation vials.

3 Methods

Deionized water (18 MΩ/cm, Millipore Milli-Q) was exclusively used for all rinsing procedures. All procedures are conducted at room temperature unless specified otherwise.

3.1 Expression of CTPR Proteins

1. Transform the CTPR3 gene in pPROEx vector into *E. Coli* BL21(DE3) cells and culture it overnight at 37 °C on agar plates with 100 μg/mL ampicillin.
2. Select a single colony and incubate overnight in 50 mL of LB media containing 100 μg/mL of ampicillin.
3. Dispense 10 mL of the overnight culture into 1 L of fresh LB media supplemented with 100 μg/mL of ampicillin. Grow the cells in an incubator-shaker (250 rpm) at 37 °C until the optical density (OD_{600}) reaches 0.8.
4. Add IPTG solution to a final concentration of 1 mM to induce the expression of CTPR3. Grow the cultures for 5 h at 37 °C with continuous shaking (250 rpm).
5. Harvest the cells by centrifugation at 3600 × *g* for 20 min and freeze the pellets at −80 °C until purification.

3.2 Purification of CTPR Proteins

1. Resuspend cell pellets in lysis buffer.
2. Sonicate the cells for 2 min at 30% power using an ultrasonic homogenizer.
3. Centrifuge the lysed cells at 17000 × *g* for 30 min, save the supernatant.
4. Purify the protein from the supernatant using a standard Ni-NTA affinity purification protocol.
5. To cleave the N-terminal hexahistidine tag from the CTPR proteins, incubate the protein sample at 4 °C in the presence of 1 mM DTT and TEV protease (30–60 μg protease to 1 mg protein).
6. Purify the collected protein by size exclusion chromatography on the FPLC using Superdex 75 16/600 for CTPR3 (*see* **Note 3**).
7. Dialyze the aqueous solution of protein against dialysis buffer three times at 4 °C using a dialysis cassette with molecular weight cutoff of 3.5 kDa for CTPR3 (*see* **Note 4**) [35].
8. The concentration of purified protein is determined by Nanodrop spectrophotometer. The molar extinction coefficient (ε) of CTPR3 is 4.93×10^4 m^{-1} cm^{-1}, measured at 280 nm in water.
9. The cell pellets can be kept at −20 °C for 3 months whereas the purified protein should be kept at 4 °C for short-term storage (1–2 days).

3.3 Preparation of CTPR-Stabilized Ag Seeds

1. Make an aqueous solution containing 5 mL of citrate solution, 0.3 mL $NaBH_4$ solution, and 0.125 mL of 0.1–2 mg/mL CTPR3 protein in a glass vial (Fig. 2) (*see* **Note 5**).
2. Inject 5 mL of $AgNO_3$ solution into the above solution at the rate of 2 mL/min using a syringe pump under vigorous agitation (*see* **Notes 5–8**).

3.4 Seeded Growth of Ag NPSMs

1. Add 20 μL of CTPR3-stabilized Ag seed solution as prepared in Subheading 3.3 into 5 mL of ascorbic acid solution, followed by dropwise addition of 3 mL of $AgNO_3$ solution at the rate of 1 mL/min under stirring (Fig. 2a) (*see* **Notes 9–17**).
2. Centrifuge the as-prepared Ag NPSMs solution at 1500 × *g* for 20 min and decant the supernatant.
3. To remove residual ascorbic acid, rinse the resultant pellet with DI water by adding water into the centrifuge tube, shaking it up and then centrifuge and decant the supernatant as in **step 2** above.
4. Repeat **step 3** above two more times.
5. Redisperse Ag NPSMs into 5 mL of DI water for further use (Fig. 3) (*see* **Note 18**).

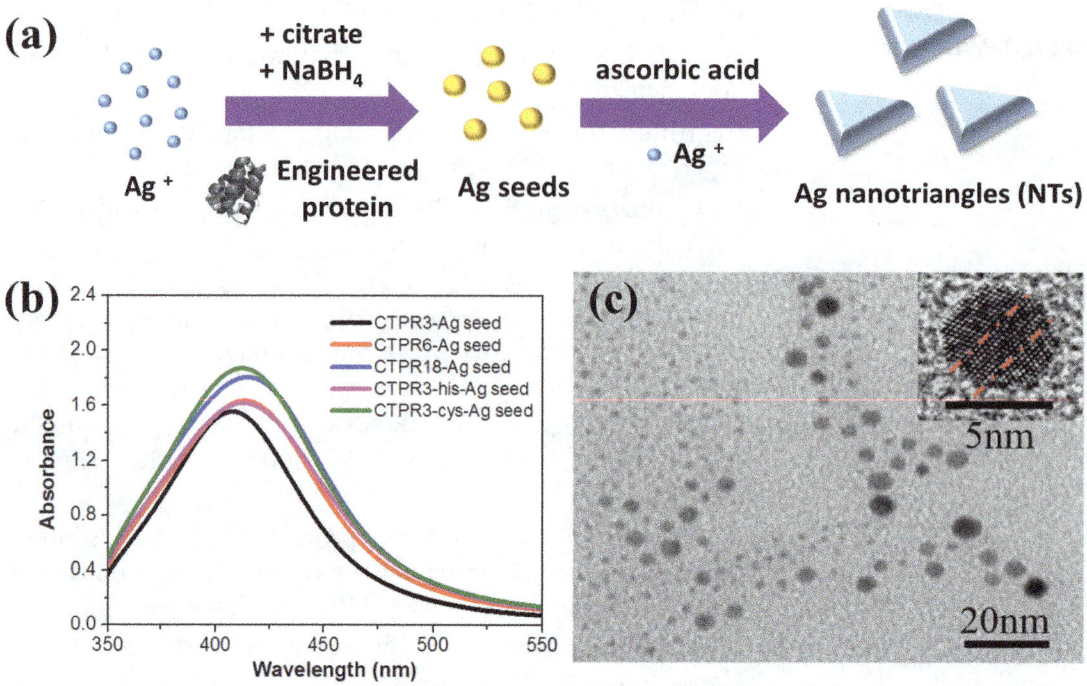

Fig. 2 (**a**) Schematic illustration of synthetic procedures for CTPR-stabilized Ag seeds and seeded growth of Ag nanoprisms. (**b**) UV-Vis spectra and (**c**) TEM images of CTPR3-Ag seeds (inset: HR-TEM of twinning nanocrystal) (Modified from Geng et al. (2016) with permission from the Royal Society of Chemistry)

3.5 Seeded Biomineralization of Au NPSMs

1. Mix 0.1 mL of 80 μM CTPR3 (prepared in Subheading 3.2) with 0.85 mL of DI water and 40 μL of $HAuCl_4$ solution, followed by adding 0.2 mL of NaBr solution and 2.5–50 μL of CTPR3-stabilized Ag seed from Subheading 3.3 (*see* **Notes 19–21**).
2. Inject 0.8 mL of MOPS buffer into the above solution. Shake gently for a few seconds until the color changes from pale yellow to clear (Au(III) → Au(I)). Leave this solution undisturbed overnight (Fig. 4a) (*see* **Notes 22** and **23**).
3. Centrifuge the as-prepared Au NPSMs solution at 1500 × *g* for 10 min and decant the supernatant.
4. To remove residual NaBr, CTPR3, and MOPS, rinse the resultant pellet with DI water by adding water into the centrifuge tube, shaking it up and then centrifuge and decant the supernatant as in **step 3** above.
5. Repeat above **step 4** three times.
6. Redisperse the Au NPSMs into 2 mL of DI water for further use (*see* **Notes 24–26**).
7. Characterize the Ag and Au NPSMs (*see* **Notes 27–30**).

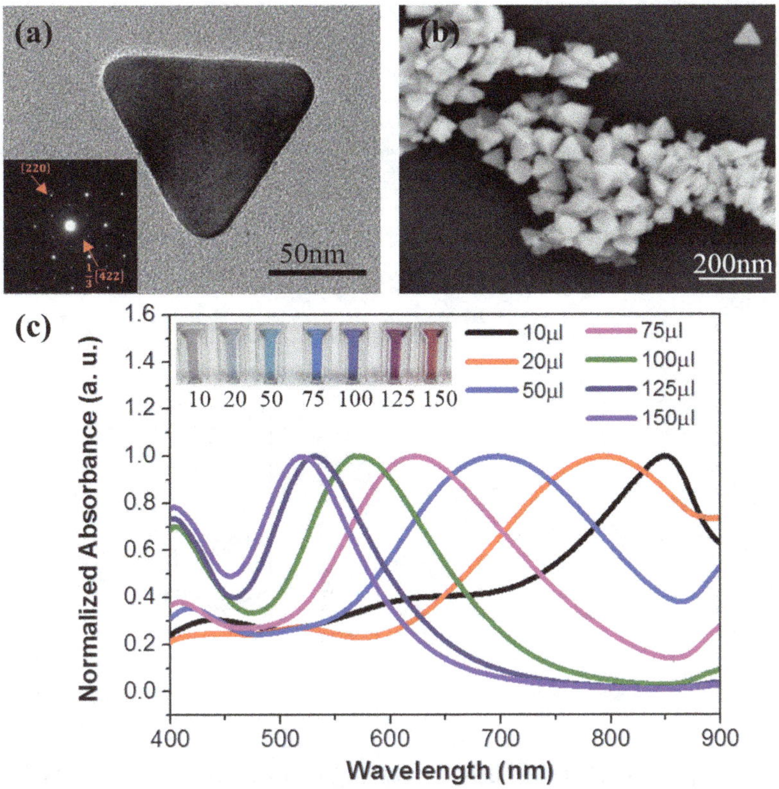

Fig. 3 (**a**) High resolution TEM image and selective area diffraction (SAED) pattern of individual Ag NPSM (inset). (**b**) SEM image of Ag NPSMs prepared using 20 μL CTPR3-Ag seed solution. (**c**) UV-Vis spectra and photographic images of the collodial Ag NPSMs obtained using different volumes of seed solution (Modified from Geng et al. (2016) with permission from the Royal Society of Chemistry)

4 Notes

1. The synthetic gene encoding CTPR3 (three tandem CTPR repeats; amino acid sequence of one CTPR repeat is shown in Fig. 1) was digested with BamHI and HindIII and subcloned into the in-house pProEx-HTAm vector. This vector was derived from pProEx-HTA vector (Invitrogen; this vector is no longer commercially available.) so that the BamHI cloning site is in frame. The vector results in a gene that codes for N-terminal His_6-tagged protein. Both gene and vector were gifts from the Regan Lab at Yale University. Any expression vector for *E. coli* (such as pET vectors) can be used, preferably including a tag (e.g., His_6) for ease of protein purification.
2. CTPR3 has 107 amino acids, MW of 12.5 kDa and an aspect ratio of 1.1. We have also made CTPR6 (213 amino acids, 25 kDa, aspect ratio 1.9) and CTPR18 (629 amino acids, 74.1 kDa, aspect ratio 4.6).

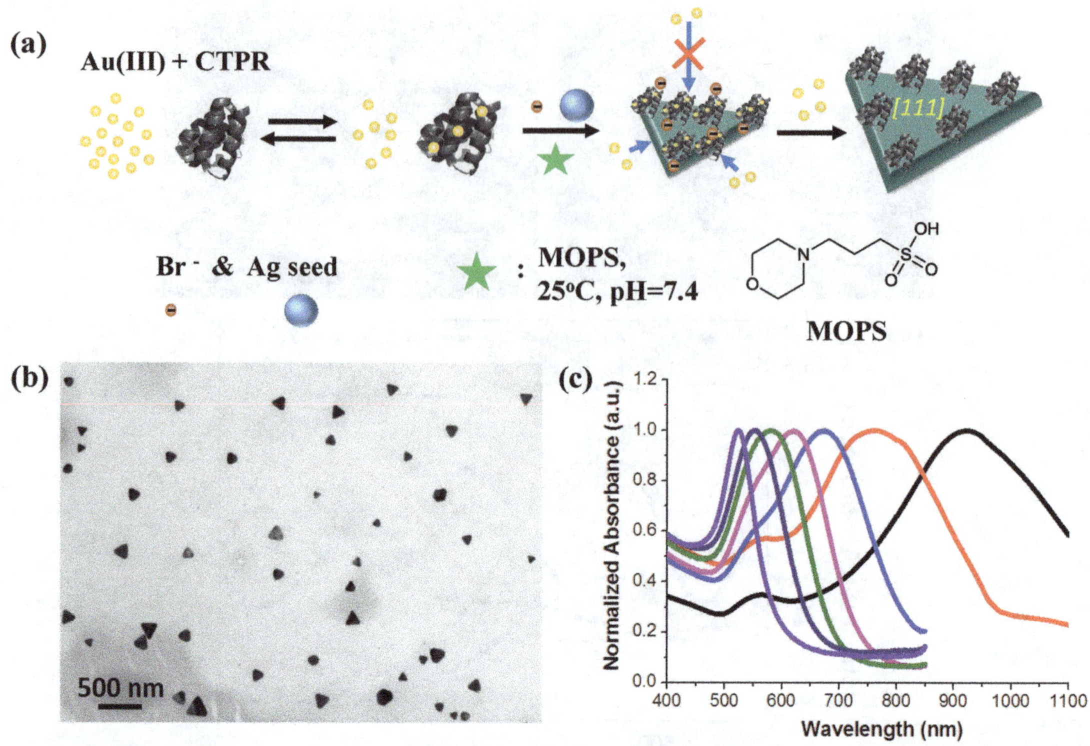

Fig. 4 (**a**) Schematic illustration of seeded biomineralization of triangular Au nanoprisms. (**b**) Representative TEM image of the as-prepared Au NPSMs (**c**) UV-Vis spectra of Au NPSMs prepared in the presence of 2.5 μL (black), 5 μL (red), 10 μL (light blue), 15 μL (magenta), 20 μL (green), 25 μL (dark blue), 50 μL (violet) of Ag seeds ($[Ag^0]$ = 0.24 mM) (Modified from Ref. [33] with permission from the Royal Society of Chemistry)

3. For other CTPRs, use Superdex 200 16/600 Prep Grade column for CTPR6 and CTPR18 in chromatography buffer with a flow rate of 0.5 mL/min.
4. Use a 10 kDa dialysis membrane for CTPR6 and CTPR18.
5. For Ag NPSMs synthesis, the syringes, vials, and pipettes should be cleaned carefully with acetone and DI to get rid of all the contaminants. Otherwise, the chemical residues and dust in the glassware may interfere with the nucleation and the subsequent seeded growth process.
6. $NaBH_4$ (10 mM) and $AgNO_3$ (0.5 mM) solutions should be freshly made.
7. The as-prepared Ag seeds should be aged for at least 3 h to circumvent interference such as homogeneous nucleation arising from the residual $NaBH_4$.
8. The as-prepared Ag seeds should be kept away from light until further use. This can be achieved by wrapping all vials with aluminum foil. The quality of seeds is retained for 3 weeks when stored at room temperature in the dark.

9. In the seeded growth step, fast seed addition and dropwise $AgNO_3$ injection into the reaction mixture enables the rapid growth of high-quality anisotropic NPSMs. It is noted that the twinning-crystal structure Ag seed could be disrupted when exposed to excessive amounts of ascorbic acid.
10. Both CTPR3 and citrate are indispensable for the high-yield production of Ag NPSMs through seed-mediated process. The absence of either one will result in poor yield of Ag NPSMs and Ag NPs with diverse shapes formed concomitantly.
11. The morphology of Ag NPs and the size of Ag NPSMs can be tuned by adjusting the concentration of protein stabilizer, ascorbic acid and the volume of CTPR3-stabilized Ag seeds
12. The ratio of [ascorbic acid]/[Ag^+] is suggested to be within the range of 0.5–1 for the high-yield of well-defined Ag NPSMs. Overabundant ascorbic acid has a detrimental effect on the morphology of resulting Ag NPSMs, as evidenced by the obvious defects such as dislocations and stacking faults.
13. In addition to CTPR3, a series of CTPR proteins including CTPR3-His, CTPR6 and CTPR18 all work as potent stabilizers and shape directing agents to facilitate the growth of Ag NPSMs using the aforementioned protocols.
14. A set of control experiments has been conducted to indicate that lysozyme, bovine serum albumin (BSA), and CTPR3-Cys are not favorable for the growth of Ag NPSMs. Thus we suggest that neither thiols nor overall positive charge are applicable in the synthetic route toward Ag NPSMs.
15. The concentration of CTPR3 plays a critical role in determining the yield and quality of final products. The ideal concentration of CTPR3 was found to be approximately 1.8 μM based on morphological and UV-Vis observations.
16. Precise control over the wavelength of LSPR is attainable when less seeds (<100 μL of solution as prepared in Subheading 3.1) are used. Otherwise, a bimodal absorbance profile is observed, attributed to the competitive growth between isotropic nanospheres and facet-selected NPSMs in the presence of a large number of nuclei.
17. The LSPR could be conveniently tuned to span a broad spectral range, from visible to NIR region (500–900 nm), through the variation of the quantities of Ag seeds used during the growth step (Fig. 3, Table 1). As the total number of seeds increases, the mean particle size declines steadily and the resultant Ag NPs undergo a progressive shape transition from triangular NPSMs to round disks.

Table 1
The optical and morphological features of Ag NPSMs prepared using varied amount of CTPR3-Ag seeds [34]

Ag seed (μL)	LSPR λ_{max} (nm)	Shape	Size (nm)
10	849	NPSMs	144.1 ± 25
20	793	NPSMs	102.8 ± 18.9
50	698	NPSMs	69 ± 7.5
100	572	NPSMs + disks	38.6 ± 8.7
150	520	NPSMs + disks	19.4 ± 6.1

18. For the purpose of long-term storage of Ag NPSMs, 0.3 mL of 25 mM citrate aqueous solution is typically added immediately after the synthesis. The colloidal Ag NPSMs are subject to etching and deformation in the absence of sufficient capping agent, which is revealed as the dramatic blue shift of absorbance wavelength within a short while.
19. For Au NPSM synthesis, all the glassware should be cleaned with aqua regia, followed by extensive rinsing with Milli-Q water. Preparation of aqua regia: In a glass beaker, mix hydrochloric acid (HCl) and nitric acid (HNO_3) in a volume ratio 3:1. The preparation should take place inside a fume hood, since the liquids and its vapors are extremely corrosive and highly oxidizing. Therefore, the use of appropriate gloves, lab coat, and chemical splash goggles is mandatory.
20. Ag seeds are of significant importance to the production of Au NPSMs. In the case of a Ag seed-free synthetic process, a mixture of Au NPSMs, fivefold penta-twinned Au NPs (decahedrons) and Au nanospheres are produced. Sizes and shapes of these structures are highly polydisperse due to inherent drawbacks such as unavoidable self-nucleation events and poor quality of initial nuclei with diverse crystal structures.
21. The addition of Ag seeds not only accelerates the reaction but also lowers the usage of CTPR protein and NaBr without compromising the final yield. The details are listed in Table 2.
22. During the synthesis of Au NPSMs, MOPS serves as a mild reducing agent while CTPR3 and NaBr were employed as the shape-directing agents. Other than MOPS, the choice of reducing agent could be further extended to other Good's buffers such as 4-(2-hydroxyethyl)-1-piperazineethanesulfonic acid (HEPES) and so forth.
23. The original synthetic process could be scaled up to 5 mL or beyond for more efficient fabrication of Au NPSMs.
24. The as-prepared Au NPSMs exhibited remarkable colloidal stability since no severe agglomeration was detected through DLS study

Table 2
Comparison between one pot and seeded growth of Au NPSMs in terms of reaction time, yield, usage of chemicals, morphological and optical features [33]

Ag seed	Reaction time	Au NPSMs yield (%)	Size (nm)	CTPR (μM)	NaBr (mM)	LSPR (nm)
No	1–2 weeks	<50	<50	10	50	550–580
Yes	Overnight	>75	50–140	2	5	500–900

Table 3
The optical and morphological features of Au NPSMs prepared using varied amount of CTPR3-Ag seed solution as prepared in Subheading 3.3 (2.5–20 μL) [33]

Ag seed (μL)	LSPR λ_{max} (nm)	Shape	Size (nm)
2.5	922	NPSMs	137 ± 33
5	762	NPSMs	99 ± 15
10	675	NPSMs + disks	79 ± 10
15	620	NPSMs + disks	61 ± 7
20	582	Disks + spheres	N/A

during the centrifugation and incubation in FBS. The excellent colloidal stability is mainly associated with the negatively charged CTPR proteins corona (Fig. 1b) with moderate binding affinity to the Au NPSMs surface [36].

25. Not only CTPR3, but a series of CTPR proteins of different length (Fig. 1a) can serve as alternative stabilizers and shape directing agents simply by following the identical protocols.
26. Analogous to the synthesis of Ag NPSMs, the LSPR of Au NPSMs can also be readily tailored from ca. 500 nm to 900 nm by simply adjusting the amount of Ag seeds added in the initial growth stage (Fig. 4, Table 3). Accordingly, the size of the Au NPSMs could be finely tailored from 61 ± 7 nm to 137 ± 33 nm.
27. In sharp contrast with the conventionally used CTAB based protocols, the depicted protocols efficiently mitigate the cytotoxicity of the Au NPSMs, which was ascertained by the 3-(4,5-dimethylthiazol-2-yl)-2,5-diphenyltetrazolium bromide (MTT) assay [33].
28. The optical properties of Au and Ag NPs are characterized by UV-Vis and UV-Vis-NIR spectroscopy.
29. The morphology of Au and Ag NPs are characterized by TEM, FESEM, HRTEM, and AFM. The size and size distribution of Au and Ag NPs are analyzed using Image-J.
30. The zeta potential and hydrodynamic size of Au NPSMs can be estimated using DLS.

Acknowledgments

The authors would like to thank Dr. Guoliang Liu for insightful comments, suggestions, and discussion on this research work. The authors acknowledge ICTAS Nanoscale Characterization and Fabrication Lab (NCFL) for the use of AFM, TEM, and SEM. This work was in part supported by the JFC ICTAS grant number 119106 to TZG.

References

1. Beeram SR, Zamborini FP (2010) Purification of gold nanoplates grown directly on surfaces for enhanced localized surface Plasmon resonance biosensing. ACS Nano 4:3633–3646
2. Pelaz B, Grazu V, Ibarra A et al (2012) Tailoring the synthesis and heating ability of gold nanoprisms for bioapplications. Langmuir 28:8965–8970
3. Millstone JE, Hurst SJ, Métraux GS et al (2009) Colloidal gold and silver triangular nanoprisms. Small 5:646–664
4. Yu H, Zhang Q, Liu H et al (2014) Thermal synthesis of silver nanoplates revisited: a modified photochemical process. ACS Nano 8:10252–10261
5. Xue B, Wang D, Zuo J et al (2015) Towards high quality triangular silver nanoprisms: improved synthesis, six-tip based hot spots and ultra-high local surface plasmon resonance sensitivity. Nanoscale 7:8048–8057
6. Jin R, Charles Cao Y, Hao E et al (2003) Controlling anisotropic nanoparticle growth through plasmon excitation. Nature 425:487–490
7. Jin R, Cao Y, Mirkin CA et al (2001) Photoinduced conversion of silver nanospheres to nanoprisms. Science 294:1901–1903
8. Xue C, Mirkin CA (2007) pH-switchable silver nanoprism growth pathways. Angew Chem Int Ed 46:2036–2038
9. Pastoriza-Santos I, Liz-Marzán LM (2002) Synthesis of silver nanoprisms in DMF. Nano Lett 2:903–905
10. Kim MH, Yoon DK, Im SH (2015) Growth pathways of silver nanoplates in kinetically controlled synthesis: bimodal versus unimodal growth. RSC Adv 5:14266–14272
11. Millstone JE, Park S, Shuford KL et al (2005) Observation of a quadrupole plasmon mode for a colloidal solution of gold nanoprisms. J Am Chem Soc 127:5312–5313
12. Millstone JE, Métraux GS, Mirkin CA (2006) Controlling the edge length of gold nanoprisms via a seed-mediated approach. Adv Funct Mater 16:1209–1214
13. Sau TK, Murphy CJ (2004) Room temperature, high-yield synthesis of multiple shapes of gold nanoparticles in aqueous solution. J Am Chem Soc 126:8648–8649
14. Huang Y, Ferhan AR, Gao Y et al (2014) High-yield synthesis of triangular gold nanoplates with improved shape uniformity, tunable edge length and thickness. Nanoscale 6:6496–6500
15. Scarabelli L, Coronado-Puchau M, Giner-Casares JJ et al (2014) Monodisperse gold nanotriangles: size control, large-scale self-assembly, and performance in surface-enhanced Raman scattering. ACS Nano 8:5833–5842
16. Chen L, Ji F, Xu Y et al (2014) High-yield seedless synthesis of triangular gold nanoplates through oxidative etching. Nano Lett 14:7201–7206
17. Millstone JE, Wei W, Jones MR et al (2008) Iodide ions control seed-mediated growth of anisotropic gold nanoparticles. Nano Lett 8:2526–2529
18. Ha TH, Koo H-J, Chung BH (2007) Shape-controlled syntheses of gold nanoprisms and nanorods influenced by specific adsorption of halide ions. J Phys Chem C 111:1123–1130
19. Jana NR, Gearheart L, Murphy CJ (2001) Seeding growth for size control of 5–40 nm diameter gold nanoparticles. Langmuir 17:6782–6786
20. Niu W, Zhang L, Xu G (2013) Seed-mediated growth of noble metal nanocrystals: crystal growth and shape control. Nanoscale 5:3172–3181
21. Personick ML, Mirkin CA (2013) Making sense of the mayhem behind shape control in the synthesis of gold nanoparticles. J Am Chem Soc 135:18238–18247
22. Lohse SE, Burrows ND, Scarabelli L et al (2014) Anisotropic noble metal nanocrystal growth: the role of halides. Chem Mater 26:34–43
23. Alkilany AM, Nagaria PK, Hexel CR et al (2009) Cellular uptake and cytotoxicity of gold nanorods: molecular origin of cytotoxicity and surface effects. Small 5:701–708

24. Alkilany AM, Murphy CJ (2010) Toxicity and cellular uptake of gold nanoparticles: what we have learned so far? J Nanopart Res 12:2313–2333
25. Geng X, Grove TZ (2015) Repeat protein mediated synthesis of gold nanoparticles: effect of protein shape on the morphological and optical properties. RSC Adv 5:2062–2069
26. Shankar SS, Rai A, Ankamwar B et al (2004) Biological synthesis of triangular gold nanoprisms. Nat Mater 3:482–488
27. Liu B, Xie J, Lee JY et al (2005) Optimization of high-yield biological synthesis of single-crystalline gold nanoplates. J Phys Chem B 109:15256–15263
28. Xie J, Lee JY, Wang DIC (2007) Synthesis of single-crystalline gold nanoplates in aqueous solutions through biomineralization by serum albumin protein. J Phys Chem C 111:10226–10232
29. Au L, Lim B, Colletti P et al (2010) Synthesis of gold microplates using bovine serum albumin as a reductant and a stabilizer. Chem Asian J 5:123–129
30. Li C, Bolisetty S, Mezzenga R (2013) Hybrid nanocomposites of gold single-crystal platelets and amyloid fibrils with tunable fluorescence, conductivity, and sensing properties. Adv Mater 25:3694–3700
31. Xie J, Lee JY, Wang DIC et al (2007) Identification of active biomolecules in the high-yield synthesis of single-crystalline gold nanoplates in algal solutions. Small 3:672–682
32. Naik RR, Stringer SJ, Agarwal G et al (2002) Biomimetic synthesis and patterning of silver nanoparticles. Nat Mater 1:169–172
33. Geng X, Roth KL, Freyman MC et al (2016) Seed-mediated biomineralizaton toward the high yield production of gold nanoprisms. Chem Commun 52:9829–9832
34. Geng X, Leng W, Carter NA et al (2016) Protein-aided formation of triangular silver nanoprisms with enhanced SERS performance. J Chem Mater B 4:4182–4190
35. Cortajarena AL, Yi F, Regan L (2008) Designed TPR modules as novel anticancer agents. ACS Chem Biol 3:161–166
36. Roth KL, Geng X, Grove TZ (2016) Bioinorganic interface: mechanistic studies of protein-directed nanomaterial synthesis. J Phys Chem C 120:10951–10960

Chapter 12

Type III Secretion Filaments as Templates for Metallic Nanostructure Synthesis

Anum Azam Glasgow and Danielle Tullman-Ercek

Abstract

Nanostructured materials can be interfaced with living cells to enable unique chemical and biological outcomes. However, it is challenging to precisely control the shape and chemical composition of submillimeter sized, cell-associated materials. In this protocol, we describe how to genetically modify and isolate a self-assembling filament protein from *Salmonella enterica*, PrgI, to bind Au nanoparticles. Au-conjugated filaments can be chemically reduced in vitro to form contiguous wires and networks that are several micrometers in length. We also describe a strategy to assemble PrgI-based filaments on live cells, which can then be sheared or remain tethered to cells for gold conjugation. These methods form the basis of a strategy for interactions between inorganic and organic systems, and could be expanded to introduce interactions with other metal nanoparticles for which peptide binding partners are known.

Key words Nanowires, Biomineralization, Protein secretion, Self-assembly, Microbial electrocatalysis

1 Introduction

As synthetic biology matures as a field, it is increasingly desirable to link the inorganic world with organic materials and organisms. For example, controlling electron transfer into and out of a cell could serve as a signaling event for a cellular process, or a cellular process could signal via such transfer that a particular cell state was reached. Alternatively, electrons could be used to power cells (or cells producing electrons could power devices), if such connections between living organisms and inorganic materials could be reliably engineered. It is crucial to precisely control the size and shape of nanostructured materials for their predictable behavior in these complex biological contexts. Although materials are engineered by traditional microfabrication and nanofabrication strategies with exquisite control over nanoscale properties [1–3], method development for the production and assembly of inorganic nanostructures for use in biological systems lags behind, due to unique complications arising from materials toxicity, precise

Andrew K. Udit (ed.), *Protein Scaffolds: Design, Synthesis, and Applications*, Methods in Molecular Biology, vol. 1798, https://doi.org/10.1007/978-1-4939-7893-9_12,

targeting of inorganic structures to biomacromolecules, and the inherent difficulty of characterizing hybrid biological/inorganic ensembles in a dynamic environment.

Naturally occurring biological materials such as lipids, DNA, and proteins have long been used as scaffolds for building three-dimensional inorganic materials at biologically relevant size scales from the bottom up and serve as an excellent starting point for creating materials that may continue to interact with the biological system [4–6]. Proteins present an especially useful engineering target because they self-assemble predictably, sample a large number of geometric conformations, and have amino acid sequence-based molecular recognition properties. In addition, protein complexes and oligomers are often more thermodynamically favorable than monomers in solution. Finally, there are already numerous bioorthogonal methods for conjugating proteins to inorganic nanostructures, ligands, membranes, and surfaces [7]. There are several examples of engineered inorganic nanostructures based on a protein scaffold, including phage coat proteins [8, 9], clathrin [10], amyloid fibers [11, 12], actin [13], and microtubules [14]. Each system has specific advantages and limitations based on the inherent chemical and geometric constraints of the template proteins and conjugation strategy.

The type III secretion system (T3SS) filament from *Salmonella enterica* forms a suitable template for constructing metal nanostructures for applications requiring display on the exterior of the cell surface or in solution in vitro. The hallmark of the *S. enterica* T3SS is a 60 nm-long homo-oligomeric protein filament or "needle" that extends into the extracellular space and is composed of ~120 copies of the PrgI monomer [15, 16]. PrgI is 80 amino acids long and has a molecular weight of 8.9 kDa. Importantly, PrgI is alpha-helical and its N-terminal amino acids form a flexible loop that extends away from the polymerized filament; this outward-facing loop is a structural anomaly in the family of T3SS filament proteins [17]. We engineered the N-terminal loop of PrgI to incorporate short metal-binding peptide motifs such as a hexahistidine tag (6XHis) without disrupting filament formation (Fig. 1), and showed that these motifs facilitate binding to nanoparticles such as Ni^{2+}-nitrilotriacetic acid-Au (gold nanoparticles decorated with NTA-Ni groups, Ni^{2+}-NTA-Au), both in vitro and in vivo on the *S. enterica* cell surface (Fig. 2) [18].

In this chapter, we detail three ways to produce Au-functionalized PrgI filaments: (1) contiguous gold structures of controllably variable lengths based on purified PrgI filaments assembled in vitro; (2) Au-bound PrgI filaments of consistent lengths sheared from *S. enterica* cells; and (3) Au-bound PrgI filaments anchored to living *S. enterica* cells (Fig. 3). For the methods involving native PrgI assembly in *S. enterica*, we use recombination-mediated genetic engineering techniques in order to preserve the secretion phenotype of the natural system. This modified T3SS can be further engineered to include peptide tags that selectively bind metals or minerals for cell-powered sensing applications for bioremediation or energy applications.

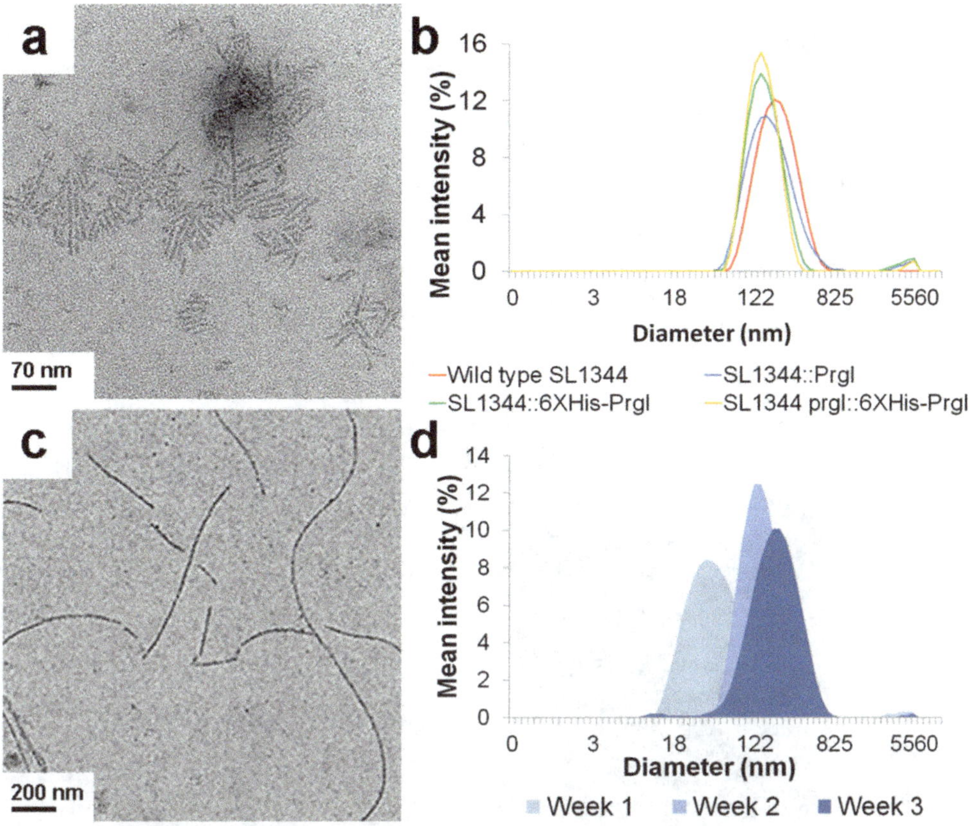

Fig. 1 Needle morphology of 6XH-PrgI is similar to wild-type PrgI both in vivo and in vitro. (**a**) TEM image of 6XHis-PrgI needles mechanically sheared from cells, which are morphologically similar to wild-type needles. (**b**) DLS traces of T3SS needles mechanically sheared from cells, indicating that the *S. enterica* wild-type strain with no plasmid, wild-type strains expressing PrgI and 6XHis-PrgI from the *pTet* promoter, and Δ*prgI::pTet-PrgI* strains all form needles. (**c**) TEM image and (**d**) DLS traces tracking needle growth of recombinant purified 6XHis-PrgI monomers into filaments over 3 weeks. Image is from the 3-week time point. Graphs use a log scale on the *x*-axis (Reproduced from Azam et al. (2016) with permission from the Royal Society Publishing)

2 Materials

Reagents should be analytical grade and can be used as received from commercial sources. In all experiments, water should be deionized using a Milli-Q system. Experiments should be conducted in a biosafety level II laboratory due to the use of the pathogen *S. enterica*.

2.1 Growth Media, Supplements, and Culture Materials

1. Lysogeny broth, Lennox (LB-Lennox): 10 g/L tryptone, 5 g/L yeast extract, 5 g/L NaCl in H_2O, autoclaved.
2. Lysogeny broth, inducing media (LB-IM): 10 g/L tryptone, 5 g/L yeast extract, 17 g/L NaCl in H_2O, autoclaved.

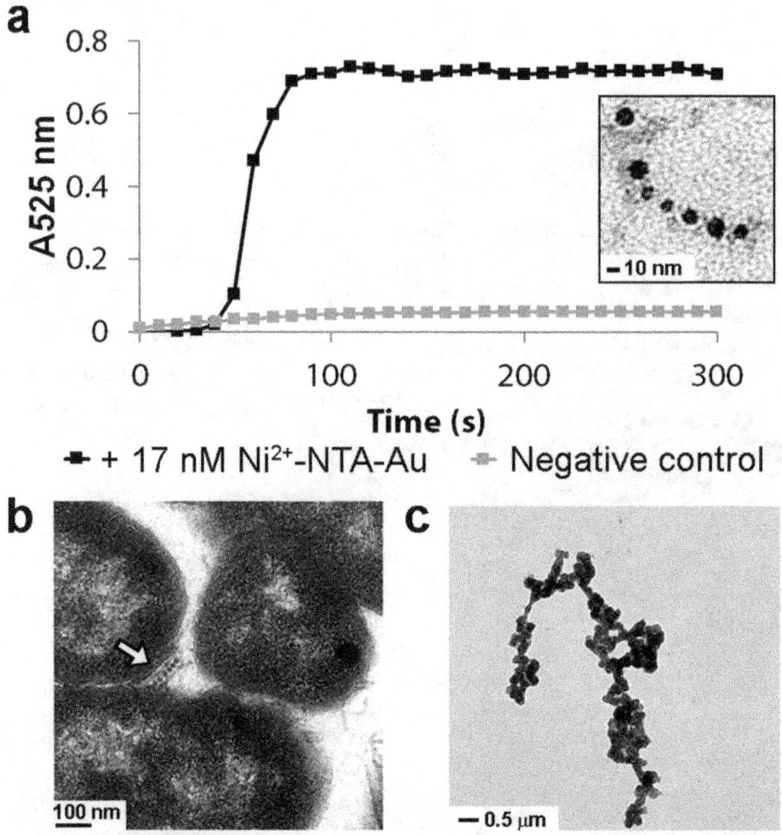

Fig. 2 Au reduction to make contiguous nanowires. (**a**) Au reduction on recombinant 6XHis-PrgI needles nucleated with Au particles can be monitored in solution by UV/Vis spectroscopy at 525 nm. Negative control indicates non-Au-conjugated 6XHis-PrgI in solution with the same reagents. Inset: 6XHis-PrgI filament seeded with Au nanoparticles that have been enlarged due to partial reduction. (**b**) TEM image of Au particles bound to T3SS filaments in vivo. (**c**) TEM image of contiguous Au structure produced in vitro (Reproduced from Azam et al. (2016) with permission from the Royal Society Publishing)

3. Terrific broth: 12 g/L tryptone, 24 g/L yeast extract, 4 mL glycerol in H_2O, autoclaved. After cooling add 10 mL/L autoclaved solution of 0.17 M KH_2PO_4, 0.72 M K_2HPO_4.
4. SOC media: 20 g/L tryptone, 5 g/L yeast extract, 0.58 g/L NaCl, 0.19 g/L KCl in H_2O, autoclaved. After cooling add 10 mL of a 0.2 μM filter-sterilized 1 M $MgCl_2$ solution, and 20 mL of a 0.2 μM filter-sterilized 1 M glucose solution.
5. 1 M isopropyl β-D-1-thiogalactopyranoside (IPTG) (1000×): 0.2 μM filter-sterilized, made in H_2O and stored at −20 °C.
6. 100 μg/mL anhydrotetracycline (aTc) (1000×): 0.2 μM filter-sterilized, made in 50% ethanol and stored at −20 °C in a foil-covered tube.
7. 34 mg/mL chloramphenicol (1000×): 0.2 μM filter-sterilized, made in 100% ethanol and stored at −20 °C.

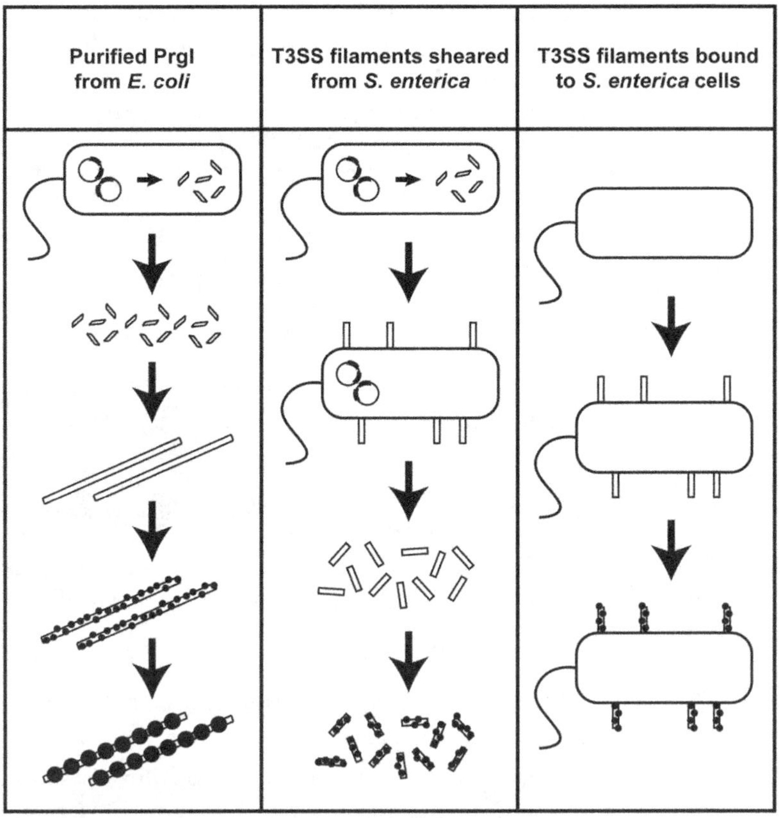

Fig. 3 Three ways to produce Au-functionalized PrgI filaments: (1) express and purify recombinant PrgI in *E. coli*, which can be assembled into contiguous Au structures of controllably variable lengths in vitro; (2) shear T3SS filaments from *S. enterica* cells and functionalize with Au; and (3) attach Au to PrgI filaments anchored to living *S. enterica* cells

8. 50 mg/mL kanamycin (1000×): 0.2 μM filter-sterilized, made in H_2O and stored at 4 °C.
9. 50 mg/mL carbenicillin (1000×): 0.2 μM filter-sterilized, made in H_2O and stored at 4 °C.
10. 1 mm electroporation cuvettes and electroporator.
11. 5 mL glass or disposable culture tubes.
12. 1 L glass baffled flasks.

2.2 Strains

All *Salmonella* strains are derived from *Salmonella enterica* serovar Typhimurium strain LT2 (ATCC® 700720™), and specific strains are available by request from the Tullman-Ercek Lab at Northwestern University. Wild type, Δ*flhCD* (do not express flagella), Δ*prgI* (not secretion-competent), and genomically His-tagged *prgI* strains are used. Additionally, three types of plasmids are used: export plasmids, PrgI overexpression plasmids in *S. enterica*, and PrgI overexpression plasmids in *E. coli*. PrgI overexpression is achieved using plasmids with the *pTet* promoter in *S. enterica*, and pET28 plasmids in *E. coli* BL21(DE3) *pLysS* cells (*see* **Notes 1–3**).

Strain	Plasmid
Wild type *S. enterica*	SptP-6XH-DH-FLAG-CmR-ColE1
Δ*flhCD S. enterica*	SptP-6XH-DH-FLAG-CmR-ColE1
Δ*prgI*Δ*flhCD S. enterica*	SptP-6XH-DH-FLAG-CmR-ColE1
*6XH-prgI*Δ*flhCD S. enterica*	SptP-6XH-DH-FLAG-CmR-ColE1
Δ*flhCD S. enterica*	pTet-6XHis-prgI-KanR-p15a
Δ*flhCD S. enterica*	pTet-prgI
BL21 *E. coli*	pET28b-6XHis-prgI$^{Q26A/K50A}$-KanR-p15a
BL21 *E. coli*	pET28b-6XH-prgI-KanR-p15a

See **Note 4** for full descriptions of plasmid abbreviations used above. These plasmids were constructed for and used in previous studies [18–21].

2.3 Genomic Modifications

1. TUC01 bacterial strain with genomic *cat-sacB* selection/counterselection cassette (D. Court lab, National Cancer Institute).
2. *pSim6* temperature-sensitive ampicillin-resistant plasmid with pSC101 origin and the λ red system (D. Court lab, National Cancer Institute).
3. LB-sucrose-agar plates: 5 g/L yeast extract, 10 g/L tryptone, 60 g/L sucrose, 15 g/L agar.
4. LB-Lennox plates: 5 g/L yeast extract, 10 g/L tryptone, 5 g/L NaCl, 15 g/L agar.

2.4 Buffers and Reagents for Protein Purification and Needle Shearing

1. Ni^{2+} column buffer: 10 mM Na_2HPO_4, 10 mM NaH_2PO_4, 20 mM imidazole, 500 mM NaCl buffer, pH to 7.3 with 1 M HCl.
2. Ni^{2+} elution buffer: 10 mM Na_2HPO_4, 10 mM NaH_2PO_4, 500 mM imidazole, 500 mM NaCl buffer, pH to 7.3 with 1 M HCl.
3. Storage buffer: 20 mM HEPES, 50 mM NaCl, pH to 7.5 with 1 M HCl.
4. TE buffer: 10 mM Tris, 1 mM EDTA, pH to 7.5 with 1 M HCl.
5. Phenylmethane sulfonyl fluoride (PMSF).
6. Sonicator.
7. 10 mL Ni-NTA affinity column.
8. 10 mL PD-10 desalting column.
9. 3 kDa molecular weight cutoff spin concentration column.
10. Dynamic light scattering instrument (e.g., Malvern Instruments Zetasizer Nano) and quartz cuvette.

11. 0.4 μM syringe filters.
12. 25-gauge needles and 10 mL syringes.

2.5 Buffers and Reagents for Au–Protein Conjugation and Reduction

1. 5 nm-diameter Ni^{2+}-NTA-Au nanoparticles stored at 0.5 μM in 50 mM MOPS, commercially available from Nanoprobes.
2. Reduction buffer: 0.01% $HAuCl_4$, 40 mM NH_2OH, and 0.05% v/v Triton X-100.
3. UV-vis spectrophotometer (e.g., NanoDrop 2000).
4. 0.5 mL NAP-5 Sephadex G-25 DNA Grade column.
5. Phosphate buffered saline (PBS): 137 mM NaCl, 2.7 mM KCl, 10 mM Na_2HPO_4, 1.8 mM KH_2PO_4, pH to 7.4 with 1 M HCl, autoclaved in deionized water.

2.6 Buffers for SDS-PAGE and Western Blotting

1. Laemmli buffer (4×): 62.5 mM Tris, 10% v/v glycerol, 1% w/v sodium dodecyl sulfate, 10% v/v beta-mercaptoethanol, 0.005% w/v Bromophenol Blue in H_2O, pH to 6.8 with 1 M HCl.
2. Running buffer: 3 g/L Tris, 14.4 g/L glycine, 1 g/L sodium dodecyl sulfate in H_2O.
3. Transfer buffer: 3 g/L Tris, 14.4 g/L glycine, 20% v/v methanol in H_2O.
4. TBST: 2.4 g/L Tris, 8.8 g/L NaCl, 0.05% v/v Tween 20 in H_2O, pH to 7.4 with 1 M HCl.
5. Blocking buffer: 5% w/w nonfat powdered milk in TBST.
6. Primary antibodies: Horseradish peroxidase-conjugated anti-Flag and anti-histidine antibodies to detect secreted proteins.
7. Secondary antibody: 0.4 μg/mL anti-mouse IgG-HRP antibody.
8. 0.22 μM syringe filters.
9. SuperSignal West Pico Chemiluminescent Substrate (ThermoFisher).
10. Polyvinylidene fluoride (PVDF) membranes.
11. Western blotting wet transfer system.
12. Chemiluminescence-sensitive gel imaging system (e.g., ChemiDoc Imager).
13. Coomassie stain solution: 0.1% Coomassie Blue R250 in 10% acetic acid, 50% methanol, 40% H_2O, v/v.
14. Coomassie destain solution: 10% acetic acid, 50% methanol, 40% H_2O, v/v.
15. Chamber and power source for running protein gels.
16. 12.5% polyacrylamide gels.
17. Protein ladder with size range 5 kDa–220 kDa.

2.7 Transmission Electron Microscopy

1. Formvar-coated copper mesh grids.
2. 0.5 mL NAP-5 Sephadex G-25 DNA grade column.
3. 100 kDa molecular weight cutoff spin concentration column.
4. Stain solution: 1% $UO_2(OAc)_2$ v/v stain in 70% methanol (*see* **Note 5**).
5. Primary fix solution: 25% v/v glutaraldehyde in fresh LB media.
6. Cacodylate buffer: 0.2 M $Na(CH_3)_2AsO_2$ in H_2O.
7. Transmission electron microscope (e.g., FEI Technai 12 TEM).

3 Methods

Carry out all procedures at room temperature except for those involving PrgI variants, which must be kept at 4 °C, unless otherwise stated. For the construction of plasmids, use standard restriction enzyme-based molecular cloning techniques and confirm by sequencing. We constructed three types of plasmids to be used in wild-type and genomically modified strains: (1) *prgI* overexpression plasmids for purification from *E. coli* BL21(DE3) *pLysS* cells; (2) plasmids for the inducible expression of *prgI* in *S. enterica*; and (3) non-*prgI* plasmids that encode proteins to be secreted by *S. enterica* to test secretion by engineered strains (*see* **Notes 6–8**). Useful plasmids and strains are listed in Subheading 2.2 [18–21].

3.1 Creation of Δ*flhCD* and Δ*prgI* Strains and Genomic Integration of 6XHis-*prgI*

1. Use the method of Thomason et al. [22] to make a homologous recombination-based *flhCD* and *prgI* knockout strain, which does not express flagella or the T3SS filament.
   (a) The Δ*flhCD* strain should be created first, and subsequent modifications to remove and replace the *prgI* gene can follow (*see* **Note 7**).
   (b) To make the knockout, PCR amplify the *cat-sacB* cassette from the TUC01 plasmid, transform cells with the temperature-sensitive *pSim6* plasmid, and perform another transformation of cells with the *cat-sacB* amplicon to introduce the cassette in the knockout region.
2. Introduce the mutant gene *6XHis-prgI* in the *prgI* genetic locus [22] (*see* **Notes 8–13**).
   (a) Knock the genes back in via PCR amplification of the gene of interest, following the method of Thomason et al. until the second transformation step. In this step, use the amplicon as the gene of interest.
   (b) Correct insertion should be confirmed by sequencing in the genetic locus.

3.2 Expression, Purification, and Assembly of 6XH-PrgI

Following this procedure, purified recombinant PrgI monomers should be monitored as they spontaneously assemble into filaments.

1. Grow 5 mL of *E. coli* BL21(DE3) *pLysS* cultures harboring *pET28b(+)-6XH-PrgI* overnight in LB-Lennox with 1:1000 dilutions of chloramphenicol and kanamycin stocks.
2. Subculture 1:100 from the overnight culture in 500 mL of Terrific Broth in a 1 L baffled flask. Grow at 225 rpm and 30 °C for 6 h, inducing the cells by adding IPTG to a final concentration of 1 mM after 2 h.
3. Centrifuge cells at 6000 × *g* for 20 min at 4 °C.
4. Thoroughly resuspend the cell pellets by pipetting in 30 mL of Ni^{2+} column buffer.
5. Lyse the cells by sonication on ice. Adjust power and time settings to avoid foaming.
6. Pellet the insoluble fraction by centrifugation at 17,000 × *g* for 1 h. Decant the supernatant and discard the pellet.
7. Pass the supernatant through a Ni-NTA affinity column at room temperature. Wash the column with 60 mL of Ni^{2+} column buffer, then elute in 7.5 mL of Ni^{2+} elution buffer.
8. Buffer-exchange the protein solution into 10.5 mL of storage buffer using a PD-10 desalting column (*see* **Note 14**).
9. Concentrate protein solution to ~1 mg/mL using 3 kDa molecular weight cutoff spin concentration columns, determining concentration by measuring the absorbance at 280 nm and using an extinction coefficient of 11,460 M^{-1} cm^{-1}.
10. Allow purified 6XH-PrgI monomers to self-assemble into filamentous structures at 4 °C over 3 weeks. Monitor daily using dynamic light scattering (*see* Subheading 3.9) (*see* **Note 15**).
11. Recombinant filaments will be 1–2 μm in length (*see* **Note 16**).

3.3 Ni^{2+}-NTA-Au Conjugation with Filaments

1. Determine the concentration of 6XH-PrgI filaments by measuring the absorbance at 280 nm and using an extinction coefficient of 11,460 M^{-1} cm^{-1}.
2. Incubate 6XH-PrgI filaments with an eightfold excess of 5 nm-diameter Ni^{2+}-NTA-Au nanoparticles at 4 °C for 16 h in storage buffer, rocking gently.
3. Remove excess nanoparticles by size exclusion using a benchtop NAP-5 column (*see* **Note 17**).
4. Store Au-labeled filaments in 2.5 mL of PBS at 4 °C for 2 weeks. Monitor daily for an increase in filament bundling and aggregation using DLS and transmission electron microscopy (*see* Subheadings 3.9 and 3.10).

3.4 Growth of Contiguous Au Structures

1. Incubate 1 mL of labeled filament solution (*see* Subheading 3.3) with reduction buffer at room temperature for 5 min in a cuvette (*see* **Note 18**).
2. Measure the absorbance at 525 nm over 5 min. The formation of crude nanowire networks and large Au-containing structures will cause the emergence of a surface plasmon resonance at 525 nm (Fig. 2).

3.5 Isolation of T3SS Needle Structures

Following this procedure, T3SS filaments are separated from cells and collected using a scaled-up culture and high-speed centrifugation method adapted from Kubori et al. [23].

1. Grow a 1 mL overnight culture of the *ΔprgI::pTet-6XHis-prgI* strain in LB-Lennox media.
2. Subculture 1:100 in 30 mL of LB-IM. Add 1:1000 dilution of aTc stock after growing for 2 h in T3SS-inducing conditions: 37 °C and shaking at 120 rpm.
3. Harvest the cells during the late log phase (about 5 h total in T3SS-inducing growth conditions). Pellet the cells at 5000 × *g* for 20 min, and resuspend the pellet in 10 mL of TE buffer containing 1 mM phenylmethylsulfonyl fluoride (PMSF).
4. Vortex the suspension 3 times for 1.5 min at maximum power.
5. Pass the cell solution through a 25-gauge needle five times to shear the needle filaments from the cells.
6. Remove intact cells by centrifugation at 9000 × *g* for 20 min at 4 °C, keeping the supernatant.
7. Collect the macromolecular needle structures from the supernatant by high-speed centrifugation at 100,000 × *g* for 1 h at 4 °C. There will be a very small and fluffy white pellet (*see* **Note 19**).
8. Resuspend the pellet in 30 mL of TE buffer.
9. Repeat **steps 6** through **8** two more times to wash the needles and remove cell debris, resuspending the final pellet in 50 μL of TE buffer (*see* **Note 20**).
10. Characterize needle size and morphology by TEM (*see* Subheading 3.10).

3.6 SDS-PAGE and Western Blotting

1. Prepare protein samples for sodium dodecyl sulfate polyacrylamide electrophoresis (SDS-PAGE) by boiling for 6 min in 4× Laemmli buffer. Mix three parts sample with one part buffer.
2. Load samples onto 12.5% polyacrylamide gels alongside a protein ladder and electrophorese at 130 V for 70 min in running buffer.
3. If the gel is not to be used for western blotting, it can be Coomassie stained according to standard methods.

4. For gels to be analyzed by western blotting, store the gels in transfer buffer with blotting papers and a polyvinylidene fluoride (PVDF) membrane cut to the size of the gels for 10 min, shaking gently. Blot the samples from the gels to the PVDF membranes at 100 V for 1 h at 4 °C in transfer buffer, using a wet transfer system.
5. Store the membranes in 10 mL of blocking buffer for 1 h, shaking gently.
6. Add a 1:1000 dilution of the primary antibodies against the incorporated N-terminal 6XHis tags on all proteins of interest to the 10 mL of blocking buffer. Continue shaking gently for 1 h (*see* **Note 21**).
7. Wash the membranes three times for 10 min each in TBST.
8. Incubate the membranes in 10 mL of TBST with 10 μL of secondary antibody.
9. Wash membranes three times for 10 min each in TBST.
10. Add SuperSignal West Pico Chemiluminescent Substrate to the membranes and monitor over a total exposure time of 10 min on a Bio-Rad Chemidoc Imager (or similar imager) to detect bands (*see* **Note 22**).

3.7 Protein Secretion Assay

1. Transform wild type *S. enterica* strain with protein export plasmid (e.g., SptP-6XH-DH-FLAG-CmR-ColE1) by electroporation in a 1 mm electroporation cuvette at 1900 V for 5 ms, and grow colonies on an LB-agar plate with appropriate antibiotics overnight at 37 °C.
2. Grow *S. enterica* cultures from colonies with and without T3SS-inducing conditions for 6–8 h (*see* **Notes 23** and **24**).
3. Centrifuge cultures at 4000 × *g* for 10 min to collect supernatants. Filter supernatants using 0.2 μm syringe filters.
4. Probe supernatants for the presence of secreted proteins by western blotting with an anti-FLAG antibody as described in Subheading 3.6 (*see* **Note 25**).

3.8 Dynamic Light Scattering (DLS)

1. Before analysis, filter samples through 0.4 μm filters.
2. Take DLS measurements in storage buffer at 25 °C in backscattering mode with 10 mm × 10 mm quartz cuvettes, using 1 mL sample at 0.2 mg/mL concentration (*see* **Note 26**).

3.9 Transmission Electron Microscopy

1. Desalt samples using NAP-5 desalting columns.
2. Concentrate samples to approximately 50 μM as determined by absorbance at 280 nm (extinction coefficient of 11,460 M^{-1} cm^{-1}) using 100 kDa MWCO spin filters.
3. Incubate samples on Formvar-coated copper mesh grids for 5 min; wick off excess fluid using filter paper and air-dry for 1 min.

4. Wash samples with water by gently pipetting directly on top of the grids, and immediately wick and air-dry again.
5. Stain samples with 1% $UO_2(OAc)_2$ for 2 min, then wick and dry again (*see* **Note 27**). Follow by washing and wicking with water.
6. Take TEM images using an FEI Technai 12 transmission electron microscope, or similar instrument, with an accelerating voltage of 120 kV.

3.10 Au Labeling and Primary Fixation of Live S. enterica

1. Grow a 1 mL overnight culture of the *ΔprgI::pTet-6XHis-prgI* strain in LB-Lennox media.
2. Subculture 1:100 in 5 mL LB-IM and grow in T3SS-inducing conditions for 5 h, inducing with a 1:1000 dilution of the aTc stock solution after 2 h.
3. Add Ni^{2+}-NTA-Au directly to the culture to a final concentration of 17 nM.
4. Allow the culture to grow for 10 min, and then centrifuge it at 4000 × *g* for 10 min.
5. Resuspend the pellet in 7 mL of LB-Lennox and mix with primary fix solution for a final concentration of 4% glutaraldehyde.
6. Mix well and rock for 15 min at room temperature.
7. Centrifuge the samples at 4000 × *g* for 5 min at 4 °C. Decant most of the supernatant, leaving only a thin layer to keep the cells wet.
8. Resuspend the pellet in 1 mL of 2.5% glutaraldehyde in cacodylate buffer (diluted from the primary fix solution). Rock for 15 min at room temperature, then store at 4 °C for secondary fixation and cell sectioning by standard methods.

4 Notes

1. Export plasmids encode proteins to be secreted. Our model protein is the human Dbl homology domain of intersection (DH) with a C-terminal FLAG epitope and an N-terminal 6XHis tag. It is C-terminally fused to the SptP protein, which is a native *S. enterica* effector protein, a requirement for type III secretion.
2. PrgI overexpression plasmids for use in *S. enterica* include the *pTet* promoter for sensitivity to anhydrotetracycline.
3. PrgI overexpression plasmids for protein production and purification in *E. coli* BL21 cells include the T7 promoter for sensitivity to the IPTG-inducible T7 RNA polymerase transcription system.
4. Full description of abbreviations used in plasmid descriptions: SptP, first 167 amino acids of native *S. enterica* effector protein SptP; 6XH, hexahistidine tag; DH, human Dbl homology domain of intersectin; FLAG, FLAG octapeptide epitope for western blot-

ting; CmR, chloramphenicol resistance cassette; ColE1, ColE1 origin of replication; pTet, *pTet* promoter for sensitivity to anhydrotetracycline; prgI, *S. enterica prgI* needle protein; KanR, kanamycin resistance cassette; p15a, p15a origin of replication; pET28a, IPTG-inducible pET28a vector.

5. 1% $UO_2(OAc)_2$ stain is light-sensitive. For best results, store at 4 °C in a foil-covered Falcon tube.
6. Plasmid constructs for the inducible expression of *prgI* in *S. enterica* and plasmid constructs for non-*prgI* plasmids that encode heterologous proteins to be secreted by *S. enterica* can be coexpressed in *S. enterica* strains. Plasmids that encode modified *prgI* variants should be expressed in a Δ*prgI* strain if one wishes to only express genetically modified PrgI proteins, and no wild type PrgI. However, Δ*prgI* strains are not secretion-competent.
7. We recommend construction of the Δ*flhCD* strain first because even though flagella appear morphologically distinct from secretion filaments by TEM and DLS, protein aggregation and the fragmentation of protein assemblies through sample incubation and processing can make the characterization more difficult. Once a Δ*flhCD* Δ*prgI* strain with the *cat-sacB* cassette in the *prgI* genomic locus has been constructed, it is much easier to replace native *prgI* with variants.
8. Complementation of a plasmid-borne *prgI* in the Δ*prgI* strain does not rescue the secretion phenotype, even though complemented cells construct T3SS needles that are morphologically indistinguishable from wild-type needles by TEM. Therefore, to restore the secretion phenotype in *S. enterica* while maintaining its ability to produce metal-binding needles, we used a recombination-mediated genetic engineering strategy to reintroduce the *6XHis-prgI* gene back into the Δ*prgI* strain at the native genomic locus for *prgI*, as described in Methods, Subheading 3.1 [18]. This strain, *S. enterica 6XHis-prgI*, was able to secrete heterologous proteins at wild-type level titers. It is intriguing that a genomic copy of *prgI* is required for secretion but a plasmid-complemented copy of the gene is sufficient for needle construction, and understanding of this feature will require further study into the genetic architecture and regulation of the *pPrgH* operon of *Salmonella* Pathogenicity Island-1.
9. The homology sequence to TUC01 that we use for all amplifications is as follows. *Cat* forward homology (Tm = 56 °C): TGTGACGGAAGATCACTTCG; *sacB* reverse homology (Tm = 57 °C): ATCAAAGGGAAAACTGTCCATAT.
10. Generally, we incorporate as much homology to the genome as can be reasonably used in the recombineering primers. Some genetic loci are trickier to incorporate than others, and *prgI* is one of them. Therefore we used 50 base pairs of homology, as the standard 40 base pairs did not work.

11. For best results using TUC01 as a template for PCR, miniprep the genome from an overnight culture, resuspend the pellet in 600 μL H_2O with 10% SDS, add lysis buffer and neutralization buffer as recommended from a commercial miniprep kit with 600 μL ddH_2O, and then proceed as usual with the miniprep.
12. There may be polar effects resulting from the removal of *prgI* that can affect other T3SS regulation. Therefore it may not always be advantageous to remove the *cat-sacB* cassette to make a clean knockout strain.
13. In our study, all engineered PrgI proteins also had the solubility-enhancing mutations V65A and V67A, which do not impact the protein's oligomerization dynamics [24]. Additionally, all *S. enterica* strains from which we sheared needles were Δ*flhCD* to avoid flagellar contamination [25].
14. At any point in **steps 7–9**, we recommend confirming the purity of 6XH-PrgI by western blotting (*see* Subheading 3.6).
15. Acidic buffer conditions slow needle oligomerization and prevent the formation of large aggregates, and storage buffer at pH as low as 5.5 does not inactivate PrgI oligomerization.
16. We recommend also testing a nonassembling 6XH-PrgI mutant, 6XH-PrgI$^{Q26A/K50A}$, as an assembly and expression negative control [26]. 6XH-PrgI$^{Q26A/K50A}$ can be made using standard site-directed mutagenesis. This mutant is unable to form a critical hydrogen bond joining the helices of the protein, and its incubation does not result in observable filaments.
17. Several methods to remove excess nanoparticles after filament binding were tested. Although we observed unbound particles by TEM in the protein sample after running it through a desalting column, the binding interaction can be reversed, so particles may rebind to filaments while other particles may become unbound. Spin concentration and ammonium sulfate precipitation are not recommended as purification strategies for fully formed filaments, especially after the introduction of Au particles, as these techniques can break the filaments.
18. The surfactant concentration can be increased up to 0.1% to reduce aggregation [27].
19. A 36% w/v CsCl density gradient, followed by overnight centrifugation and additional washing, is recommended at this stage in Kubori et al. [23]. However, we found it to be unnecessary for our purposes because we do not observe many other structures by TEM. Occasionally we observe a ring-shaped structure that closely corresponds to the size of the outer membrane ring that anchors the T3SS filament in the membrane.
20. If the pellet is too small, it may not be visible, but that does not mean that there are no isolated needles.

21. Either antibody incubation step can be done overnight.
22. For sheared and recombinant 6XHis-tagged PrgI needles, we expect to see single bands close to 10 kDa that correspond to PrgI.
23. Improved secretion is observed when starting a culture directly from a freshly transformed *S. enterica* colony, grown overnight on an LB-agar plate for 12–16 h, rather than from a frozen stock. *S. enterica* transformations are most efficient using electroporation rather than chemical transformation protocols. We transform *S. enterica* by concentrating cells from 1 L to approximately 1 mL, making 100 μL aliquots, pipetting each aliquot into ice-cold 1 mm electroporation cuvettes, and electroporating at 1900 V for 5 ms.
24. *S. enterica* grown in T3SS-inducing conditions harboring protein export plasmids with C-terminal FLAG tags fused to target proteins should yield a positive signal in the secretion assay, and strains grown in non-T3SS-inducing conditions or without a protein export plasmid should yield a negative signal in the secretion assay.
25. Following this western blotting procedure on secretion fractions should result in single bands that correspond to the expected size of the target protein plus 18.6 kDa, the size of the SptP secretion tag. T3SS-induced cells secrete about 10% of the total expressed protein. This comparison should be made by western blotting cell culture fractions alongside secretion fractions. OD-normalization and dilution of cell culture fractions by tenfold is recommended to account for differences in growth rate and avoid overwhelming the signal from the secretion fraction samples. Additionally, we recommend running the following negative controls: secretion and whole culture fractions from (1) cells harboring protein export plasmids that were grown in non-T3SS growth conditions (37 °C/225 rpm), (2) Δ*prgI* cells harboring protein export plasmids that were grown in T3SS-inducing conditions, and (3) wild type *S. enterica* cells that do not harbor a protein export plasmid. These strains should not yield any bands in the secretion fraction, and only the first two controls should have bands at the expected size in the whole culture fraction.
26. Aggregated proteins and nascent filaments will be distinguishable by DLS shortly following protein purification. We have not observed a completely monomeric solution. Additionally, DLS is used qualitatively only; the hydrodynamic radius reported by DLS does not correspond exactly to the length of the filaments, as the technique is optimized for spherical particles.
27. Avoid excessive salt buildup from the sample buffer and the staining solution by using a low-salt buffer, filtering the protein solution before applying to the grid, and washing the grid after sample application, because it can cause TEM artifacts that appear as large, regularly shaped rods.

References

1. Nugroho FA, Iandolo B, Wagner JB et al (2016) Bottom-up nanofabrication of supported noble metal alloy nanoparticle arrays for plasmonics. ACS Nano 23:2871–2879. https://doi.org/10.1021/acsnano.5b08057
2. Kuroda Y, Kuroda K (2010) Morphosynthesis of nanostructured gold crystals by utilizing interstices in periodically arranged silica nanoparticles as a flexible reaction field. Angew Chem Int Ed 17:6993–6997. https://doi.org/10.1002/anie.201002430
3. Lee D, Yoon S (2015) Gold nanocube–nanosphere dimers: preparation, plasmon coupling, and surface-enhanced Raman scattering. J Phys Chem C 119:7873–7882. https://doi.org/10.1021/acs.jpcc.5b00314
4. Xu G-K, Li Y, Li B et al (2009) Self-assembled lipid nanostructures encapsulating nanoparticles in aqueous solution. Soft Matt 5:3977–3983. https://doi.org/10.1039/B906918F
5. Yin P, Choi HM, Calvert CR et al (2008) Programming biomolecular self-assembly pathways. Nature 451:318–322. https://doi.org/10.1038/nature06451
6. Gradišar H, Jerala R (2014) Self-assembled bionanostructures: proteins following the lead of DNA nanostructures. J Nanobiotech 12:4. https://doi.org/10.1186/1477-3155-12-4
7. Chen CL, Rosi NL (2010) Peptide-based methods for the preparation of nanostructured inorganic materials. Angew Chem Int Ed 49:1924–1942. https://doi.org/10.1002/anie.200903572
8. Lee SW, Mao C, Flynn CE et al (2002) Ordering of quantum dots using genetically engineered viruses. Science 296:892–895. https://doi.org/10.1126/science.1068054
9. Shenton W, Douglas T, Young M et al (1999) Inorganic–organicnanotubecompositesfromtemplate mineralization of tobacco mosaic virus. Adv Mater 11:253–256. https://doi.org/10.1002/(SICI)1521-4095(199903)11:3<253::AID-ADMA253>3.0.CO;2-7
10. Schoen AP, Schoen DT, Huggins KN et al (2011) Template engineering through epitope recognition: a modular, biomimetic strategy for inorganic nanomaterial synthesis. J Am Chem Soc 133:18202–18207. https://doi.org/10.1021/ja204732n
11. Scheibel T, Parthasarathy R, Sawicki G et al (2003) Conducting nanowires built by controlled self-assembly of amyloid fibers and selective metal deposition. Proc Natl Acad Sci U S A 100:4527–4532. https://doi.org/10.1073/pnas.0431081100
12. Acar H, Garifullin R, Guler MO (2011) Self-assembled template-directed synthesis of one-dimensional silica and titania nanostructures. Langmuir 27:1079–1084. https://doi.org/10.1021/la104518g
13. Henry E, Dif A, Schmutz M, Legoff L et al (2011) Crystallization of fluorescent quantum dots within a three-dimensional bio-organic template of actin filaments and lipid membranes. Nano Lett 11:5443–5448. https://doi.org/10.1021/nl203216q
14. Boal AK, Headley TJ, Tissot RG et al (2003) Microtubule-templated biomimetic mineralization of lepidocrocite. Adv Func Mater 14:19–24. https://doi.org/10.1002/adfm.200304435
15. Burkinshaw BJ, Strynadka NC (2014) Assembly and structure of the T3SS. Biochim Biophys Acta 1843:1649–1663. https://doi.org/10.1016/j.bbamcr.2014.01.035
16. Rathinavelan T, Lara-Tejero M, Lefebre M et al (2014) NMR model of PrgI-SipD interaction and its implications in the needle-tip assembly of the *Salmonella* type III secretion system. J Mol Biol 426:2958–2969. https://doi.org/10.1016/j.jmb.2014.06.009
17. Loquet A, Sgourakis NG, Gupta R et al (2012) Atomic model of the type III secretion system needle. Nature 486:276–279. https://doi.org/10.1038/nature11079
18. Azam A, Tullman-Ercek D (2016) Type-III secretion filaments as scaffolds for inorganic nanostructures. J R Soc Interface 13:20150938. https://doi.org/10.1098/rsif.2015.0938
19. Azam A, Li C, Metcalf K et al (2015) Type III secretion as a generalizable strategy for the production of full-length biopolymer-forming proteins. Biotechnol Bioeng 113:2313–2320. https://doi.org/10.1002/bit.25656
20. Glasgow AA, Wong HT, Tullman-Ercek D (2017) A secretion-amplification role for *Salmonella enterica* translocon protein SipD. *ACS Synth Biol* 6(6):1006–1015. https://doi.org/10.1021/acssynbio.6b00335
21. Metcalf KJ, Finnerty C, Azam A et al (2014) Using transcriptional control to increase titer of secreted heterologous proteins by the type III secretion system. Appl Environ Microbiol 80:5927–5934. https://doi.org/10.1128/AEM.01330-14

22. Thomason L, Court DL, Bubunenko M et al (2007) Recombineering: genetic engineering in bacteria using homologous recombination. Curr Protoc Mol Biol 78:1.16.1–1.16.24. https://doi.org/10.1002/0471142727.mb0116s78
23. Kubori T, Sukhan A, Aizawa SI et al (2000) Molecular characterization and assembly of the needle complex of the *Salmonella typhimurium* type III protein secretion system. Proc Natl Acad Sci U S A 97:10225–10230. https://doi.org/10.1073/pnas.170128997
24. Poyraz O, Schmidt H, Seidel K et al (2010) Protein refolding is required for assembly of the type three secretion needle. Nat Struct Mol Biol 17:788–792. https://doi.org/10.1038/nsmb.1822
25. Ikebe T, Iyoda S, Kutsukake K (1999) Promoter analysis of the class 2 flagellar operons of *Salmonella*. Genes Genet Syst 74:179–183. https://doi.org/10.1266/ggs.74.179
26. Wang Y, Ouellette AN, Egan CW et al (2007) Differences in the electrostatic surfaces of the type III secretion needle proteins PrgI, BsaL, and MxiH. J Mol Biol 371:1304–1314. https://doi.org/10.1016/j.jmb.2007.06.034
27. Brown KR, Natan MJ (1998) Hydroxylamine seeding of colloidal Au nanoparticles in solution and on surfaces. Langmuir 14:726–728. https://doi.org/10.1021/la970982u

Chapter 13

Incorporation of Non-Canonical Amino Acids into Proteins by Global Reassignment of Sense Codons

Katharine Y. Fang, Seth A. Lieblich, and David A. Tirrell

Abstract

Non-canonical amino acids are finding increasing use in basic and applied research. Proteins that evolved naturally for biological function did so by exploiting the chemistries of the canonical amino acids; however, when proteins are repurposed for biomedical and pharmacological applications, they are often subject to conditions different from those characteristic of their original biological environments. Non-canonical amino acids can impart properties that are inaccessible within canonical protein sequence space, and can thereby lead to improved or new functionality. We describe simple methods for global replacement of canonical amino acids by their non-canonical counterparts in recombinant proteins made in high yield in bacterial expression hosts. These methods can be used to engineer both chemical and physical properties of recombinant proteins.

Key words Non-canonical amino acids, Protein engineering, Proteomics, Recombinant protein

1 Introduction

The functions of naturally occurring proteins are determined by evolutionary constraints that are defined in large part by the chemistries of the canonical amino acids (cAAs). Because of this, exploiting such proteins as therapeutics ultimately subjects them to conditions that create additional challenges for practical application. For example, humans normally produce and use insulin within days, while diabetic patients are treated with insulin formulations that are stored in consumer packaging and used months later [1]. Prolonged storage of insulin and other protein therapeutics can lead to aggregation and altered pharmacokinetics [2].

The introduction of chemically distinct amino acids into recombinant proteins can impart properties that cannot be accessed within canonical protein sequence space. Over the past decade [3–5], there has been substantial progress in the development of methods to engineer proteins by introducing non-canonical amino acids (ncAAs) (Fig. 1). Non-canonical amino acids

Andrew K. Udit (ed.), *Protein Scaffolds: Design, Synthesis, and Applications*, Methods in Molecular Biology, vol. 1798,
https://doi.org/10.1007/978-1-4939-7893-9_13,

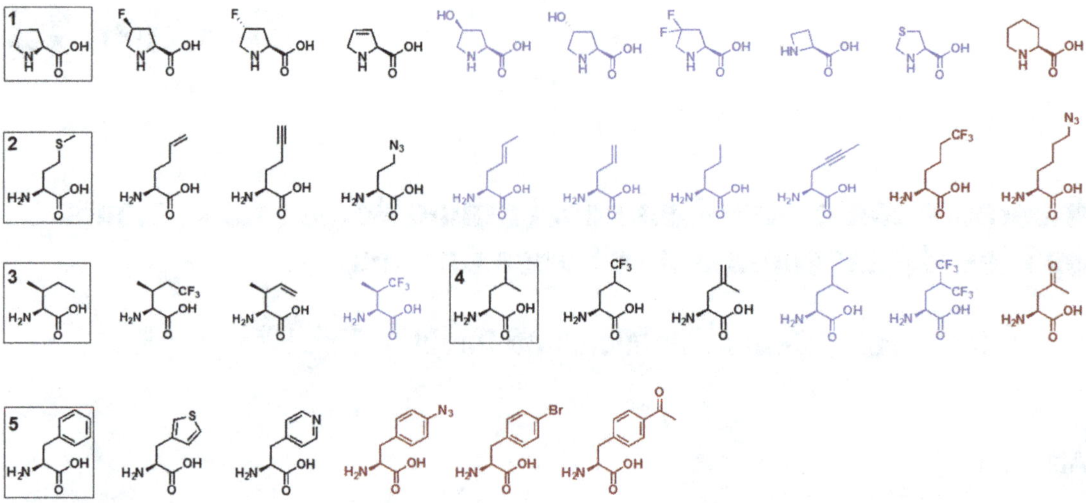

Fig. 1 Representative set of non-canonical amino acids that have been used for global codon reassignment [10, 11, 35, 36]. Structures of L-proline (1) and analogs [32], L-methionine (2) and analogs [37–40], L-isoleucine (3) and analogs [41, 42], L-leucine (4) and analogs [30, 43–45], and L-phenylalanine (5) and analogs [46, 47]. Color code denotes aminoacyl-tRNA synthetase required for efficient amino acid replacement: *Black* Endogenous synthetase, *Light Blue* Overexpressed wild-type synthetase, *Red* Overexpressed mutant synthetase

have enabled advances in protein engineering and modification, proteomics, and biomaterials. For example, our laboratory has used ncAAs to modulate the biophysical properties of insulin [6], to track the injection of pathogen proteins into a host [7], and to control chain mobility in protein hydrogels [8].

ncAA mutagenesis is most often accomplished via one of two complementary methods: nonsense suppression or global codon reassignment. Nonsense suppression methods [9, 10] use stop codons (most frequently the amber codon) and orthogonal tRNA/synthetase pairs to direct incorporation of new amino acids into proteins, usually at a single site. Global codon reassignment [11, 12] involves proteome-wide replacement of cAAs by their non-canonical counterparts. The two amino acids compete for the codons of interest, with the level of ncAA incorporation determined by the concentrations of the amino acids and the selectivity of the cognate synthetases. The two methods are complementary; nonsense suppression is generally preferred for the preparation of proteins modified at a single site for biophysical or pharmaceutical purposes, while global reassignment offers advantages in materials design and proteomic analysis.

The introduction of ncAAs into cellular proteins has been demonstrated in a wide variety of organisms and cultured cells, including *Escherichia coli*, *Staphylococcus aureus*, *Pseudomonas aeruginosa* [13], *Yersinia enterocolitica* [7], *Vibrio harveyi* [14], *Saccharomyces cerevisiae* [15, 16], *Danio rerio* [17], *Caenorhabditis elegans* [18], *Mus musculus* [19, 20], *Cricetulus griseus* (e.g., Chinese hamster ovary cells; CHO) [21], *Homo sapiens* (e.g., Hela cells) [22, 23], *Drosophila melanogaster* [24], and *Arabidopsis* [25]. The incorporation of ncAAs in cell culture and in vivo has been shown to be effective in a variety of experimental conditions such as planktonic culture, agar plates, bioreactors, biofilms, in vivo and ex vivo infection models, cocultures, and animal models [7, 13–25].

Herein we describe a recommended protocol for global codon reassignment in *E. coli,* the most common expression system for recombinant protein production.

2 Materials

All solutions are prepared using double distilled water at room temperature, and pH is adjusted by addition of HCl and/or NaOH.

2.1 Plasmid Construction

1. Plasmid pQE80L.
2. Forward and reverse primers to amplify the region of DNA encoding the aminoacyl-tRNA synthetase (aaRS) of interest (*see* **Note 1**). The primers should contain appropriate sequences as needed for cloning (e.g., restriction enzyme recognition sites).
3. Gene fragment encoding protein of interest (POI) (*see* **Note 2**).
4. DNA polymerase with reaction buffer for PCR.
5. Restriction endonucleases and respective buffers (*see* **Note 3**).
6. DNA purification kits (in-gel and in-solution).
7. Genomic DNA purification kit.
8. 0.5× TBE buffer: 40 mM Tris–HCl, 45 mM boric acid, 1 mM EDTA, pH 8 (*see* **Note 4**).
9. DNA ligation kit.
10. Chemically Z-competent *E. coli*.
11. LB-agar plates: 5 g/L NaCl, 10 g/L tryptone, 5 g/L yeast extract with 1.5% Bacto agar, sterilized by autoclave.
12. LB liquid medium: 5 g/L NaCl, 10 g/L tryptone, 5 g/L yeast extract, sterilized by autoclave.
13. Antibiotic stock solution: 200 mg/L ampicillin, sterile-filtered.
14. Apparatus and gels for DNA electrophoresis.
15. NanoDrop spectrophotometer for DNA quantification.

2.2 Protein Expression

1. Chemically competent *E. coli* strain which is auxotrophic for cAA counterpart of ncAA.
2. Electroporation apparatus
3. LB liquid medium: 5 g/L NaCl, 10 g/L tryptone, 5 g/L yeast extract, sterilized by autoclave.

4. SOC liquid medium: SOB medium (20 g/L tryptone, 5 g/L yeast extract, 10 mM NaCl, 2.5 mM KCl, 10 mM $MgCl_2$, 10 mM $MgSO_4$, sterilized by autoclave) with 20 mM glucose (added to SOB medium from a 10× filter-sterilized glucose solution), sterile.
5. Amino acid stock solutions: 19aa solution, cAA solution, and ncAA solution, all sterile filtered (*see* **Note 5**).
6. 1× M9, 20aa medium: 1× M9 salts (8.5 mM NaCl, 18.7 mM NH_4Cl, 22 mM KH_2PO_4, 47.8 mM Na_2HPO_4), 0.1 mM $CaCl_2$, 1 mM $MgSO_4$, 5 mg/L $FeSO_4$, 1 μg/L trace metals (Cu^{2+}, Mn^{2+}, Zn^{2+}, MoO_4^{2-}), 35 mg/L thiamine hydrochloride, 10 mg/L biotin with 1× 20aa solution (50 mg/L of each L-amino acid) (*see* **Note 6**).
7. 1.25× M9, 19aa medium: 1× M9 salts (8.5 mM NaCl, 18.7 mM NH_4Cl, 22 mM KH_2PO_4, 47.8 mM Na_2HPO_4), 0.1 mM $CaCl_2$, 1 mM $MgSO_4$, 5 mg/L $FeSO_4$, 1 μg/L trace metals (Cu^{2+}, Mn^{2+}, Zn^{2+}, MoO_4^{2-}), 35 mg/L thiamine hydrochloride, 10 mg/L biotin with 1× 19aa solution (50 mg/L of each L-amino acid, minus cAA of interest) (*see* **Note 6**).
8. Antibiotic stock solution: 200 mg/L ampicillin, sterile-filtered.
9. Isopropyl-β-thiogalactoside (IPTG) stock solution: 1 M IPTG, sterile filtered.
10. Sterile saline solution: 0.9 wt% NaCl (sterilized by autoclave), chilled to 4 °C.
11. 1.5 M NaCl solution (*see* **Note 7**).
12. UV-Vis spectrophotometer.

2.3 Protein Purification

1. Lysis buffer: 1× B-PER (Bacterial Protein Extraction Reagent, Thermo Fisher), 0.5 mg/mL lysozyme, 50 U/mL benzonase nuclease.
2. Ni-NTA binding buffer: 8 M urea, 300 mM NaCl, 50 mM NaH_2PO_4, pH 8.0 (*see* **Note 8**).
3. Ni-NTA wash buffer 1: 8 M Urea, 300 mM NaCl, 50 mM NaH_2PO_4, 10 mM imidazole, pH 8.0 (*see* **Note 8**).
4. Ni-NTA wash buffer 2: 8 M Urea, 20 mM Tris, 5 mM imidazole, pH 7.5 (*see* **Note 8**).
5. Denaturing Ni-NTA elution buffer: 8 M UltraPure urea (Thermo Fisher), 20 mM Tris, pH 3.0 (*see* **Notes 8** and **9**).
6. Ni-NTA agarose beads.
7. SDS-PAGE materials: Protein loading buffer, NuPAGE Novex protein gels, and electrophoresis equipment.
8. BCA Protein Assay Kit.

2.4 Mass Spectrometry Validation of ncAA Incorporation

1. Denaturing buffer: 8 M UltraPure urea, 20 mM Tris, pH 8.0.
2. Dithiothreitol (DTT) stock solution: 0.25 M DTT.
3. Iodoacetamide stock solution: 0.45 M iodoacetamide.
4. Sequencing-grade protease (*see* **Note 10**).
5. Protease reaction buffer: 50 mM NH_4HCO_3, pH 8.0 or as recommended by protease manufacturer (*see* **Note 10**).
6. Ziptips containing C_{18} resin (e.g., ZTC18S960 from Millipore).
7. Mass spectrometer.

3 Method

3.1 Plasmid Construction

In some cases, even if the ncAA of interest can be incorporated into proteins by the endogenous machinery of the host, overexpression of the wild-type synthetase may improve the yield of the recombinant protein. Some ncAAs will require mutant synthetases for activation [26]. The protocol below describes the construction of a plasmid containing gene elements encoding both the aaRS, which is constitutively expressed, and the POI, which is IPTG-inducible (Fig. 2).

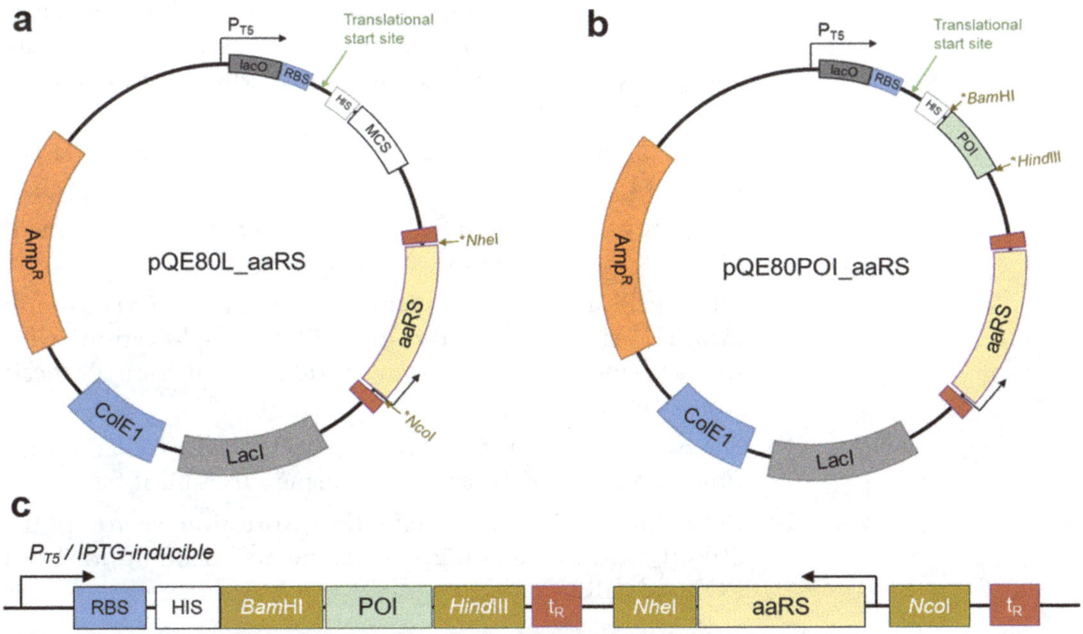

Fig. 2 General plasmid design for protein expression and components for ncAA incorporation. (**a**) Plasmid map for pQE80L_aaRS. Gene fragment aaRS is inserted into plasmid pQE80L at *Nco*I and *Nhe*I sites. (**b**) Plasmid map for pQE80POI_aaRS. Gene fragment POI is inserted into plasmid pQE80L_aaRS at *Bam*HI and *Hind*III sites to enable translation of an N-terminal hexahistidine tagged POI. (**c**) Linearized plasmid segment containing necessary components for expression with ncAAs, highlighting location of POI and aaRS with respect to recognition sites for selected endonucleases, and transcriptional initiation (denoted by black arrows) and termination sites (red t_R blocks)

1. Digest plasmid pQE80L with *Nhe*I and *Nco*I (*see* **Note 11**).
2. Gel purify the vector backbone fragment (approximately 4.2 kb).
3. PCR amplify aaRS fragment (endogenous promoter + gene encoding aaRS) using polymerase chain reaction with forward and reverse oligonucleotides coding for flanking regions upstream and downstream of aaRS fragment (*see* **Note 12**).
4. Digest PCR product (*Nco*I-aaRS gene fragment-*Nhe*I) with *Nhe*I and *Nco*I, and purify using DNA purification kit.
5. Quantify concentration of purified DNA products using NanoDrop.
6. Ligate purified PCR product and digested vector backbone at a 3:1 molar ratio using the ligation kit. Include a ligation reaction with water in place of the PCR product as a negative control for vector religation.
7. Transform the expression strain by adding 2 μL of the ligation reaction mixture to 50 μL of chemically Z-competent *E. coli* cells that were thawed on ice. Incubate for 10 min on ice (*see* **Note 13**).
8. Spread the cell mixture on a LB agar plate containing 100 μg/L ampicillin and incubate at 37 °C overnight (or until colonies become visible).
9. Pick several colonies and grow overnight in LB liquid medium with 200 μg/L ampicillin using a shaker at 37 °C. Prepare plasmid from confluent bacterial cultures and verify plasmid by DNA sequencing.
10. Digest plasmid pQE80L_aaRS with *Bam*HI and *Hind*III (*see* **Notes 11** and **14**).
11. Gel-purify the modified vector backbone fragment (4.2 kb + length of aaRS DNA fragment).
12. Obtain POI DNA fragment with flanking *Bam*HI and *Hind*III sites either through PCR amplification with forward and reverse oligonucleotides or custom synthesis (*see* **Note 15**).
13. Digest gene insert fragment (*Bam*HI-POI-*Hind*III) with *Bam*HI and *Hind*III, and purify using DNA purification kit.
14. Repeat **steps 5–8** to obtain the expression vector pQE-80POI_aaRS, containing gene elements encoding for both POI and aaRS.

3.2 Protein Expression

E. coli cells that are auxotrophic for the cAA are transformed with plasmid pQE80POI_aaRS (Fig. 2) in preparation for expression of the POI and incorporation of ncAA (Fig. 3a). The extent of replacement of cAA residues depends on the rate of activation of the ncAA by the (wild-type or mutant) synthetase (*see* **Note 16**).

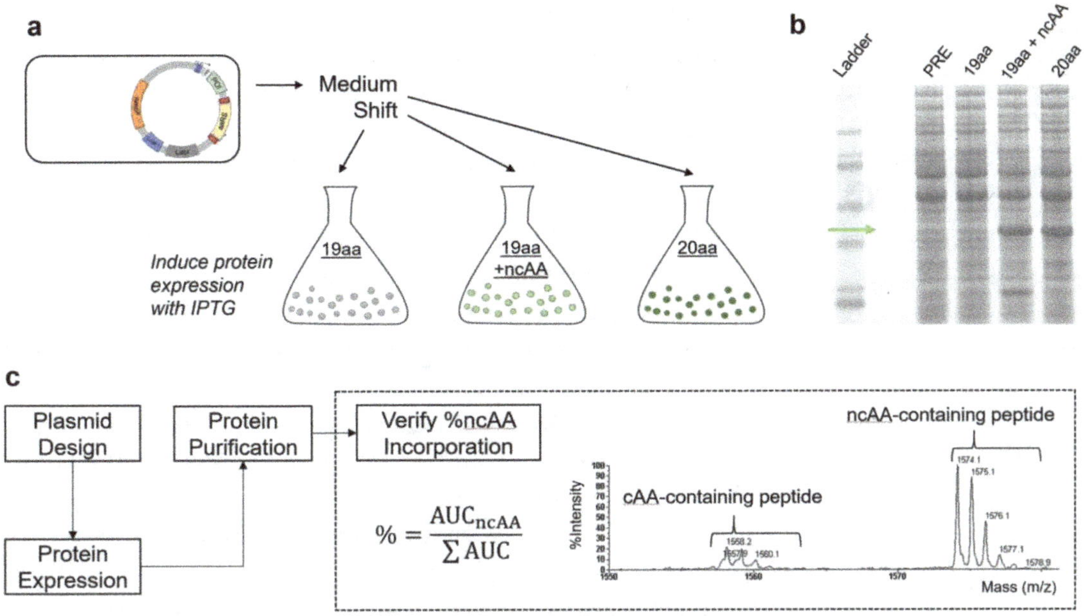

Fig. 3 Schematic for protein expression and ncAA incorporation. (**a**) Medium shift protocol with 19aa, 19aa + ncAA, and 20aa (19aa + cAA of interest) cultures. (**b**) Example protein gel electrophoresis of cell lysates with lanes labeled PRE: no induction with IPTG; 19aa: induction in 19aa medium; 19aa + ncAA: induction in 19aa + ncAA medium; 20aa: induction in 20aa medium. Note the expressed protein in the 19aa + ncAA and 20aa lanes (denoted by green arrow). (**c**) General strategy for verifying ncAA incorporation by mass spectrometry. The extent of amino acid replacement is estimated through analysis of the area under the curve (AUC) of the signals corresponding to the ncAA (AUC_{ncAA}) and cAA forms of peptides generated by proteolysis

1. Transform plasmid pQE80POI_aaRS into cAA auxotrophic *E. coli* strain by electroporation and grow bacteria in SOC medium for 1 h at 37 °C (*see* **Note 17**).
2. Spread cells on LB agar plate containing 100 μg/L of ampicillin and allow to grow overnight.
3. Pick a single colony with a sterile inoculating loop and transfer to a sufficient volume of LB liquid medium containing 200 μg/L ampicillin and grow overnight in a shaker at 37 °C (e.g., for a subsequent 1:40 dilution into 40 mL of expression culture, a volume of 1 mL would be appropriate).
4. Dilute the overnight culture at 1:40 into fresh 1× M9, 20aa medium (e.g., 1 L of M9 for a highly expressed protein).
5. Grow for 3–5 h in a shaker at 37 °C to mid-log phase.
6. At an optical density of approximately 0.9 at 600 nm, initiate a medium shift by first centrifuging the cells at 5000 × *g*, 4 °C for 10 min.
7. Discard the supernatant, resuspend the cell pellet in a volume equal to the culture volume of cold, sterile saline solution, and centrifuge the cells again at 5000 × *g*, 4 °C for 10 min.

8. Repeat **step 7** two more times for a total of three washes.
9. Discard the saline, resuspend the cell pellet in 1.25× M9, 19aa medium and return the culture to the 37 °C shaker for 45 min (*see* **Note 18**).
10. A 1 mL sample of bacterial culture should be taken at this step as a preinduction and postdepletion control.
11. Add 5× ncAA solution to the culture and return to the 37 °C shaker for 15 min (*see* **Notes 5** and **19**).
12. Add IPTG stock solution to a final concentration of 1 mM to induce protein expression for 2 h at 37 °C with agitation (*see* **Note 20**).
13. Obtain 1 mL samples of 19aa, 19aa + ncAA, and 20aa cultures to verify protein expression by protein electrophoresis (*see* Fig. 3b, **Note 21**).
14. Harvest the cells by centrifugation at 5000 × *g*, 4 °C for 10 min.
15. Cell pellets can be stored at −80 °C until further use.

3.3 Protein Purification

Below we detail a denaturing Ni-NTA based purification of the recombinant protein expressed in Subheading 3.2. There is no inherent limitation in ncAA mutagenesis that requires Ni-NTA or denaturing purification. Other purification methods (e.g., affinity, size exclusion, ion exchange) can all be successfully utilized to produce purified recombinant protein.

1. Thaw the cell pellet on ice for 1 h.
2. Resuspend the cell pellet in lysis buffer (e.g., 10 mL of lysis buffer for a cell pellet obtained from 1 L of expression), and gently agitate at room temperature for 45 min.
3. Separate soluble and insoluble fractions by centrifugation at 10,000 × *g* for 30 min.
4. Redissolve the insoluble fraction in Ni-NTA binding buffer, (e.g., 5 mL of buffer per 10 mL of lysis buffer). This insoluble fraction is commonly referred to as inclusion bodies (IBs).
5. Determine the location of the protein by running protein gel electrophoresis on both the soluble and insoluble fractions.

   Many recombinant proteins expressed in *E. coli* are found in the insoluble fraction [27]; therefore, this method will proceed with purification under denaturing conditions. For purification of soluble protein, omit urea from Ni-NTA buffers, include protease inhibitors and perform purification at 4 °C to minimize protease activity (*see* **Notes 8** and **9**).
6. Equilibrate Ni-NTA agarose beads (e.g., 5 mL of Ni-NTA agarose beads for a protein solution obtained from 1 L of expression culture with average expression levels of 100 mg of protein per liter) with Ni-NTA binding buffer.

7. Incubate IB suspension and equilibrated Ni-NTA agarose beads for 30–45 min at RT with gentle agitation.
8. Collect flow-through from suspension of Ni-NTA agarose beads and IBs.
9. Wash Ni-NTA resin using five times the volume of Ni-NTA beads with each of Ni-NTA wash buffers 1 and 2.
10. Elute purified protein from Ni-NTA resin using denaturing Ni-NTA elution buffer.
11. Verify purification by protein electrophoresis. Evaluate purity and estimate protein concentration using a BCA Protein Assay Kit (*see* **Note 21**).

3.4 Mass Spectrometry Validation of ncAA Incorporation

1. Perform bioinformatic peptide cleavage on protein of interest for protease selection (*see* **Note 22**).
2. Normalize protein samples to approximately 0.5 mg/mL by dilution with denaturing buffer (*see* **Note 23**).
3. Add DTT stock solution to 5 mM and incubate at 55 °C, 20 min to reduce disulfides.
4. Add iodoacetamide stock solution to 15 mM and incubate for 20 min in the dark at room temperature to block free cysteines.
5. Dilute the protein sample solution 1:10 into protease reaction buffer for a total volume of 100 μL.
6. According to manufacturer's recommendations, add protease and allow sample to digest.
7. Desalt the sample using Ziptip cleanup as per manufacturer's recommendations (*see* **Note 24**).
8. Estimate ncAA incorporation level (e.g., Fig. 3c) by mass spectrometry (*see* **Notes 22** and **25**).

4 Notes

1. Typically, the synthetase is sourced from the expression host but this is not a requirement. The need for an overexpressed synthetase is dependent on the ncAA. In our experience, even if the ncAA can be incorporated using the endogenous machinery, it may be beneficial to overexpress the wild-type synthetase to increase protein yield or incorporation level.
2. In this protocol, we describe production of a typical recombinant protein in *E. coli*. We anticipate that this protocol can be adapted to other organisms (e.g., yeast) and to production of many recombinant proteins.

3. *Bam*HI, *Hind*III, *Nhe*I, and *Nco*I are used throughout the chapter due to their inclusion in the plasmid chosen to illustrate the technique. We recommend using these respective enzymes because they are located at convenient locations on plasmid pQE80L. Enzymes should be chosen with careful consideration of the expression system and protein of interest (POI).
4. TAE buffer can be used in place of TBE.
5. The 19aa solution contains all of the canonical amino acids (excluding the cAA of interest) at 50 mg/L. Generally, the pH of the solution will be titrated with sodium hydroxide to the basic region (i.e., pH 9–10.5) for complete dissolution. The solution can be brought down to the neutral region (i.e., pH 7–8) afterward with concentrated hydrochloric acid. Separate solutions are made for the cAA of interest and replacement ncAA at concentrations of 100–300 mM; these amino acid solutions can be diluted accordingly to the 5× working concentration of cAA or ncAA in the expression medium (e.g., if ncAA in medium is at 1 mM, 5× ncAA solution would be at 5 mM).
6. For convenience, a 10× M9/additives stock solution can be made and diluted as necessary.
7. Certain analogs require hypertonic medium for efficient cellular uptake (e.g., proline [28]). Hyperosmotic conditions can be achieved with the addition of NaCl at 100–600 mM [29].
8. If the POI partitions into the soluble fraction, buffers for Ni-NTA purification should be made in the absence of 8 M urea, and denaturing Ni-NTA elution buffer should be replaced with 300 mM NaCl, 50 mM NaH_2PO_4, 250 mM imidazole, pH 8.0. We also suggest adding protease inhibitors to the purification buffers and performing the purification step at 4 °C to reduce protein degradation by cellular proteases.
9. The imidazole concentration may need to be optimized for successful elution of the POI.
10. Standard proteases such as trypsin, GluC, and several others can be used depending on desired peptide fragment containing cAA/ncAA residue of interest.
11. We recommend vector dephosphorylation. Antarctic phosphatase and 10× dephosphorylation buffer can be added directly to the digestion mixture.
12. *E. coli* aaRS genes can be amplified from genomic DNA.
13. If using cells that are not Z-competent, use the transformation protocol appropriate for the nature of the chemical competence (e.g., heat shock).

14. We recommend using *Bam*HI and *Hind*III sites due to the N-terminal histidine tag present in the pQE80 vector. Ensure that the aaRS DNA sequence does not contain *Bam*HI and/ or *Hind*III; if this cannot be avoided, other restriction enzyme sites in the MCS of pQE80L_aaRS may be used instead.
15. For POIs with DNA sequences that are <1 kb, we recommend custom synthesis for ease of cloning.
16. If incorporation levels are too low, we suggest the following steps that may enhance incorporation: (1) changes in plasmid design by (a) placing the aaRS under a stronger promoter or (b) placing the aaRS on a separate plasmid that has a higher copy number; or (2) changes in the expression conditions, including (a) increasing the length of the cAA depletion step or (b) increasing the concentration of ncAA in the expression medium.
17. We highly recommend, prior to expression with ncAAs, transforming the plasmid of interest into prototrophic *E. coli* (i.e., BL21), expressing the protein of interest in rich medium, and purifying the product to determine the expression yield. Generally, though not always, volumetric expression yields of proteins containing ncAAs will be less than those of the corresponding wild-type proteins.
18. We suggest that the time allowed for depletion of the intracellular cAA of interest be approximately the same as the doubling time of the expression strain. Instructions for specific ncAAs are available in the literature [30–34].
19. To ensure proper depletion of the intracellular cAA, be sure to prepare negative and positive control samples: 19aa (add sterile water or 1.5 M NaCl to 1× M9, 19aa solution) and 20aa (add 5× cAA solution made in either water or 1.5 M NaCl), respectively.
20. Alternative growth times (1–20 h) or growth temperatures (16–37 °C) may result in enhanced expression levels for some proteins. If 37 °C for 2 h does not result in sufficient expression, we suggest performing time-course expressions at 20 °C and 37 °C by taking samples every hour post-induction for the first 6 h and another sample at 20 h.
21. If further purification is required, adjust the pH of the elution fractions to 8.0 and apply the protein to an additional chromatography column chosen on the basis of the physical properties of the protein. Refer to the manufacturer's protocol for the additional chromatography steps.
22. The protease most useful for preparing peptides for analysis by mass spectrometry will depend on the sequence of the protein. PeptideMass on ExPASy is a simple, open-source

bioinformatic tool to obtain a list of peptide masses generated by candidate proteases.

23. We recommend denaturing the protein prior to protease digestion. If the protein is purified from the soluble fraction, we recommend precipitating it from the elution buffer using standard acetone or chloroform–methanol precipitation methods.

24. For example: Add 10 μL of 5% TFA to protease–protein solution to quench protease digestion. Vortex and centrifuge to remove bubbles. Wet Ziptip with 10 μL of 50% ACN and discard, twice. Equilibrate Ziptip with 10 μL of 0.1% TFA and discard, once. Bind peptides to the Ziptip's C_{18} column by pipetting up and down in the quenched protease–protein solution, ten times. Wash Ziptip with 10 μL of 0.1% TFA and discard, 15 times. Elute peptides in 3 μL of 50% ACN, 0.1% TFA. Store at −20 °C until further use.

25. For example, we commonly use matrix-assisted laser desorption/ionization mass spectrometry (MALDI-MS). Briefly, our protocol: Dissolve 10 mg of α-CN into 1 mL of 50% ACN, 0.1% TFA. Remove undissolved α-CN by centrifugation. Dilute peptide at 1:3 ratio into α-CN solution, and spot and dry peptides onto matrix plate. Perform MALDI-MS on peptides and analyze spectra (Fig. 3c).

Acknowledgments

We thank past and present members of the Tirrell laboratory for their contributions to the development and use of ncAAs in protein science and engineering.

References

1. Zaykov AN, Mayer JP, DiMarchi RD (2016) Pursuit of a perfect insulin. Nat Rev Drug Discov 15:425–439
2. Shire SJ (2009) Formulation and manufacturability of biologics. Curr Opin Biotechnol 20(6):708–714
3. Zhang WH, Otting G, Jackson CJ (2013) Protein engineering with unnatural amino acids. Curr Opin Struct Biol 23(4):581–587
4. Ravikumar Y, Nadarajan SP, Yoo TH et al (2015) Unnatural amino acid mutagenesis-based enzyme engineering. Trends Biotechnol 33(8):462–470
5. Neumann-Staubitz P, Neumann H (2016) The use of unnatural amino acids to study and engineer protein function. Curr Opin Struct Biol 38:119–128
6. Lieblich SA, Fang KY, Cahn JKB et al (2017) 4*S*-Hydroxylation of insulin at ProB28 accelerates hexamer dissociation and delays fibrillation. J Am Chem Soc 139(25):8384–8387
7. Mahdavi A, Szychowski J, Ngo JT et al (2014) Identification of secreted bacterial proteins by noncanonical amino acid tagging. Proc Natl Acad Sci U S A 111(1):433–438
8. Rapp PB, Omar AK, Shen JJ et al (2017) Analysis and control of chain mobility in protein hydrogels. J Am Chem Soc 139(10):3796–3804
9. Liu DR, Magliery TJ, Pastrnak M et al (1997) Engineering a tRNA and aminoacyl-tRNA

synthetase for the site-specific incorporation of unnatural amino acids into proteins *in vivo*. Proc Natl Acad Sci U S A 94(19):10092–10097

10. Liu CC, Schultz PG (2010) Adding new chemistries to the genetic code. Annu Rev Biochem 79:413–444
11. Johnson JA, Lu YY, Van Deventer JA et al (2010) Residue-specific incorporation of non-canonical amino acids into proteins: recent developments and applications. Curr Opin Chem Biol 14(6):774–780
12. Budisa N (2004) Prolegomena to future experimental efforts on genetic code engineering by expanding its amino acid repertoire. Angew Chem Int Ed Engl 43(47):6426–6463
13. Babin BM, Bergkessel M, Sweredoski MJ et al (2016) SutA is a bacterial transcription factor expressed during slow growth in Pseudomonas aeruginosa. Proc Natl Acad Sci U S A 113(5):E597–E605
14. Bagert JD, van Kessel JC, Sweredoski MJ et al (2016) Time-resolved proteomic analysis of quorum sensing in Vibrio harveyi. Chem Sci 7(3):1797–1806
15. Van Deventer JA, Le DN, Zhao J et al (2016) A platform for constructing, evaluating, and screening bioconjugates on the yeast surface. Protein Eng Des Sel. [Epub ahead of print]
16. Wiltschi B (2016) Incorporation of non-canonical amino acids into proteins in yeast. Fungal Genet Biol 89:137–156
17. Hinz FI, Dieterich DC, Tirrell DA et al (2012) Noncanonical amino acid labeling in vivo to visualize and affinity purify newly synthesized proteins in larval zebrafish. ACS Chem Nerosci 3(1):40–49
18. Yuet KP, Doma MK, Ngo JT et al (2015) Cell-specific proteomic analysis in Caenorhabditis elegans. Proc Natl Acad Sci U S A 112(9):2705–2710
19. Han S, Yang A, Lee S et al (2017) Expanding the genetic code of Mus musculus. Nat Commun 8:14568
20. Hinz FI, Dieterich DC, Schuman EM (2013) Teaching old NCATs new tricks: using non-canonical amino acid tagging to study neuronal plasticity. Curr Opin Chem Biol 17(5):738–746
21. Axup JY, Bajjuri KM, Ritland M et al (2012) Synthesis of site-specific antibody-drug conjugates using unnatural amino acids. Proc Natl Acad Sci U S A 109(40):16101–16106
22. Lu YY, Sweredoski MJ, Huss D et al (2014) Pro-metastatic GPCR CD97 is a direct target of tumor suppressor microRNA-126. ACS Chem Biol 9(2):334–338
23. Mahdavi A, Hamblin GD, Jindal GA et al (2016) Engineered aminoacyl-tRNA synthetase for cell-selective analysis of mammalian protein synthesis. J Am Chem Soc 138(13):4278–4281
24. Erdmann I, Marter K, Kobler O et al (2015) Cell-selective labelling of proteomes in *Drosophila melanogaster*. Nat Commun 6:7521
25. Glenn WS, Stone SE, Ho SH et al (2017) Bioorthogonal noncanonical amino acid tagging (BONCAT) enables time-resolved analysis of protein synthesis in native plant tissue. Plant Physiol 173(3):1543–1553
26. Fan Y, Evans CR, Ling J (2017) Rewiring protein synthesis: from natural to synthetic amino acids. Biochim Biophys Acta 1861:3024–3029
27. Burgess RR (2009) Refolding solubilized inclusion body proteins. Methods Enzymol 463:259–282, (Elsevier)
28. Grothe S, Krogsrud RL, McClellan DJ et al (1986) Proline transport and osmotic stress response in *Escherichia coli* K-12. J Bacteriol 166(1):253–259
29. Weber A, Kögl SA, Jung K (2006) Time-dependent proteome alterations under osmotic stress during aerobic and anaerobic growth in *Escherichia coli*. J Bacteriol 188(20):7165–7175
30. Van Deventer JA, Fisk JD, Tirrell DA (2011) Homoisoleucine: a translationally active leucine surrogate of expanded hydrophobic surface area. ChemBioChem 12(5):700–702
31. Yoo TH, Tirrell DA (2007) High-throughput screening for methionyl-tRNA synthetases that enable residue-specific incorporation of noncanonical amino acids into recombinant proteins in bacterial cells. Angew Chem Int Ed Engl 46(28):5340–5343
32. Kim W, George A, Evans M et al (2004) Cotranslational incorporation of a structurally diverse series of proline analogues in an *Escherichia coli* expression system. ChemBioChem 5(7):928–936
33. Tanrikulu IC, Schmitt E, Mechulam Y et al (2009) Discovery of *Escherichia coli* methionyl-tRNA synthetase mutants for efficient labeling of proteins with azidonorleucine in vivo. Proc Natl Acad Sci U S A 106(36):15285–15290
34. Yoo TH, Link AJ, Tirrell DA (2007) Evolution of a fluorinated green fluorescent protein. Proc Natl Acad Sci U S A 104(35):13887–13890
35. Link AJ, Mock ML, Tirrell DA (2003) Non-canonical amino acids in protein engineering. Curr Opin Biotechnol 14(6):603–609

36. Davis L, Chin JW (2012) Designer proteins: applications of genetic code expansion in cell biology. Nat Rev Mol Cell Biol 13(3):168–182
37. Kiick KL, van Hest JCM, Tirrell DA (2000) Expanding the scope of protein biosynthesis by altering the methionyl-tRNA synthetase activity of a bacterial expression host. Angew Chem Int Ed Engl 39(12):2148–2152
38. van Hest JCM, Kiick KL, Tirrell DA (2000) Efficient incorporation of unsaturated methionine analogues into proteins *in vivo*. J Am Chem Soc 122(7):1282–1288
39. Kiick KL, Saxon E, Tirrell DA et al (2002) Incorporation of azides into recombinant proteins for chemoselective modification by the Staudinger ligation. Proc Natl Acad Sci U S A 99(1):19–24
40. Truong F, Yoo TH, Lampo TJ et al (2012) Two-strain, cell-selective protein labeling in mixed bacterial cultures. J Am Chem Soc 134(20):8551–8556
41. Mock ML, Michon T, van Hest JCM et al (2006) Stereoselective incorporation of an unsaturated isoleucine analogue into a protein expressed in *E. coli*. ChemBioChem 7(1):83–87
42. Wang P, Tang Y, Tirrell DA (2003) Incorporation of trifluoroisoleucine into proteins in vivo. J Am Chem Soc 125(23):6900–6906
43. Tang Y, Wang P, Van Deventer JA et al (2009) Introduction of an aliphatic ketone into recombinant proteins in a bacterial strain that overexpresses an editing-impaired leucyl-tRNA synthetase. ChemBioChem 10(13):2188–2190
44. Montclare JK, Son S, Clark GA et al (2009) Biosynthesis and stability of coiled-coil peptides containing (2S,4R)-5,5,5-trifluoroleucine and (2S,4S)-5,5,5-trifluoroleucine. ChemBioChem 10(1):84–86
45. Tang Y, Ghirlanda G, Vaidehi N et al (2001) Stabilization of coiled-coil peptide domains by introduction of trifluoroleucine. Biochemistry 40(9):2790–2796
46. Kirshenbaum K, Carrico IS, Tirrell DA (2002) Biosynthesis of proteins incorporating a versatile set of phenylalanine analogues. ChemBioChem 3(2–3):235–237
47. Datta D, Wang P, Carrico IS et al (2002) A designed phenylalanyl-tRNA synthetase variant allows efficient in vivo incorporation of aryl ketone functionality into proteins. J Am Chem Soc 124(20):5652–5653

Chapter 14

Modification of Protein Scaffolds via Copper-Catalyzed Azide–Alkyne Cycloaddition

Stanislav Presolski

Abstract

The copper-catalyzed azide–alkyne cycloaddition (CuAAC) has proven to be a reliable, high-efficiency method for modification of protein scaffolds. This "click" reaction offers specificity and nearly quantitative yields even at low reagent concentrations. While robust, CuAAC still requires proper setup to achieve the high efficiency characteristic of this reaction, as well as to avoid degradation of sensitive substrates. Detailed herein is a generic CuAAC protocol for protein modification. Key features include the use of DMSO and triazole-based accelerating ligands for protection against reactive oxygen species, as well as aminoguanidine for intercepting deleterious ascorbate by-products formed during the bioconjugation.

Key words Click chemistry, CuAAC, Bioconjugation, Protein modification, Azide–alkyne cycloaddition

1 Introduction

The copper-catalyzed azide–alkyne cycloaddition (CuAAC) reaction, in which the eponymous small and fairly unreactive functional groups are ligated to form a stable triazole ring (Fig. 1), has been used in the functionalization of peptides since its discovery in 2002 [1, 2]. It has become synonymous with "click chemistry" [3], because its selectivity and biocompatibility has led to its wide adoption in the fields of molecular biology, medicinal chemistry, and materials science [4–6]. While the exact nature of all the species involved in the catalytic cycle is dependent on the reaction conditions [7], all proposed mechanisms have now converged on identifying two Cu^+ ions as being responsible for the remarkable rate acceleration and regioselectivity [7–9]. Thus, in the majority of cases when oxygen is not excluded from the CuAAC media, mild reducing agents are employed to generate in situ and then maintain the copper ions in their active +1 oxidation state. Using an aqueous solution of sodium ascorbate (NaAsc) and $CuSO_4$ is typically all that is necessary to obtain excellent yields of small molecule

Andrew K. Udit (ed.), *Protein Scaffolds: Design, Synthesis, and Applications*, Methods in Molecular Biology, vol. 1798, https://doi.org/10.1007/978-1-4939-7893-9_14,

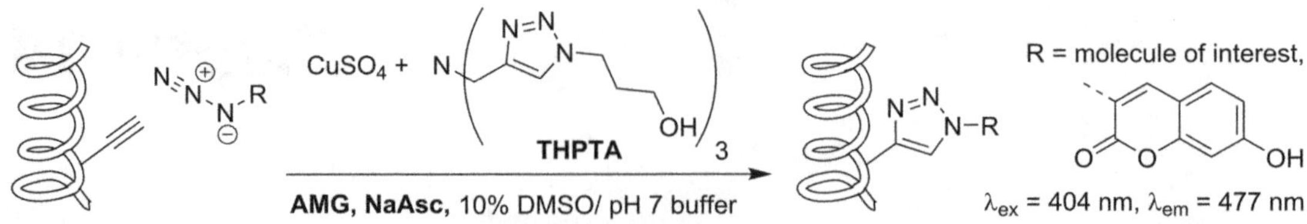

Fig. 1 CuAAC modification of protein scaffolds

triazoles; however care must be taken to avoid degradation of biological substrates. Proteins are especially susceptible to oxidative damage and crosslinking by the aldehyde by-products of ascorbate decomposition, so the use of sacrificial reagents is crucial for effective CuAAC bioconjugation [10]. Here we describe an optimized procedure for protein scaffold modifications employing the copper-chelating ligand tris(3-hydroxypropyltriazolylmethyl) amine (THPTA), which serves the dual purpose of accelerating the cycloaddition and intercepting reactive oxygen species formed during the reaction [11]. Addition of aminoguanidine (AMG) and dimethyl sulfoxide (DMSO) affords even further protection from harmful dehydroascorbate and reactive oxygen species, respectively [12]. This robust CuAAC method can even be successfully applied to labeling live cells [13, 14].

2 Materials

Prepare stock solutions in 18 MΩ·cm ultrapure water and reagent grade DMSO and store at room temperature unless otherwise indicated. Coumarin azide can be synthesized according to Wang [15] or purchased from Glen Research (http://www.glenresearch.com/ProductFiles/product.php?item=50-2004). THPTA is available through Sigma-Aldrich and TCI, but can also be prepared following the accounts of Finn [10, 16].

1. Phosphate buffer: 100 mM potassium phosphate pH 7 (*see* **Note 1**).
2. Protein-alkyne: 65 μM total alkyne on protein of interest in phosphate buffer (*see* **Note 2**).
3. Protein solution: Protein of interest in phosphate buffer that has not been labeled with alkyne, but at the same concentration as the protein-alkyne above (*see* **Note 3**).
4. Azide solution: 1 mM solution of azide of interest in DMSO (or buffer, *see* **Note 4**).
5. 10 mM $CuSO_4$: Dissolve 25 mg of copper sulfate pentahydrate in 10 mL of water.

6. 50 mM THPTA: Dissolve 44 mg of THPTA in 2 mL of water (*see* **Note 5**).
7. 100 mM AMG: Dissolve 11 mg of aminoguanidine hydrochloride in 1 mL of H_2O.
8. 100 mM NaAsc: Dissolve 20 mg of sodium ascorbate in 1 mL of H_2O. Prepare fresh just before use and store for not more than a few hours (*see* **Note 6**).
9. EDTA solution: Add 10 g of ethylenediaminetetraacetic acid disodium salt dihydrate to 100 mL of water to prepare a saturated EDTA stock solution.
10. 658 μM propargyl alcohol: Pipette 2 μL of propargyl alcohol into 50 mL of phosphate buffer.
11. 1 mM coumarin azide: Dissolve 2 mg of coumarin azide in 10 mL of DMSO (*see* **Note 7**).
12. 2 mL microcentrifuge tubes.
13. Rotisserie for microcentrifuge tubes.
14. Fluorimeter with appropriate cuvette.

3 Methods

The procedures below should be carried out at room temperature. The exact volumes and stock solution concentrations can be altered to some extent (*see* **Note 8**), but the order of addition is critical and must be diligently followed.

3.1 Protein Modification

The final concentrations in this 500 μL reaction are 50 μM alkyne, 100 μM azide, 100 μM $CuSO_4$, 500 μM THPTA, 5 mM AMG, 5 mM NaAsc in phosphate buffer with 10% DMSO (*see* **Note 9**).

1. Add 390 μL of protein-alkyne to a 2 mL microcentrifuge tube.
2. Pipette 50 μL of the azide solution into the same microcentrifuge tube.
3. In a separate tube, premix 5 μL of 10 mM $CuSO_4$ with 5 μL of 50 mM THPTA, then add the resulting mixture to the reaction tube containing azide and alkyne (*see* **Note 10**).
4. Add 25 μL of 100 mM AMG.
5. Initiate the reaction by pipetting in 25 μL of 100 mM NaAsc into the tube, close the tube immediately and attach to an end-over-end rotator or rotisserie, causing the solution to tumble inside once every 2–3 s (*see* **Note 11**).
6. After 1 h, stop the reaction by the addition of 500 μL of EDTA (*see* **Note 12**) and purify the conjugate in accordance with the methods required for the specific protein (e.g., dialysis, precipitation, gel permeation chromatography).

3.2 Determination of CuAAC Efficiency

In order to determine CuAAC efficacy, and troubleshoot if necessary, prepare the following fluorescence assay. The final reagent concentrations in this 500 μL reaction are: 500 μM propargyl alkyne, 100 μM coumarin azide, 200 μM $CuSO_4$, 1 mM THPTA, 5 mM AMG, 5 mM NaAsc, and 10% DMSO in phosphate buffer.

1. Add 380 μL of 658 μM propargyl alcohol to a 2 mL microcentrifuge tube.
2. Pipette 50 μL of 1 mM coumarin azide into the same tube.
3. In a separate tube, premix 10 μL of 10 mM $CuSO_4$ with 10 μL of 50 mM THPTA, then add the resulting mixture to the reaction tube containing propargyl alcohol and coumarin azide.
4. Add 25 μL of 100 mM AMG.
5. Initiate the reaction by pipetting in 25 μL of 100 mM NaAsc, close the tube immediately and attach to an end-over-end rotator/rotisserie, causing the solution to tumble inside once every 2–3 s.
6. After 1 h quench the reaction mixture by adding 720 μL of EDTA solution.
7. Add 780 μL of the protein of interest (without alkyne) in phosphate buffer at the same concentration as the protein in Subheading 3.1 **step 1**. The concentration of fluorescent triazole formed will be 25 μM.
8. In a different microcentrifuge tube set up a test reaction as outlined in Subheading 3.1, using coumarin azide for the azide solution and the protein-alkyne. After quenching with EDTA, the combined concentration of alkyne and triazole will be 25 μM (*see* **Note 13**).
9. Compare the fluorescence intensity at 477 nm (404 nm excitation) between the test reaction (**step 8**) and the fluorescence standard solution (**steps 1**–**7**) to determine the extent of protein modification.
10. If necessary, optimize the reaction conditions (*see* **Note 14**).

4 Notes

1. Triazole-based ligands such as THPTA have an optimal pH around 7, but the CuAAC reaction can accommodate a wide acid-base range established by phosphate, acetate, MOPS and HEPES buffers. Reaction media containing copper-binding species like Tris should be avoided, as well as those with chloride concentration greater than 200 mM.
2. The reaction can also be performed in the reverse orientation, i.e., with a protein-azide and a small molecule alkyne. However,

in that case the test reaction must be performed with an appropriate fluorogenic alkyne (*see* Uttamapinant et al. [14]).

3. If the protein is precious, bovine serum albumin could be used as a surrogate instead.
4. We recommend DMSO since it is an excellent solvent for most small molecules of interest, and it provides some protection against reactive oxygen species. DMF and *N*-methyl-2-pyrrolidone (NMP) are suitable alternatives, and aqueous buffer can also be used. Do not substitute with acetonitrile, which has been shown to inhibit CuAAC under bioconjugation reaction conditions due to its strong copper-chelating ability. Aqueous DMSO, DMF and NMP below 30 v/v% are well tolerated, but above that amount the THPTA is outcompeted for the copper by these coordinating solvents. Different types of more strongly donating ligands can provide appropriate metal chelation, rate acceleration, and most importantly protection from reactive oxygen species (*see* Presolski et al. [7]).
5. THPTA can be substituted with other water-soluble CuAAC ligands, as long as they do not contain heterocycles such as pyridines and imidazoles that are more strongly chelating than triazole (*see* Uttamapinant et al. [14]).
6. NaAsc stock solution should be prepared immediately before setting up the CuAAC reaction to ensure maximum efficacy of the reducing agent and minimal interference from its oxidized by-product dehydroascorbate, which can cause undesired arginine modification. If, however, the protein scaffolds show excessive degradation by dehydroascorbate despite the use of excess amount of aminoguanidine, 10 mM hydroxylamine can be used instead of NaAsc.
7. Coumarin azide and its stock solutions should be protected from light and refrigerated when not in use.
8. It is advisable to use an excess amount of copper (and never below 50 μM) to ensure that CuAAC is complete before the reducing agent has been depleted. If only low micromolar concentrations of a particular substrate are available, its partner should be used in excess to speed up the reaction to appreciable levels.
9. The concentration of dissolved O_2 in water is approximately 250 μM, so its copper-catalyzed reduction to H_2O requires a minimum of 500 μM NaAsc. Therefore, most of the ascorbate is consumed not in reducing the Cu^{2+} ions, but in removing the oxygen from the solution. And since the reaction mixture is in equilibrium with the air above it where O_2 is 8 mM, it is essential to cap the vial/tube tightly and minimize the headspace volume by choosing the smallest possible container.

10. THPTA should be used in excess, because it sacrificially reacts with the radicals and peroxides generated by the Cu^{+} reduction of dissolved O_2 during the early stages of the reaction. Thus, it is best if it is premixed with the $CuSO_4$ solution before NaAsc is added in. The weak donor ability of the triazole-based THPTA precludes it from forming inhibitory complexes even at ligand–Cu ratios greater than 5:1 (*see* ref. [7]).
11. If the biomolecule substrates are particularly susceptible to oxidation, they can be added as a very concentrated stock solution to the reaction mixture 30 min after the THPTA, $CuSO_4$ and NaAsc have been premixed. At that point in time, the concentration of reactive oxygen species will be substantially lower. Alternatively, CuOAc, CuOTf, or CuBr (alongside freshly cleaned Cu wire as a comproportionation partner in the eventuality of some Cu^{2+} formation) can be used with rigorous exclusion of oxygen in degassed solvent.
12. The EDTA quench can be omitted if purification is done immediately and the protein is not very sensitive to reactive oxygen species or Cu^{2+}-catalyzed hydrolysis.
13. Before quenching the test reaction with EDTA, expose it to a handheld long-wavelength UV lamp. If the solution is not visibly fluorescent, add a drop of neat propargyl alcohol and allow the reaction to proceed for another hour to determine whether the catalyst is active under the particular conditions. If there is increased fluorescence signal, then the alkyne groups have either been unsuccessfully incorporated into the biomolecule, or have become inaccessible by hydrophobic collapse. Otherwise, the catalyst must have been inactivated by the protein, such as one containing many histidines or free cysteine residues.
14. Hydrophobic collapse could bury reactive groups, so appropriate measures must be taken, such as DMSO addition and/or higher reaction temperature. Heating the reaction to 37 °C and above will accelerate it and might provide the necessary flexibility in the protein scaffold to expose the alkynes or azides it contains. Histidine and cysteine-rich Cu-binding motifs might sequester catalytically active copper, so if they cannot be avoided, premix with Zn^{2+} and use much greater copper excess.

Acknowledgments

The author would like to acknowledge Dr. Vu P. Hong, Dr. Alexander Kislukhin, and Prof. M.G. Finn for many fruitful discussions on the mechanism and the optimization of the ligand-accelerated CuAAC reaction, as well as Yale-NUS start-up grant for funding.

References

1. Tornøe CW, Christensen C, Meldal M (2002) Peptidotriazoles on solid phase: [1,2,3]-triazoles by regiospecific copper(I)-catalyzed 1,3-dipolar cycloadditions of terminal alkynes to azides. J Org Chem 67:3057–3064
2. Rostovtsev VV, Green LG, Fokin VV et al (2002) A stepwise Huisgen cycloaddition process: copper(I)-catalyzed regioselective ligation of azides and terminal alkynes. Angew Chem Int Ed 41:2596–2599
3. Kolb HC, Finn MG, Sharpless KB (2001) Click chemistry: diverse chemical function from a few good reactions. Angew Chem Int Ed 40:2004–2021
4. Meldal M, Tornøe CW (2008) Cu-catalyzed azide–alkyne cycloaddition. Chem Rev 108: 2952–3015
5. McKay CS, Finn MG (2014) Click chemistry in complex mixtures: bioorthogonal bioconjugation. Chem Bio 21:1075–1101
6. Haldón E, Nicasio MC, Pérez PJ (2015) Copper-catalysed azide–alkyne cycloadditions (CuAAC): an update. Org Biomol Chem 13:9528–9550
7. Presolski SI, Hong V, Cho S-H et al (2010) Tailored ligand acceleration of the Cu-catalyzed azide–alkyne cycloaddition reaction: practical and mechanistic implications. J Am Chem Soc 132:14570–14576
8. Worrell BT, Malik JA, Fokin VV (2013) Direct evidence of a dinuclear copper intermediate in Cu(I)-catalyzed azide–alkyne cycloadditions. Science 340:457–460
9. Jin L, Tolentino DR, Melaimi M et al (2015) Isolation of bis(copper) key intermediates in Cu-catalyzed azide–alkyne “click reaction”. Sci Adv 1:e1500304
10. Hong V, Presolski SI, Ma C et al (2009) Analysis and optimization of copper-catalyzed azide–alkyne cycloaddition for bioconjugation. Angew Chem Int Ed 48:9879–9883
11. Presolski SI, Hong VP, Finn MG (2011) Copper-catalyzed azide–alkyne click chemistry for bioconjugation. Curr Protoc Chem Biol 3:153–162
12. Abel GR Jr, Calabrese ZA, Ayco J et al (2016) Measuring and suppressing the oxidative damage to DNA during Cu(I)-catalyzed azide−alkyne cycloaddition. Bioconj Chem 27:698–704
13. Hong V, Steinmetz NF, Manchester M et al (2010) Labeling Live cells by copper-catalyzed alkyne–azide click chemistry. Bioconj Chem 21:1912–1916
14. Uttamapinant C, Tangpeerachaikul A, Grecian S et al (2012) Fast, cell-compatible click chemistry with copper-chelating azides for biomolecular labeling. Angew Chem Int Ed 51:5852–5856
15. Sivakumar K, Xie F, Cash BM et al (2004) A fluorogenic 1,3-dipolar cycloaddition reaction of 3-azidocoumarins and acetylenes. Org Lett 6:4603–4606
16. Kislukhin AA, Hong VP, Breitenkamp KE et al (2013) Relative performance of alkynes in copper-catalyzed azide−alkyne cycloaddition. Bioconj Chem 24:684–689

Chapter 15

Antibody Modification of *p*-Aminophenylalanine-Containing Proteins

Adel M. ElSohly, Chawita Netirojjanakul, and Matthew B. Francis

Abstract

The use of antibody conjugates for biomedical applications has garnered increased attention due to the ability of antibodies to specifically engage targets of interest. Despite these appealing qualities, the preparation of antibody–protein conjugates remains challenging. Here we detail an approach to attaching targeting antibodies to proteins of interest that combines advances in genetic code expansion and an efficient bioconjugation strategy. As an example, we prepare bacteriophage MS2 viral capsids bearing antibodies on their surfaces for applications in molecular targeting. This technique provides a modular framework to easily prepare antibody–MS2 conjugates in an efficient manner, even at low concentrations of the reacting biomolecules.

Key words Bacteriophage MS2, Antibodies, Bioconjugation, Oxidative coupling, Genetic code expansion, Noncanonical amino acids, *p*-aminophenylalanine

1 Introduction

Here we detail a bioorthogonal method to chemically modify the external surface of proteins with antibodies (Abs) [1]. This strategy relies on the incorporation of a noncanonical amino acid (*p*-aminophenylalanine, *p*aF) using Amber stop codon suppression techniques [2, 3] in conjunction with an efficient oxidative coupling reaction of *ortho*-aminophenols in the presence of sodium periodate [4]. To illustrate the technique, we specifically describe the modification of bacteriophage MS2 capsids. Nanoscale carriers derived from self-assembling biomolecular complexes have shown great promise for the development of novel platforms for imaging and drug delivery [5–12]. Viral capsids in particular are attractive due to their ability to self-assemble into homogeneous particles that contain both interior and exterior surfaces that can be orthogonally modified for various applications. The large number of coat protein monomers allows for the attachment of many cargo molecules, such as therapeutics or imaging agents, which have the

Andrew K. Udit (ed.), *Protein Scaffolds: Design, Synthesis, and Applications*, Methods in Molecular Biology, vol. 1798, https://doi.org/10.1007/978-1-4939-7893-9_15,

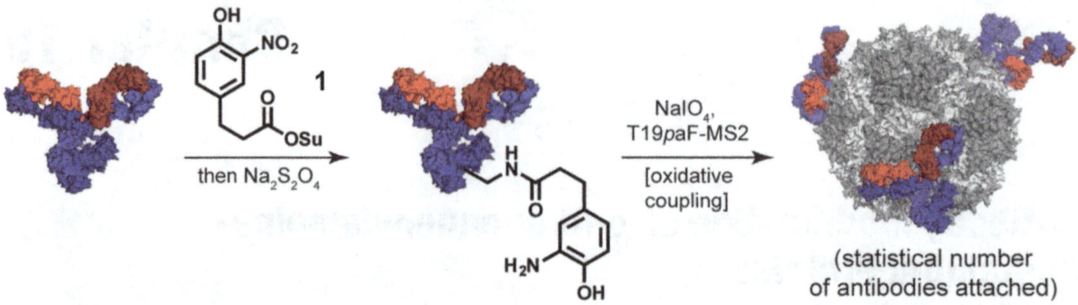

Fig. 1 Schematic of two-step protocol for conjugation of antibodies to T19*p*aF MS2 viral capsids. Lysine residues on the antibody are nonspecifically labeled with a nitrophenol N-hydroxysuccinimide (NHS) ester reagent **1**. Subsequent treatment with sodium dithionite ($Na_2S_2O_4$) results in chemoselective reduction to the corresponding antibody–aminophenol conjugate (Ab-AP). Treatment of the Ab-AP with sodium periodate ($NaIO_4$) in the presence of T19*p*aF MS2 capsids results in efficient, covalent attachment of the antibodies to the exterior surface of the capsid

potential to improve therapeutic index or signal contrast, respectively. In each of these cases, however, active targeting of a region of interest is required to achieve the desired outcome [13]. Indeed, particularly for tumor targeting, many studies have demonstrated the importance of active targeting in achieving the desired intratumoral localization ([13] and references therein).

This technique results in rapid and quantitative attachment of Abs to the capsid when a 3:1 ratio of antibody to capsid is used. The modular nature of this protocol facilitates the preparation of diverse Ab-targeted MS2 conjugates for a variety of applications. The overall approach is summarized in Fig. 1.

2 Materials

All aqueous solutions are prepared using ultrapure water (deionized to a resistivity of 18 MΩ·cm at 23 °C). Buffer solutions are stored at ambient temperature in the absence of azide. Stock solutions in DMSO are prepared from anhydrous DMSO and stored at −20 °C. All chemicals are used as received from their respective commercial sources. Antibodies can be obtained from commercial sources.

2.1 Synthesis of Ab-Aminophenol (Ab-AP)

1. Nitrophenol NHS ester: 5 mM stock solution in anhydrous DMSO. Accurately weigh out 10 mg of nitrophenol NHS ester (*see* **Note 1**) in a 1.7 mL Eppendorf tube and dissolve in 648 μL of DMSO to give a 50 mM solution. Dilute tenfold to give a final concentration of 5 mM (*see* **Note 2**).
2. Antibody: Antibodies from commercial sources at 1 mg/mL (~6.6 μM) in phosphate buffered saline are used as received. Stock solutions are stored at 4 °C (*see* **Note 3**).

3. Antibody pH adjustment buffer: 100 mM sodium phosphate, pH 8.0 in water. Add 0.690 g of sodium phosphate (monobasic monohydrate) to a 50 mL conical centrifuge tube and dissolve in 30 mL of water. Adjust the pH to 8.0 with solutions of NaOH and HCl and bring to a final volume of 50 mL.
4. Dithionite dissolution buffer: 100 mM sodium phosphate, pH 6.5 in water. Add 0.690 g of sodium phosphate (monobasic monohydrate) to a 50 mL conical centrifuge tube and dissolve in 30 mL of water. Adjust the pH to 6.5 with solutions of NaOH and HCl and bring to a final volume of 50 mL.
5. Sodium dithionite solution: 100 mM in dithionite dissolution buffer. Accurately weigh 10 mg of sodium dithionite into a 1.7 mL Eppendorf tube and dissolve with 574 μL of dithionite dissolution buffer (*see* **Note 4**).
6. Washing buffer: 10 mM sodium phosphate, pH 7.0 in water. Add 0.690 g of sodium phosphate (monobasic) to a 50 mL conical centrifuge tube and dissolve in 30 mL of water. Adjust the pH to 7.0 with solutions of NaOH and HCl and bring to a final volume of 50 mL to give a 100 mM solution. Dilute this stock tenfold to give the final 10 mM buffer solution.
7. Microcentrifuge spin concentrators with 30 kDa MWCO membranes.
8. UV-Vis spectrometer.
9. Microcentrifuge.

2.2 Conjugation of Ab-AP to T19paF MS2 Capsids

1. T19*pa*F Bacteriophage MS2: Stock solutions are maintained at 100 μM (coat protein monomer concentration) in 20 mM phosphate buffer, pH 7.0 in the absence of azide and stored at 4 °C (*see* **Note 5**).
2. Reaction and periodate dissolution buffer: 10 mM sodium phosphate, pH 7.0 in water. Add 0.690 g of sodium phosphate (monobasic) to a 50 mL conical centrifuge tube and dissolve in 30 mL of water. Adjust the pH to 7.0 with solutions of NaOH and HCl and bring to a final volume of 50 mL to give a 100 mM solution. Dilute this stock tenfold to give the final 10 mM buffer solution.
3. Sodium periodate solution: 10 mM stock solution in 10 mM sodium phosphate, pH 7.0. Accurately weigh 5 mg of sodium periodate into a 1 dram glass vial, and add 2.34 mL of 10 mM sodium phosphate, pH 7.0. Cap the vial and vortex to give the final solution (*see* **Note 6**).
4. Microcentrifuge spin concentrators with 100 kDa MWCO membranes.
5. SDS-PAGE gel electrophoresis equipment and reagents.

3 Methods

All operations were conducted at ambient temperature using standard pipetting techniques. The protocols described below can be scaled according to the desired amount of conjugate to be prepared.

3.1 Synthesis of Ab-AP Conjugate

1. Sequentially mix 100 μL of antibody stock solution, 33.3 μL of antibody pH adjustment buffer, and 0.66 μL of nitrophenol NHS ester (five equivalents with respect to the antibody) in a clean Eppendorf tube. The reaction is allowed to proceed for 1 h at ambient temperature.
2. Add 13.3 μL of sodium dithionite solution directly to the reaction mixture (final concentration ~10 mM). The pale-yellow solution should rapidly turn colorless. Allow the reduction to proceed for 15 min at room temperature.
3. Remove excess dithionite and small molecules as well as buffer exchange by spin concentrating against a 30 kDa MWCO membrane with washing buffer to a final volume of ~50 μL. Repeat this three times.
4. Recover the Ab-AP conjugate from the spin concentrator into a clean Eppendorf tube and measure the protein concentration by UV-Vis spectroscopy using Abs_{280} of 1.4 yielding 1 mg/mL (~6.6 μM) as a conversion factor (*see* **Notes 7–9**).

3.2 Conjugation of Ab-AP to T19paF MS2 Capsids

For the following, a 50% recovery in the Ab-AP conjugation step above is assumed (i.e., 50 μL of 6.6 μM conjugate). Concentrations and volumes should be adjusted based on actual recoveries.

1. Add 198 μL of T19*pa*F MS2 solution to the Ab-AP solution obtained in Subheading 3.1 to establish a 3:1 ratio of antibody to capsid (i.e., 1/180 of the coat protein monomer concentration). Add 135 μL of reaction buffer to establish a final MS2 coat protein concentration of ~50 μM (*see* **Note 10**).
2. Add 13.1 μL of sodium periodate solution to the reaction medium to reach a final concentration of 500 μM. Allow the reaction to proceed for 5 min at room temperature.
3. Remove any residual unconjugated antibody and periodate by spin concentrating against a 100 kDa MWCO membrane with washing buffer. Repeat three times.
4. Recover the pure antibody–MS2 conjugate in a clean Eppendorf tube. Store the conjugates at 4 °C until ready for use (*see* **Notes 11** and **12**).
5. Confirm attachment of the antibodies to the capsid by SDS-PAGE under reducing conditions and/or SEC chromatography. Representative data are shown in Fig. 2.

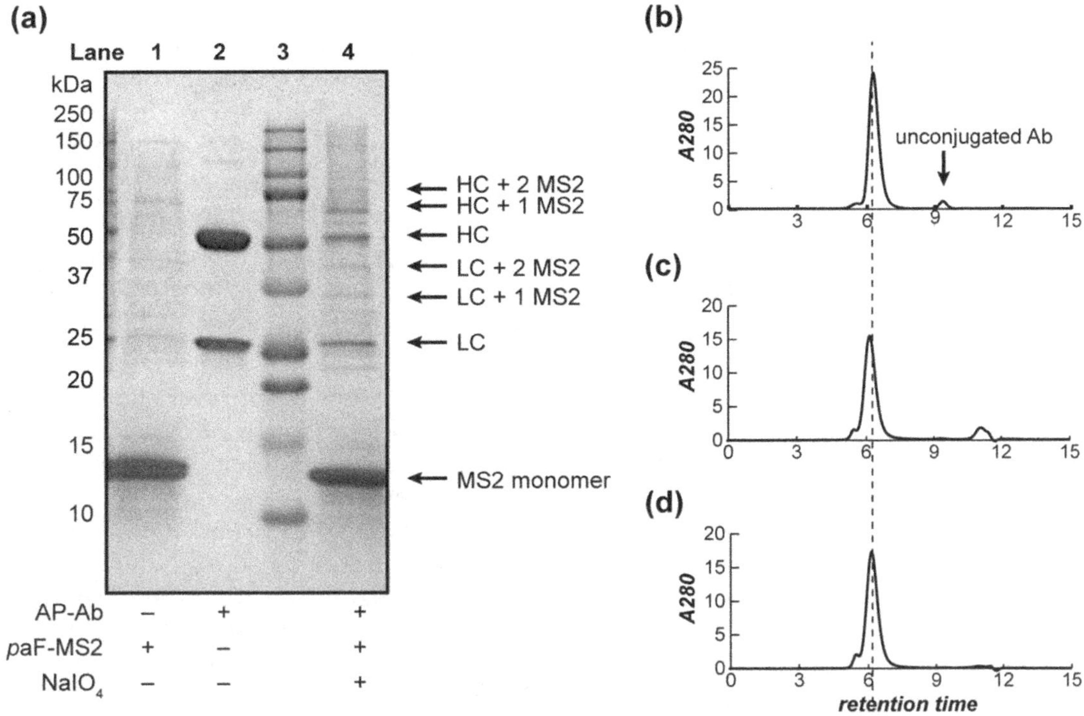

Fig. 2 Representative characterization of antibody–MS2 conjugates. (**a**) SDS-PAGE analysis under reducing conditions demonstrate clear modification of both the light and heavy chains (LC and HC) with MS2 coat protein monomers. (**b**) SEC-HPLC on a Phenomenex BioSep S4000 column of a 3:1 mixture of antibody to capsid prior to the addition of sodium periodate shows clear separation of the species. (**c**) After 5 min of reaction, SEC analysis shows complete consumption of the free antibody and a slight retention time shift of the MS2 peak. (**d**) After spin concentration against a 100 kDa MWCO membrane, pure conjugate is obtained

4 Notes

1. The nitrophenol NHS ester is prepared according to Behrens et al. [4]. Briefly, commercially available 3-(4-hydroxyphenyl) propionic acid is nitrated with concentrated nitric acid in acetic acid with cooling and precipitated by pouring into ice-water slurry. This material can be purified by crystallization or column chromatography. The NHS ester is prepared by reaction with *N*-hydroxysuccinimide and *N*-(3-Dimethylaminopropyl)-*N*′-ethylcarbodiimide hydrochloride (EDC) in dichloromethane and can be purified by flash chromatography.
2. For larger scale preparations, more concentrated nitrophenol NHS ester stock solutions can be used with no significant alteration in the reaction.
3. The protocols described here have been successfully demonstrated with humanized, human, and murine antibodies (all of

which maintained binding to their desired epitopes post-conjugation), demonstrating that the protocol is quite general. In some instances, antibodies are supplied in Tris buffer from commercial sources. In these cases, buffer exchange is essential to prevent unwanted competitive NHS ester reactions with the buffer.

4. Sodium dithionite decomposes over time in neutral to acidic aqueous solutions. In order to achieve the most consistent results, this solution should be prepared immediately prior to use. The solid should be stored in a tightly closed bottle in a desiccator.

5. While this particular example uses a self-assembling capsid, monomeric proteins can be expressed with a *p*aF residue using the same techniques without any significant alterations. MS2 capsids with the T19*p*aF mutation on each coat protein are expressed and purified according to Carrico et al. [3]. For applications where interior surface functionalization of the capsid are desired, the T19*p*aF/N87C double mutant (described by ElSohly et al. [1]) should be used instead with the caveat that the interior cysteines should be labeled prior to the oxidative coupling reaction.

6. Although sodium periodate can be dissolved in water directly, the resultant solutions can be quite acidic with high variability between reagent lots. Dissolution in phosphate buffer neutralizes the acid and provides more consistent results, and protects the proteins from denaturation. Solutions prepared in this fashion should be used within 4 h of preparation.

7. While the concentrations detailed in Subheading 3.1 are typical for this type of system, lower concentrations can also be used successfully in this protocol.

8. We have found that the recoveries in these experiments are antibody-dependent but highly reproducible for a particular antibody. Typical recoveries are between 50% and 80%.

9. The Ab-AP conjugate can be prone to aerial oxidation. For optimal results, the conjugate should be used immediately after isolation. If a large batch of material is desired, excellent results have been obtained by stopping after **step 1** (i.e., at the antibody–nitrophenol conjugate) and then using aliquots in the subsequent steps.

10. Lower concentrations have also been used successfully without any other alteration to the protocol; however, having a final coat protein concentration of 50 μM facilitates subsequent analyses.

11. Typical recoveries are 60–85%. These conjugates tend to be quite stable when stored at 4 °C, and binding to specific targets

with several antibody-MS2 conjugates was observed several months after initial preparation.

12. In order to calculate the concentration of the conjugate (i.e., with respect to capsid and not with respect to coat protein monomer), an effective extinction coefficient for the agent is 11.3/μM/cm.

Acknowledgments

These studies were funded by the DoD Breast Cancer Research Program (Grants BC061995 and W81XWH-14-0400 to M.B.F. and A.M.E., respectively). C.N. was supported by a Howard Hughes Medical Institute International Student Research Fellowship. LC-MS instrumentation was acquired with NIH Grant 1S10RR022393-01.

References

1. ElSohly AM, Netirojjanakul C, Aanei IL et al (2015) Synthetically modified viral capsids as versatile carriers for use in antibody-based cell targeting. Bioconjug Chem 26:1590–1596
2. Mehl RA, Anderson JC, Santoro SW et al (2003) Generation of a bacterium with a 21 amino acid genetic code. J Am Chem Soc 125:935–939
3. Carrico ZM, Romanini DW, Mehl RA et al (2008) Oxidative coupling of peptides to a virus capsid containing unnatural amino acids. Chem Commun:1205–1207
4. Behrens CR, Hooker JM, Obermeyer AC et al (2011) Rapid chemoselective bioconjugation through oxidative coupling of anilines and aminophenols. J Am Chem Soc 133:16,398–16,401
5. Flenniken ML, Willits DA, Harmsen AL et al (2006) Melanoma and lymphocyte cell-specific targeting incorporated into a heat shock protein cage architecture. Chem Biol 13:161–170
6. Suci P, Kang S, Gmür R et al (2010) Targeted delivery of a photosensitizer to aggregatibacter actino-mycetemcomitans biofilm. Antimicrob Agents Chemother 54:2489–2496
7. Wang Q, Kaltgrad E, Lin T et al (2002) Natural supramolecular building blocks: wild-type cowpea mosaic virus. Chem Biol 9:805–811
8. Douglas T, Young M (2006) Viruses: making friends with old foes. Science 312:873–875
9. Kovacs EW, Hooker JM, Romanini DW et al (2007) Dual-surface-modified bacteriophage MS2 as an ideal scaffold for a viral capsid-based drug delivery system. Bioconjug Chem 18:1140–1147
10. Tong GJ, Hsiao SC, Carrico ZM et al (2009) Viral capsid DNA aptamer conjugates as multivalent cell-targeting vehicles. J Am Chem Soc 131:11,174–11,178
11. Wu W, Hsiao SC, Carrico ZM et al (2009) Genome-free viral capsids as multivalent carriers for taxol delivery. Angew Chem Int Ed 48:9493–9497
12. Garimella PD, Datta A, Romanini DW et al (2011) Multivalent, high-relaxivity MRI contrast agents using rigid cysteine-reactive gadolinium complexes. J Am Chem Soc 133:14,704–14,709
13. Bertrand N, Wu J, Xu X et al (2014) Cancer nanotechnology: the impact of passive and active targeting in the era of modern cancer biology. Adv Drug Deliv Rev 66:2–25

Chapter 16

Cross-Linked Collagen Gels Using Gold Nanoparticles

Katarzyna Slowinska

Abstract

Cross-linking collagen results in the formation of a collagen matrix that is one of the most widely used biomaterials for applications in drug delivery, tissue engineering, molecular scaffolds, and materials for wound dressing. The properties of the collagen matrix depend on the cross-linking strategy employed. Here we describe the preparation and characterization of a nonfibrillar collagen matrix stabilized by cross-linking with tiopronin-protected, 3–4 nm polyvalent gold nanoparticles.

Key words Collagen, Nonfibrillar collagen, Gold nanoparticles, Polyvalent cross-linking agents, Collagen matrix

1 Introduction

Collagen is one of the most common biomaterials used for biomedical applications [1, 2]. The medicinal applications of collagen originate from its ease of extraction, purification, and processing [3]. Decomposition of the matrix and its porosity strongly depend on how the matrix is prepared [4]. Cross-linking is the most widely used stabilization method to improve gel longevity and allows the decomposition time to be varied between 24 h and 6 months [5]. The majority of cross-linking agents are linear and their use results in a similar morphology of the gels regardless of the cross-linking agent. Gold nanoparticles are polyvalent cross-linking agents and are able to form, on average, eight bonds per crystal within the matrix [6]; the other polyvalent cross-linking agents used are dendrimers [7]. Modifying collagen matrices with polyvalent cross-linkers result in biocompatible matrices with improved thermal and enzymatic stability [6–8].

Additionally, the method of collagen preparation before cross-linking is performed. The form of collagen before gel formation will influence the morphology of the cross-linked collagen matrix. To achieve a homogeneous matrix containing small pores suitable for drug delivery applications and a small size dispersion of the

Andrew K. Udit (ed.), *Protein Scaffolds: Design, Synthesis, and Applications*, Methods in Molecular Biology, vol. 1798, https://doi.org/10.1007/978-1-4939-7893-9_16,

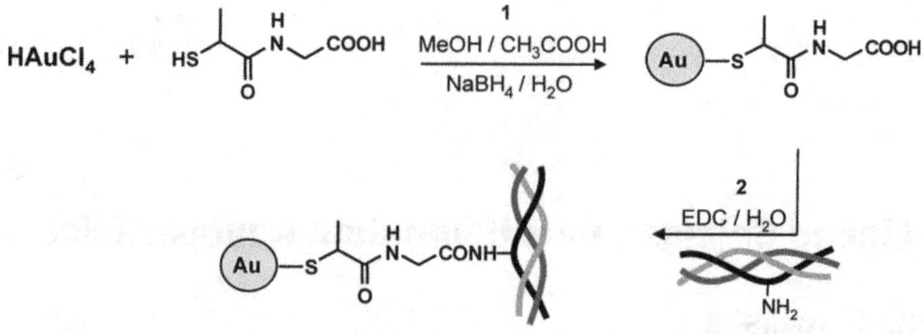

Scheme 1 Synthesis of Au nanoparticles, and cross-linking of a collagen gel. Reproduced from Castaneda et al. 2008 with permission from American Chemical Society

pores, the entire fibrillar component of the collagen has to be removed and only molecular collagen can be used for matrix preparation [9].

The synthesis of a collagen matrix cross-linked with gold nanoparticles described herein has three parts: (1) preparation of a molecular collagen solution via extraction and purification of collagen type I from rat tail tendons, (2) synthesis of tiopronin-protected gold nanoparticles (2–3 nm), and (3) cross-linking collagen type I with gold nanoparticles via formation of covalent bonds between the carboxyl group of tiopronin and primary amine groups from collagen lysine residues. The general scheme of the process is outlined in Scheme 1.

2 Materials

All solutions should be prepared with ultrapure water, referred to here as D.I. water (prepared by feeding deionized water into a Milli-Q system to attain a resistivity of 18 MΩ cm at 25 °C) and analytical grade reagents. Prepare and store all reagents at room temperature unless indicated otherwise. Follow all waste disposal regulations when disposing of waste materials.

2.1 Monomeric Collagen Type I

1. Three rat tails.
2. Blade.
3. Refrigerated centrifuge.
4. 0.01 M HCl (500 mL).
5. 3% (v/v) Acetic acid (1 L).
6. 3.5 kDa cutoff dialysis bags or tubing.
7. Millipore filtration system (glass vacuum filter holders) with 20, 5, and 0.65 μm nitrocellose filter disks.
8. Solid NaCl (50 g).

9. Phosphate buffer: 0.02 M phosphate buffer, pH 7.8, prepared by mixing 1.020 g of anhydrous potassium phosphate monobasic and 24.545 g of sodium phosphate dibasic heptahydrate, make up to 5 L with water, store at 4 °C.
10. 50 mL Falcon tubes [5–10].
11. 1 L beaker.
12. Acetone (1 L).
13. Dry ice (200 g).
14. Lyophilizer.
15. Parafilm.
16. Mini Protean System for protein gels.
17. SDS PAGE precast gel (4–20% polyacrylamide).
18. Low range SDS PAGE MW standard stained with Coomassie R-250.

2.2 Synthesis of Tiopronin-Gold Nanoparticles

1. Tetrachloroauric acid (500 mg).
2. *N*-(2-mercaptopropionyl) glycine (1 g).
3. Fresh $NaBH_4$ (1 g).
4. Methanol (100 mL).
5. Acetic acid (100 mL).
6. 50 mL of 6:1 (v/v) methanol–acetic acid mixture.
7. Concentrated HCl.
8. Stir plate.
9. 100 mL round bottom flask.
10. Rotary evaporator.
11. 3.5 kDa cutoff dialysis bags or tubing.
12. TEM.
13. Formvar-coated copper TEM grids.

2.3 Cross-Linking Collagen Type I

1. 3 mL glass vials.
2. 1.5 mL microcentrifuge tubes.
3. Fresh EDC (1-(3-dimethyl-aminopropyl)-3-ethyl carbodiimide) (1 g).

2.4 TNBS Assay to Quantify Cross-Linking

1. 0.1 M sodium bicarbonate pH 8.5 (10 mL).
2. 0.1% TNBS: 5 mL 0.1% w/v 2,4,6-trinitrobenzene sulfonic acid in 0.1 M sodium bicarbonate pH 8.5 (10 mL).
3. 12 M HCl (10 mL).
4. Plastic cuvette.
5. Spectrophotometer.

2.5 TEM of Cross-Linked Collagen Matrix

1. Araldite 502 resin (20 g) (Ted Pella).
2. 70% ethanol (50 mL).
3. 95% ethanol (50 mL).
4. 100% ethanol (50 mL).
5. Pasteur pipettes.
6. DDSA Dodecenyl succinic anhydride (DDSA) (20 g).
7. 2,4,6-tris(dimethylaminomethyl)phenol (DPM-30) (1 g).
8. Propylene oxide (50 mL).
9. Ultratome.
10. TEM.
11. Formvar-coated copper TEM grids.

3 Methods

3.1 Monomeric Collagen Type I

3.1.1 Extraction

1. Extract tendon fibers from a rat tail with a blade (*see* **Note 1**).
2. Dissolve 1 g of rat tail tendon in 100 mL of 0.01 M HCl and stir for 4 h at room temperature.

3.1.2 Purification

1. Centrifuge the solubilized tendon for 30 min at 10,000 rpm (14,000 × *g*) and 4 °C. Discard the pellet and filter the supernatant sequentially through 20, 5, and 0.65 μm filters using the Millipore system (*see* **Note 2**).
2. Obtain a salt concentration of 0.7 M by adding the appropriate amount of NaCl to the filtrate and stir for 1 h.
3. Centrifuge the mixture for 1 h at 10,000 rpm (14,000 × *g*) and 4 °C. Collect the pellet.
4. Dissolve the pellet in 100 mL of 0.01 M HCl overnight.
5. Dialyze the sample against 1 L of phosphate buffer for at least 8 h at room temperature.
6. Perform another dialysis against 1 L of fresh phosphate buffer for at least 4 h at 4 °C.
7. Centrifuge the dialysate for 1 h at 10,000 rpm (14,000 × *g*) at 4 °C.
8. Resuspend the pellet in 15 mL of D.I. water and perform dialysis against 3% acetic acid overnight.
9. Perform another dialysis against D.I. water for 24–48 h, changing the water every 8 h, at 4°C.
10. The collagen solution may be stored at 4 °C.

3.1.3 Lyophilization

1. Pour approximately 5 mL of collagen solution into 50 mL Falcon tubes. Prepare at most five tubes for each run.
2. Pour acetone onto dry ice in a 1 L beaker and place the tubes in there for 20–30 min to completely freeze the collagen solutions.
3. Remove the tubes and dry thoroughly. Unscrew the caps and replace with parafilm. Punch tiny holes into the parafilm to allow for suction and to prevent sample loss during lyophilization.
4. Place the tubes in the lyophilzer jar and leave in the lyophilizer for two nights.
5. Store lyophilized collagen in the freezer (−20 °C).
6. To confirm purity, dissolve collagen in 0.01 M HCl (overnight) and perform standard SDS-PAGE analysis (Fig. 1).

3.2 Synthesis of Tiopronin-Gold Nanoparticles (Au/Tp; Size 2.4 nm)

This protocol follows a synthesis originally published by Brust et al. [10]. The ratio of tetrachloroauric acid to *N*-(2-mercaptopropionyl) glycine determines the size of the nanoparticles (*see* **Note 3**). The protocol below yields 2.4 nm nanoparticles. Perform all steps at room temperature.

1. Codissolve tetrachloroauric acid (0.31 g, 0.80 mmol) and *N*-(2-mercaptopropionyl) glycine (0.39 g, 2.4 mmol) in 35 mL of 6:1 methanol–acetic acid.
2. Dissolve $NaBH_4$ (0.6 g, 16 mmol) in 15 mL of water.
3. Add $NaBH_4$ dropwise to tetrachloroauric acid with rapid stirring and leave stirring for 15 min (*see* **Note 4**).
4. Stir for an additional 30 min and remove the solvent under vacuum with rotary evaporator without exceeding 40 °C (*see* **Note 5**).
5. Purify the black, crude tiopronin-protected gold nanoparticles by dialysis. Dissolve 150 mg of crude product in 75 mL of water. Adjust the pH of the solution to 1.0 by adding dropwise concentrated HCl. Load this solution to 8 inch segments of dialysis tubing (3.5 kDa cutoff) and place in a 10 L beaker of water with slow and continuous stirring for 72 h, changing the water every 8 h.
6. Collect the black tiopronin-protected gold nanoparticles in a round bottom flask and remove the solvent under vacuum with rotary evaporator without exceeding 40 °C. Place the damp nanoparticles under vacuum for 24 h to completely remove any remaining water.
7. The size dispersion of the nanoparticles can be confirmed under TEM (Fig. 2). (*See* Subheading 3.5 for a detailed TEM protocol.)

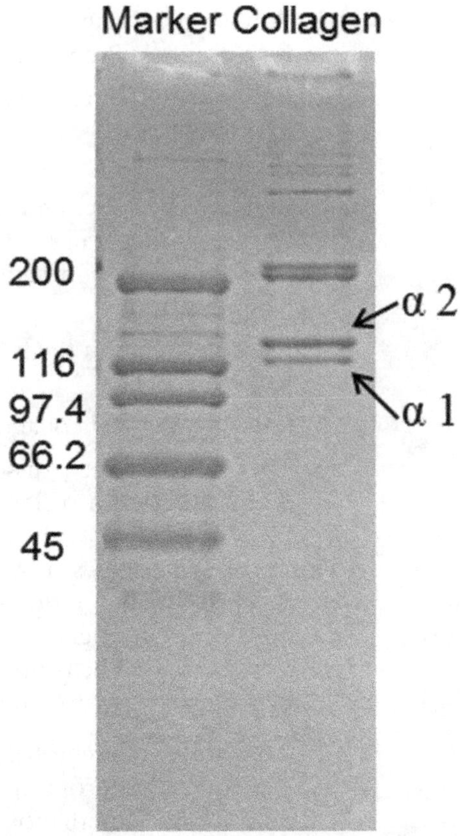

Fig. 1 SDS PAGE of pure collagen extract. The α1 band is half the size of the α2 band, reflecting a 1:2 stoichiometry of the collagen triple helix

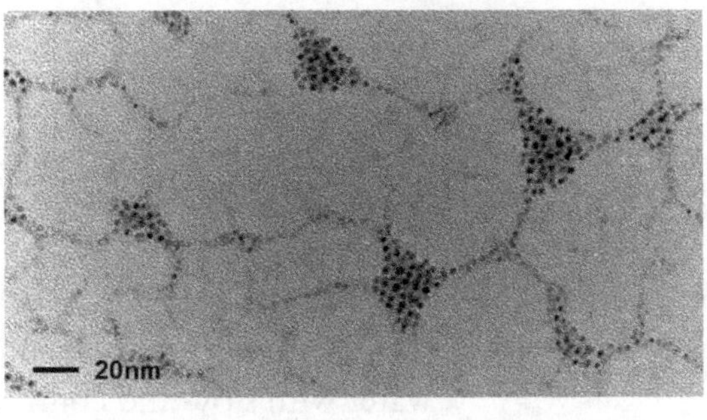

Fig. 2 TEM image of tiopronin-protected gold nanoparticles

3.3 Cross-Linking Collagen Type I

This protocol describes preparation of a molecular collagen matrix (0.4% wt/vol) cross-linked with gold nanoparticles protected by tiopronin. The cross-linking is achieved by amide bond formation between collagen primary amines (lysine; ~3% amino acid content) and the carboxyl group of tiopronin (Scheme 1). This protocol yields 1.5 mL of matrix for 3:1 collagen to nanoparticle weight ratio (*see* **Note 6**).

1. Make a 1% collagen stock solution in water by weighing out 0.2000 g of freeze-dried collagen and tearing it up into small pieces and then adding to 20 mL of water in a 50 mL Falcon tube. Vortex the solution and then let it dissolve overnight in the refrigerator at 4 °C.
2. Prepare a 30 mg/mL nanoparticle (Au/Tp) stock solution by weighing out 0.0600 g of nanoparticles and adding to 2 mL of water in a small vial. Vortex or sonicate until completely dissolved.
3. Pipette 40 μL of Au/Tp stock and 860 μL of water (for 3:1 collagen–Au/Tp ratio; for other ratios *see* **Note 6**) into a 1.5 mL microcentrifuge tube.
4. Add 30 mg of EDC to each sample (*see* **Note 7**).
5. Add 600 μL of the collagen stock solution to each sample (*see* **Note 8**).
6. Vortex the mixture in the tube to mix thoroughly (*see* **Note 9**).
7. Leave the sample overnight to fully cross-link and deactivate the EDC (*see* **Note 10**).

3.4 TNBS Assay to Quantify Cross-Linking

The degree of cross-linking can be measured as the loss of free ε-amino groups of lysines upon the addition of Au/Tp nanoparticles. The number of free amino groups can be determined using 2,4,6-trinitrobenzenesulfonic acid (TNBS) assay, as originally described by Cheung et al. and Sheu et al. [11, 12]

1. Place 100 μL of collagen treated with different amounts of Au/Tp nanoparticles in a microcentrifuge tube.
2. Add 400 μL of 0.1 M sodium bicarbonate solution and 250 μL of 0.1% TNBS.
3. Mixed well and place in a water bath at 40 °C for 2 h.
4. Add 300 μL of 12 M HCl and raise the temperature to 60 °C. Solubilization of the collagen will be complete after 60 min.
5. Transfer the samples to the cuvette, place it in the spectrophotometer and read absorbance at 425 nm (*see* **Note 11**).
6. Calculate the degree of cross-linking as a percent loss of free ε-amino groups: $(1 - (Abs_{Au}/Abs_{EDC}))100\%$, where Abs_{Au} represents the Absorbance of the Au/Tp cross-linked collagen and Abs_{EDC} represents the intensity of the collagen treated with only EDC.

3.5 TEM of Cross-Linked Collagen Matrix

1. Take the cross-linked collagen gel sample and lyophilize it.
2. Use a standard protocol for TEM tissue imaging with araldite resin. Dehydrate the sample by progressively increasing the

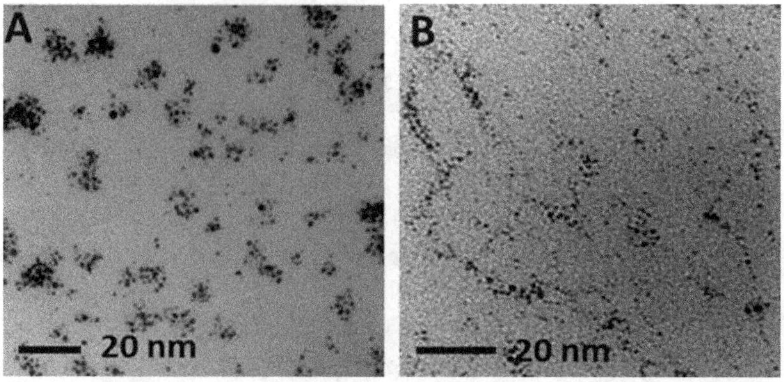

Fig. 3 TEM images of lyophilized collagen gels preserved in Araldite resin. (**a**) Collagen–Au/Tp = 3:1 and (**b**) Collagen–Au/Tp = 1:3

concentration of ethanol (70, 95, 100%), using a Pasteur pipette to add and remove the ethanol solutions to and from the sample. Keep the sample for 20 min in each solution.

3. Thoroughly mix 7.8 g of DDSA with 10.0 g Araldite 502 and 0.27 g DPM-30 and wait for at least 20 min.
4. Place the sample in propylene oxide for 1 h.
5. Place the sample in the mold to be infused with the resin. Add a 1:1 mixture of the resin solution and propylene oxide to fully immerse the sample.
6. Wait until the sample sinks and place it at 60 °C for 24–48 h.
7. Cool the sample to room temperature and cut it with Ultratome to 60–80 nm sections before imaging under TEM (Fig. 3) (*see* **Note 12**).

4 Notes

1. Cutting along the entire length gives the best results. One tail should provide 300–500 mg of crude material. If performed correctly this protocol results in 30% yield of pure monomeric collagen. Lower yields can occur if the subsequent filtration step is not performed correctly.
2. This step may require several Millipore disk filters due to clogging. Do not skip the larger pore filters, otherwise this step will take many hours and several filter disks to complete. Also make sure the vacuum is at least 100 torr to ensure speedy filtration.
3. The 3:1 ratio of tiopronin to tetrachloroauric acid (2.4–0.80 mmol) results in 2.4 nm nanoparticles; the ratio 1:12 (0.1–1.2 mmol) results in 3.8 nm nanoparticles.

4. The rapid stirring ensures a small size dispersion in the final solution. The temperature and the pH of the reaction mixture will increase and the color of the solution will turn to black. After 15 min of stirring to complete the reaction, the temperature will return to room temperature.
5. When removing the solvent do not exceed 40 °C. Higher temperature can result in partial removal of the protective monolayer of tiopronin and coalescence of gold nanoparticles. If this happens the flask will be covered with a gold mirror.
6. The ratio of collagen to nanoparticle can be adjusted by varying the dilution factor of nanoparticle solution while keeping the final collagen concentration at 0.4%. *See* table below:

Ratio (wt) Coll: Au/Tp	30 mg/mL stock Au/Tp (μL)	Water (μL)	EDC (mg)	1% stock collagen (μL)
3:1	40	860	30	600
2:1	60	840	30	600
1:1	120	780	30	600
1:2	240	660	30	600
1:3	360	540	30	600

7. 30 mg of EDC in 1.5 mL corresponds to 1.5×10^{-4} M, giving ten times excess with respect to carboxyl groups to be activated when the highest concentration of nanoparticles is used for collagen cross-linking.
8. The stock solution is very viscous and hard to pipette accurately, so it is more consistent to use the tube's 1.5 mL mark as reference to reach volume.
9. Large purple precipitates will appear. The sample can be left in the tube or placed in the dish that will be used for further tests.
10. The gels have a consistency of semiset gelatin, VERY fragile and easily broken. If one needs to prepare a matrix that is similar to hydrogels, increase the concentration of the collagen solution to 1.5–2.5% (wt/vol).
11. The reference cuvette should contain the same amount of Au/Tp nanoparticles to avoid scattering losses.
12. This protocol (collagen–Au/Tp = 3:1) results in gel a sample that exhibits the nanoparticles' organization within the gel in island-like structures. With a higher content of gold nanoparticles one should expect chain-like features in the TEM. When a higher concentration of collagen is present in a collagen gel

(above 0.5%) the TEM samples result in "shadow areas," where the collagen matrix is present, where nanoparticles are localized.

Acknowledgments

We are grateful to Dr. Douglass for his help with TEM. This work was supported by the National Institutes of Health GM099594.

References

1. Nimni ME, Cheung D, Strates B et al (1987) Chemically modified collagen: a natural biomaterial for tissue replacement. J Biomed Mater Res 21:741–771
2. Ruszczak Z (2003) Effect of collagen matrices on dermal wound healing. Adv Drug Deliv Rev 55:1595–1611
3. Ruszczak Z, Friess W (2003) Collagen as a carrier for on-site delivery of antibacterial drugs. Adv Drug Deliv Rev 55:1679–1698
4. Rault I, Frei V, Herbage D et al (1996) Evaluation of different chemical methods for cross-linking collagen gel, films and sponges. J Mater Sci Mater Med 7:215–221
5. Barbucci R (ed) (2002) Integrated biomaterials science. Kulwer Academic/ Plenum, New York
6. Castaneda L, Valle J, Yang N et al (2008) Crosslinking collagen with gold nanocrystals. Biomacromolecules 9:3383–3388
7. Duan X, Sheardown H (2005) Crosslinking of collagen with dendrimers. J Biomed Mater Res A 75:510–518
8. Castaneda L, Pluskat S, Ky D et al (2007) Novel collagen based material for local drug delivery systems. In: Naik RR, Perry CC, Shiba K, Ulijn R (eds) The nature of design—utilizing biology's portfolio. Mater. Res. Soc. Symp. Proc., 1008E, Warrendale, PA, pp T09–T08
9. Schuetz T, Richmond N, Harmon ME et al (2013) The microstructure of collagen type I gel cross-linked with gold nanoparticles. Colloids Surf B Biointerfaces 101:118–125
10. Brust M, Walker M, Bethell D et al (1994) Synthesis of thiol-derivatised gold nanoparticles in a two-phase liquid–liquid system. J Chem Soc Chem Commun:801–802
11. Cheung DT, Perelman N, Ko EC et al (1985) Mechanism of crosslinking of proteins by glutaraldehyde III: reaction with collagen in tissues. Connect Tissue Res 13:109–115
12. Sheu MT, Huang JC, Yeh GC et al (2001) Characterization of collagen gel solutions and collagen matrices for cell culture. Biomaterials 22:1713–1719

Chapter 17

Genipin Cross-Linking of Elastin and Elastin-Based Proteins

Lisa D. Muiznieks

Abstract

Genipin is a natural plant-derived compound that covalently cross-links biopolymers into lattice networks with good biocompatibility, controllable swelling, and mechanical properties. This protocol describes the genipin cross-linking of elastic proteins, including tropoelastin and elastin-based polypeptides, through steps of elastin phase-separation upon addition of salt and heat, centrifugation to rapidly concentrate the dense protein phase, and incubation. This method is applicable for the fabrication of elastic materials suitable for use as scaffolds for biomedical applications.

Key words Genipin, Elastin, Elastin-like polypeptide, Cross-link, Protein scaffold, Hydrogel

1 Introduction

Protein scaffolds have important applications in a diverse range of biomedical fields including pharmacology and tissue engineering. To maintain material integrity and stability, biopolymers must be cross-linked into three-dimensional networks. Protein cross-linkers contain reactive sites that target functional groups such as amine, thiol, and carboxylic acid groups of amino acid side chains. Commonly used covalent cross-linking agents include aldehydes, NHS-esters and genipin. Genipin is a natural plant product extracted from the fruit of *Gardenia jasmonoides* Ellis [1]. Genipin reacts with primary amines (Fig. 1), specifically the *N*-terminus (alpha amine) and lysine side chains (epsilon amine) of proteins, to covalently cross-link polypeptide chains with an efficacy similar to aldehydes but is nontoxic, whereas glutaraldehyde is cytotoxic and can lead to calcification [2, 3]. Genipin has been used as a cross-linking reagent to fix biological tissues such as pericardium, valved conduits and tendon [1, 4–6]. It can also cross-link biopolymers such as chitosan, collagen, gelatin, fibrin, elastin, and silk into materials that have good biocompatibility and controllable swelling and mechanical properties [7–12]. These materials have been tailored toward

Andrew K. Udit (ed.), *Protein Scaffolds: Design, Synthesis, and Applications*, Methods in Molecular Biology, vol. 1798, https://doi.org/10.1007/978-1-4939-7893-9_17,

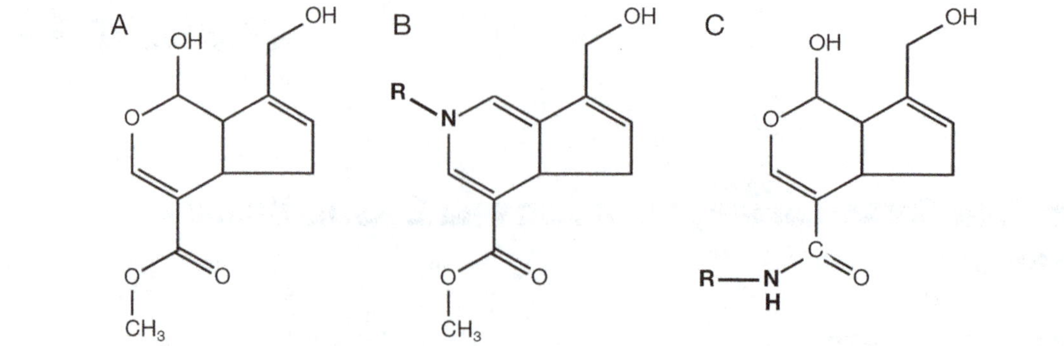

Fig. 1 Genipin molecular structure before (**a**) and after (**b**, **c**) conjugation to a biopolymer, R, via a primary amine

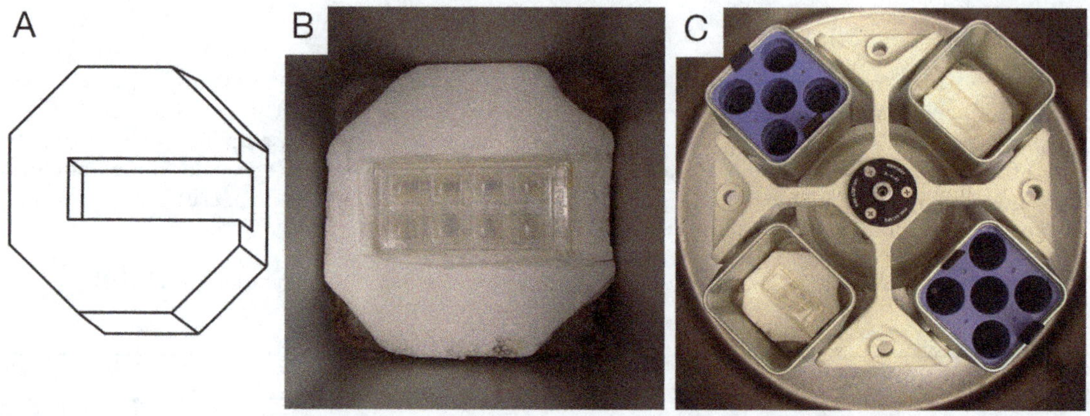

Fig. 2 Design of a styrofoam insert to support the casting mold during centrifugation. (**a**) Schematic of styrofoam support, made in-house, cut to the inner dimensions of a flat-bottomed centrifuge bucket. (**b**) Photo of the support and casting mold in a centrifuge bucket. (**c**) Setup of the sample and matching balance positioned in a swinging-bucket centrifuge

applications in drug delivery (e.g., as nanoparticles, microspheres, hydrogels, and films [13–15]), and for tissue repair and regeneration (e.g., acellular plugs for wound healing, cartilaginous sponges, and adhesive hydrogels for intervertebral disc repair [16–19]).

This protocol details the cross-linking of elastin-based proteins by genipin. Elastin provides elastic extension and recoil properties to vertebrate tissues [20]. This procedure can be applied to full-length elastin monomers (tropoelastin), smaller elastin-like polypeptides (ELPs), and elastin-containing block-copolymers, e.g., elastin-silk hybrids [7, 20]. Recombinant elastin retains similar assembly features and mechanical properties to native elastin and is biocompatible, making it suitable for use as a protein scaffold. Indeed, genipin cross-linked elastin materials elicit minimal immune response in vivo [16]. This cross-linking method exploits elastin self-assembly by inducing phase separation (corresponding to the native self-association of hydrophobic domains) via the addition of NaCl and heat, and includes a centrifugation step to concentrate phase-separated

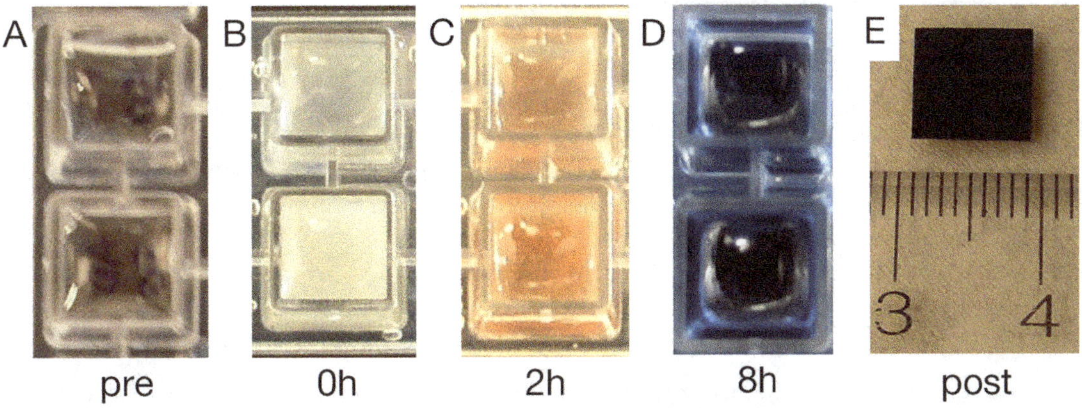

Fig. 3 Color change of the cross-linking reaction between genipin and elastin over time. (**a**) Elastin solution appears clear in the wells of a casting mold before the addition of NaCl. (**b**) Upon addition of NaCl the elastin solution turns turbid (appears white) indicating phase separation. (**c**, **d**) Solution turns from yellow/orange to dark blue as the cross-linking reaction proceeds. (**e**) Elastin scaffold cross-linked with genipin after removal from the casting mold

protein (Fig. 2) [7]. Progress of the reaction can be monitored through a color change from clear to white to yellow/orange until finally a dark blue material is formed (Fig. 3).

2 Materials

1. Elastin/ELPs, lyophilized (*see* **Note 1**).
2. Sodium borate buffer: 0.15 M sodium borate, pH 8. Weigh 5.72 g of sodium tetraborate decahydrate powder and dissolve in 90 mL of distilled water (*see* **Note 2**). Adjust pH dropwise with HCl while mixing (*see* **Note 3**). Transfer the solution to a graduated cylinder and top up the volume to 100 mL with distilled water. Store at room temperature.
3. NaCl solution, 5 M: Weigh 29.22 g of NaCl and dissolve in 100 mL of distilled water (*see* **Note 4**). Store at room temperature.
4. Genipin solution, 100 mM: Weigh 22.67 mg of genipin in a 1.5 mL eppendorf tube. Add 1 mL of ethanol to dissolve (*see* **Note 5**). Store genipin solution at 4 °C for up to 1 day.
5. Lab-Tek chamber slide and cover (*see* **Note 6**).
6. Glass scorer.
7. Temperature-controlled swinging-bucket centrifuge with flat-bottomed buckets (*see* **Note 7**).
8. Styrofoam centrifuge bucket insert (*see* **Note 8**).
9. Incubator.

3 Methods

3.1 Protein Preparation

1. Weigh 5 mg of lyophilized protein into a 1.5 mL eppendorf tube (*see* **Note 9**).
2. Add 370 μL of sodium borate buffer to each tube of weighed protein and dissolve at 4 °C for a minimum of 2 h (*see* **Note 10**).

3.2 Cross-Linking

1. Place two centrifuge buckets in an incubator set to 37 °C to warm (*see* **Note 11**).
2. Warm the interior of the centrifuge to 37 °C (*see* **Note 12**).
3. Score and snap off the overhanging glass label of a LabTek chamber slide (*see* **Note 13**).
4. Label the base of each well of the chamber slide with the sample name.
5. To the wells of the sample chamber slide, add in the following order 370 μL of elastin solution, then 80 μL of NaCl solution (*see* **Notes 14** and **15**), then 50 μL of genipin solution (*see* **Note 16**).
6. Add 500 μL of water to all remaining unused wells.
7. Prepare a centrifuge balance by scoring and breaking off the overhanging label from a second chamber slide. Fill all wells of the balance slide with 500 μL of water.
8. Carefully place the sample chamber slide and the balance slide into styrofoam supports and transfer to the warmed centrifuge buckets (*see* **Note 17**).
9. Insert the buckets into the prewarmed centrifuge and spin (2060 × *g*, 7 min, 37 °C).
10. Immediately place the sample slide into an incubator set to 37 °C and leave overnight for the reaction to run to completion (Fig. 3; *see* **Note 18**).

3.3 Material Handling

1. Once the cross-linking reaction is complete, visually inspect the materials and solution in the chamber wells: the solution in the wells should be relatively clear when viewed from the side, and the glass slide base should be uniformly coated with a dark blue material, corresponding to covalently cross-linked elastin/ELPs (*see* **Note 19**).
2. Remove the solution by inverting and gently flicking the chamber slide over a waste collection vessel. Gently rinse the material several times with distilled water (*see* **Note 20**).
3. The materials can either be used in-place or removed:

   (a) To use the materials in-place: Leave the chamber slide intact and use the materials in the casting mold, e.g., seeded with cells.

(b) To remove the materials from the chamber slides: Remove the plastic wells from the glass slide base by lifting from one corner in one smooth motion (*see* **Note 21**). Remove the gasket in the same manner, gripping firmly and slowly lifting from one corner. Materials can be removed from the glass by gently flushing with distilled water (*see* **Note 22**).

4 Notes

1. The number, availability (solvent exposure) and spacing of primary amine groups will influence the cross-link density and structural integrity of the final material, also affecting morphological, biological, and mechanical properties of materials, e.g., material dimensions, porosity, swelling capacity, cell interactions, protease susceptibility, stiffness, strength and extensibility [21]. To produce materials that remain intact (required for applications in which the material will be removed from the casting mold), the protein should contain multiple, separated sites for cross-linking so that a sufficient number of inter-chain cross-links are made. For reference, ELPs should contain the equivalent of at least two cross-linking domains (each containing two lysine residues) separated by a hydrophobic sequence (of approximately 30 residues or greater) [22]. If the material is to be used in the casting mold, it may be sufficient to use ELPs with fewer primary amines, although less robust materials will result.
2. Do not use boric acid or a mix of sodium tetraborate/boric acid to prepare the buffer, or boric acid to pH the sodium tetraborate solution. Tris-based buffers are not suitable for this reaction as they contain primary amines.
3. We recommend using concentrated (6–12 N) HCl to pH the solution because a 0.15 M stock of sodium tetraborate is close to the solubility limit, so only a small volume will be available for adjusting the pH. More dilute HCl, e.g., 1 N, may be used to finely adjust the pH once it is close to the desired value.
4. This concentration of NaCl should dissolve at room temperature with stirring for 2 h, but as it is close to the solubility limit (about 6 M) gentle heating may be required.
5. Gently mix the solution on a rocking platform for 1 h at room temperature to dissolve fully. Genipin should be dissolved in an organic solvent (e.g., ethanol, DMSO) as it has low solubility in aqueous solutions. Higher genipin stock concentrations can be reached when dissolving in DMSO compared to in ethanol.
6. We describe the preparation of cross-linked elastin materials using 8-well chamber slides of dimensions 10 × 10 mm per

well, but any suitable mold can be used. LabTek chamber slides comprise plastic wells sealed by a gasket to a glass microscope slide, i.e., the well walls are designed to be removable. We find protein materials are easier to remove from a glass support than from plastic, particularly if they have a large area in contact with the casting surface. Sample volumes and the amount of protein cross-linked can be readily scaled according to the final dimensions of the material desired. However, care should be taken to use a suitable cover for small reaction volumes to prevent solution from evaporating during cross-linking. We describe here the construction of a flat scaffold (sheet), but discs, sponges, and solid or hollow tubes can be cast using an appropriate mold.

7. Heat is used to induce elastin phase separation, and centrifugation concentrates the protein at the base of the casting mold, resulting in denser and more consistently cross-linked materials.
8. To stabilize the casting mold in a centrifuge bucket during centrifugation a support was fashioned from a piece of styrofoam (Fig. 2). The support was cut to fit the square base of the centrifuge bucket and the corners were trimmed to provide handholds to enable insertion and removal. A rectangular slot was carved from the top of the styrofoam support to snuggly fit the casting mold.
9. Weigh the protein precisely using an analytical balance. This starting amount is sufficient to cross-link full-length elastin or ELPs as small as 10 kDa containing a sufficient number of well-spaced lysine residues (*see* **Note 1**).
10. The most consistent results are obtained when samples are dissolved at 4 °C overnight.
11. From room temperature it takes at least 15 min to sufficiently warm metal buckets. The buckets must be warm enough to maintain elastin/ELPs in the phase-separated state, i.e., visibly turbid during centrifugation.
12. It is sufficient to set the centrifuge to 37 °C and run it empty for 10 min.
13. This step is only required to enable the slide to fit into the centrifuge bucket. Check to ensure the gasket still makes a tight seal around the wells nearest the cut edge.
14. Take care not to add any air bubbles to the wells while dispensing the solution as this could result in bubbles in the final material. Place the pipette tip under the surface of the solution when adding NaCl and genipin. This enables gentle mixing of reagents and reduces the chance of introducing air bubbles.

15. This is equivalent to a final NaCl concentration of 0.8 M. NaCl and heat are used to induce elastin/ELP phase separation. The temperature at which phase separation (T_c) occurs for full-length elastin or a given ELP is dependent largely on NaCl concentration and to a lesser degree on protein/peptide concentration and pH [23, 24]. T_c can be readily determined for a given concentration of elastin/ELP and NaCl by monitoring the solution turbidity of a 1 mL sample (e.g., dissolved in 50 mM phosphate buffer, pH 7.4) using a spectrophotometer at 440 nm upon increasing the temperature at 1 °C/min [25]. The amount of NaCl described in this protocol is sufficient to immediately turn the solution of tropoelastin at this concentration into a turbid suspension at room temperature. If the solution does not become turbid upon adding NaCl the temperature should be increased. Alternatively, a higher concentration of NaCl (e.g., 1–1.2 M) or amount of ELP (e.g., 5.2–5.4 mg) may be used. Difficulty in getting the solution to become turbid is likely to occur only for ELPs with a higher phase separation temperature, e.g., ELPs smaller than 10 kDa or those containing mutations that lower the hydrophobicity compared to native elastin (for reference, *see* Miao et al. [20] and Reichheld et al. [25]).
16. This is equivalent to a final genipin concentration of 10 mM, which is a molar excess over primary amines in 5 mg of full-length tropoelastin and of a well-characterized elastin-based peptide, ELP_3 [26]. Amino acid analysis can be used to calculate the number of amine groups on lysine side-chains that have been modified in the final material to estimate the percentage of lysines cross-linked. However, this does not confirm that functional cross-links are made, or the identity of covalently linked lysine residue pairs. Nor does it indicate the potential extent of genipin polymerization, which is dependent on pH, and results in the blue color of materials [10]. A likely reaction mechanism at pH 8 (weakly basic conditions) involves nucleophilic attack by a primary amine on the C3 carbon atom of genipin, opening the dihydropyran ring, followed by attack on the resulting aldehyde group by the newly formed secondary amine (*see* Butler et al. [11]; the heterocyclic structure is shown in Fig. 1b). The ring-opened intermediate of genipin can undergo oxygen radical-induced polymerization into short (dimer, trimer, and tetramer) heterocyclic bridges (longer chains ($n \sim 7$–88) of polymerized genipin are likely to be formed only under strongly basic conditions; *see* Mi et al. [10]). An additional, slower cross-linking reaction requiring acid catalysis involves a nucleophilic substitution of the ester group of genipin by a primary amine (the result of the reaction is shown in Fig. 1c) [11].

17. Care must be taken not to spill the contents or introduce air bubbles into the solution upon inserting the slides and supports into the centrifuge buckets. It is recommended to practice inserting styrofoam supports containing a casting mold filled with water first to check there is sufficient finger space cut out to enable smooth insertion.
18. Incubation (heat) is required to maintain elastin in the phase-separated state during cross-linking. The casting mold may be left in the styrofoam support during incubation to reduce sample handling at this stage. Progress of the reaction can be monitored through a color change from clear to white to yellow/orange until finally a dark blue material is formed (Fig. 3). Care should be taken to keep the solution adequately covered to prevent evaporation, especially if the reaction volume is small.
19. If this is not the case, it is likely that elastin/ELP phase separation needs to be optimized (i.e., increased solution turbidity is required; *see* **Note 15**). Alternatively, poorly cross-linked materials may indicate too few lysine residues are available to sufficiently connect polymer chains. Consider the number, spacing/location, and solvent accessibility of lysine residues when designing ELPs (*see* **Note 1**).
20. Elastin materials should be kept solvated at all times and not left to dry out, otherwise they will become hard and brittle.
21. A razor blade can be helpful at this point to loosen one corner, but take care not to tear the elastin material.
22. In some cases, firmly scraping a razor blade across the glass surface may be required to assist removal. If materials roll tightly after being removed from the support a drop of 5 M NaCl can be added to help flatten them.

Acknowledgments

The author thanks Fred Keeley and Sean Reichheld for useful discussions.

References

1. Sung HW, Huang RN, Huang LL et al (1998) Feasibility study of a natural crosslinking reagent for biological tissue fixation. J Biomed Mater Res 42:560–567
2. Levy RJ, Schoen FJ, Sherman FS et al (1986) Calcification of subcutaneously implanted type I collagen sponges. Effects of formaldehyde and glutaraldehyde pretreatments. Am J Pathol 122:71–82
3. Chang Y, Tsai CC, Liang HC et al (2002) In vivo evaluation of cellular and acellular bovine pericardia fixed with a naturally occurring crosslinking agent (genipin). Biomaterials 23:2447–2457

4. Huang LL, Sung HW, Tsai CC et al (1998) Biocompatibility study of a biological tissue fixed with a naturally occurring crosslinking reagent. J Biomed Mater Res 42:568–576
5. Fessel G, Cadby J, Wunderli S et al (2014) Dose- and time-dependent effects of genipin crosslinking on cell viability and tissue mechanics—toward clinical application for tendon repair. Acta Biomater 10:1897–1906
6. Chang Y, Tsai CC, Liang HC et al (2001) Reconstruction of the right ventricular outflow tract with a bovine jugular vein graft fixed with a naturally occurring crosslinking agent (genipin) in a canine model. J Thorac Cardiovasc Surg 122:1208–1218
7. Muiznieks LD, Keeley FW (2016) Phase separation and mechanical properties of an elastomeric biomaterial from spider wrapping silk and elastin block copolymers. Biopolymers 105:693–703
8. Silva SS, Maniglio D, Motta A et al (2008) Genipin-modified silk-fibroin nanometric nets. Macromol Biosci 8:766–774
9. Vieth S, Bellingham CM, Keeley FW et al (2007) Microstructural and tensile properties of elastin-based polypeptides crosslinked with genipin and pyrroloquinoline quinone. Biopolymers 85:199–206
10. Mi F-L, Shyu S-S, Peng C-K (2004) Characterization of ring-opening polymerization of genipin and ph-dependent cross-linking reactions between chitosan and genipin. J Polym Sci A Polym Chem 43:1985–2000
11. Butler MF, Ng Y-F, Pudney PDA (2003) Mechanism and kinetics of the crosslinking reaction between biopolymers containing primary amine groups and genipin. J Polym Sci A Polym Chem 41:3941–3953
12. Mekhail M, Wong KK, Padavan DT et al (2011) Genipin-cross-linked electrospun collagen fibers. J Biomater Sci Polym Ed 22:2241–2259
13. Martinez AW, Caves JM, Ravi S et al (2014) Effects of crosslinking on the mechanical properties, drug release and cytocompatibility of protein polymers. Acta Biomater 10:26–33
14. Arteche Pujana M, Perez-Alvarez L, Cesteros Iturbe LC et al (2013) Biodegradable chitosan nanogels crosslinked with genipin. Carbohydr Polym 94:836–842
15. Harris R, Lecumberri E, Heras A (2010) Chitosan-genipin microspheres for the controlled release of drugs: clarithromycin, tramadol and heparin. Mar Drugs 8:1750–1762
16. Hrabchak C, Rouleau J, Moss I et al (2010) Assessment of biocompatibility and initial evaluation of genipin cross-linked elastin-like polypeptides in the treatment of an osteochondral knee defect in rabbits. Acta Biomater 6:2108–2115
17. Schek RM, Michalek AJ, Iatridis JC (2011) Genipin-crosslinked fibrin hydrogels as a potential adhesive to augment intervertebral disc annulus repair. Eur Cell Mater 21:373–383
18. Lien SM, Ko LY, Huang TJ (2009) Effect of pore size on ECM secretion and cell growth in gelatin scaffold for articular cartilage tissue engineering. Acta Biomater 5:670–679
19. Silva SS, Motta A, Rodrigues MT et al (2008) Novel genipin-cross-linked chitosan/silk fibroin sponges for cartilage engineering strategies. Biomacromolecules 9:2764–2774
20. Miao M, Sitarz E, Bellingham CM et al (2013) Sequence and domain arrangements influence mechanical properties of elastin-like polymeric elastomers. Biopolymers 99:392–407
21. Muiznieks LD, Keeley FW (2017) Biomechanical design of elastic protein biomaterials: a balance of protein structure and conformational disorder. ACS Biomater Sci Eng 3:661–679
22. Bellingham CM, Lillie MA, Gosline JM et al (2003) Recombinant human elastin polypeptides self-assemble into biomaterials with elastin-like properties. Biopolymers 70:445–455
23. Bellingham CM, Woodhouse KA, Robson P et al (2001) Self-aggregation characteristics of recombinantly expressed human elastin polypeptides. Biochim Biophys Acta 1550:6–19
24. Vrhovski B, Jensen S, Weiss AS (1997) Coacervation characteristics of recombinant human tropoelastin. Eur J Biochem 250:92–98
25. Reichheld SE, Muiznieks LD, Stahl R et al (2014) Conformational transitions of the cross-linking domains of elastin during self-assembly. J Biol Chem 289:10057–10068
26. Reichheld SE, Muiznieks LD, Keeley FW et al (2017) Direct observation of structure and dynamics during phase separation of an elastomeric protein. Proc Natl Acad Sci U S A 114:E4408–E4415

Chapter 18

Characterization of Self-Assembled Protein Scaffolds from Collagen-Mimetic Peptides

Devyani Shukla, Jayeeta Lahiri, and Avanish Singh Parmar

Abstract

Collagen-mimetic peptides have been utilized to form structures of different morphology for various biomedical and nanotechnology applications. This chapter describes the characterization of collagen-mimetic peptide self-assembled structures formed by tuning the interactions between peptides. Inclusion of varying hydrophobicity, electrostatic forces, and stereoselectivity was mainly employed in CMP designs discussed herein. The role of these forces can be assessed using multiple characterization techniques. Light scattering techniques have been employed to study the aggregation kinetics of self-assembled nanostructures and to investigate the net charge distribution of peptides. Spectroscopy techniques like circular dichroism, fluorescence, and absorption spectroscopy have been utilized to decipher the secondary structures of peptide and binding of the peptides with dyes. Imaging techniques helped in resolving the morphology of the self-assembled structures. Confocal fluorescence microscopy and differential scanning calorimetry helped in indirect assessment of hydrophobicity and X-ray studies to determine the inter-helical spacing between the triple helical peptides of the higher order structures.

Key words Self-assembly, Collagen mimetic peptide, Triple helix, Nanostructures, Electrostatic interactions, Stereoselectivity, Hydrophobic interactions

1 Introduction

Collagen, a vital component of epithelial and connective tissues, has been of interest to the scientific community for decades. It is the most abundant protein in the human body and is very stable. Natural collagen is mostly derived or extracted from animals; however, certain disadvantages are involved with animal-derived collagen. Purification of collagen from animal sources is an expensive process, with a high risk of contamination and the potential to elicit an immune response. To overcome these problems, collagen-mimetic peptides (CMPs) have been designed by various groups to allow better control of mechanical, material and physical parameters [1, 2].

CMPs display a signature Gly-X-Y motif arranged in tandem throughout the sequence, which gives rise to the formation of the

Andrew K. Udit (ed.), *Protein Scaffolds: Design, Synthesis, and Applications*, Methods in Molecular Biology, vol. 1798, https://doi.org/10.1007/978-1-4939-7893-9_18,

characteristic triple helix. This triple helix is composed of three left-handed polyproline II (PPII)-like helical chains bound together to form a right-handed superstructure. Every third residue is strictly glycine in these triple helices since it is the only residue which can be accommodated inside without causing any distortion in the linear assembly; every third residue inherently lies buried near the centre of the chain, inaccessible to the solvent. Often, replacement of glycine by another amino acid results in disease [3–5]. A few amino acids have a higher probability of occupying X positions, while others are hardly ever found in the triple helix (e.g., Cys, Tyr, Trp). Residues like Phe, Leu, and Glu have greater proclivity for the X position, while a few amino acids like Arg and Lys are more commonly found in the Y position [5]. It has been hypothesized that steric hindrances, hydrophobic, and electrostatic interactions play a major role in determining the distribution of these amino acids. The hydrophobic and electrostatic forces contributed by these residues play a key role in the self-assembly of supramolecular collagen fibrils which then associate with various signaling molecules like integrins, proteoglycans, and fibronectin [5].

Understanding the fundamental molecular mechanism behind the self-assembly process of CMPs is crucial to controlling the functionality of the higher order structures. This knowledge can be used for the development of advanced biomaterials suited for applications like tissue engineering, cosmetics, biosensors and food engineering [6, 7]. Different molecular approaches have been utilized, such as hydrophobic/electrostatic interactions, hydrogen bonding, disulfide bridges, and metal mediated cross-linking, for generating higher order structures of synthetic CMPs. The size of such assemblies varies from particles in the nanometre range to fiber strands, macroscopic membranes, nanotapes, and sheets [8–13]. Despite their small size, CMPs effectively mimic the properties shown by natural collagen which has almost 1000 residues.

The presence of strong electrostatic and hydrogen bond interactions at staggered points along the chain mediates stability and specificity of the interactions to eventually form microfibrils, fibers, and sheets [13, 14]. Computational modeling used to treat electrostatic interactions range from using simple scoring models to more sophisticated and complex algorithmic calculations [15–20]. Even so, they are very difficult to extend to the fundamental atomistic mechanisms involved in protein systems [21]. Exchange of similarly charged amino acids in a peptide sequence with each other (e.g., arginine and lysine) can cause large changes in the molecular morphology, stability, and function (Fig. 1) [22]. CMPs with anionic and cationic domains show that those with arginine have a greater triple helix forming propensity than lysine. Only arginine-containing peptides participate in supramolecular assemblies in all three two-component systems (Fig. 2). Arginine is seen to be the most commonly occurring charged amino acid which

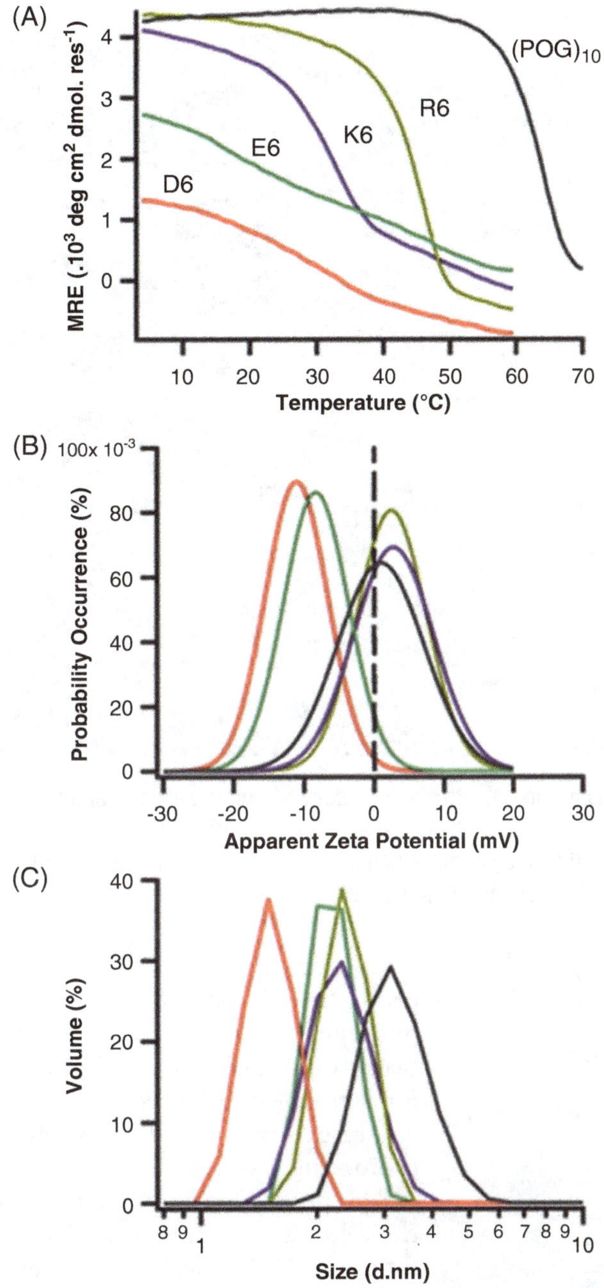

Fig. 1 Analysis of supercharged peptides. (**A**) Thermal denaturation profiles and melting temperatures measured by CD: $K_6 = 31.1$ °C, $R_6 = 44.3$ °C, $D_6 = 26.5$ °C, $E_6 = 17$ °C, $(POG)_{10}$=65 °C (O is hydroxyproline). (**B**) Zeta potential derived from electrophoretic mobility measured by DLS. (**C**) Hydrodynamic radii of peptides from 1 to 10 nm consistent with triple helical rod formation. (Reprinted (adapted) from Parmar et al. (2016) with permission from the American Chemical Society)

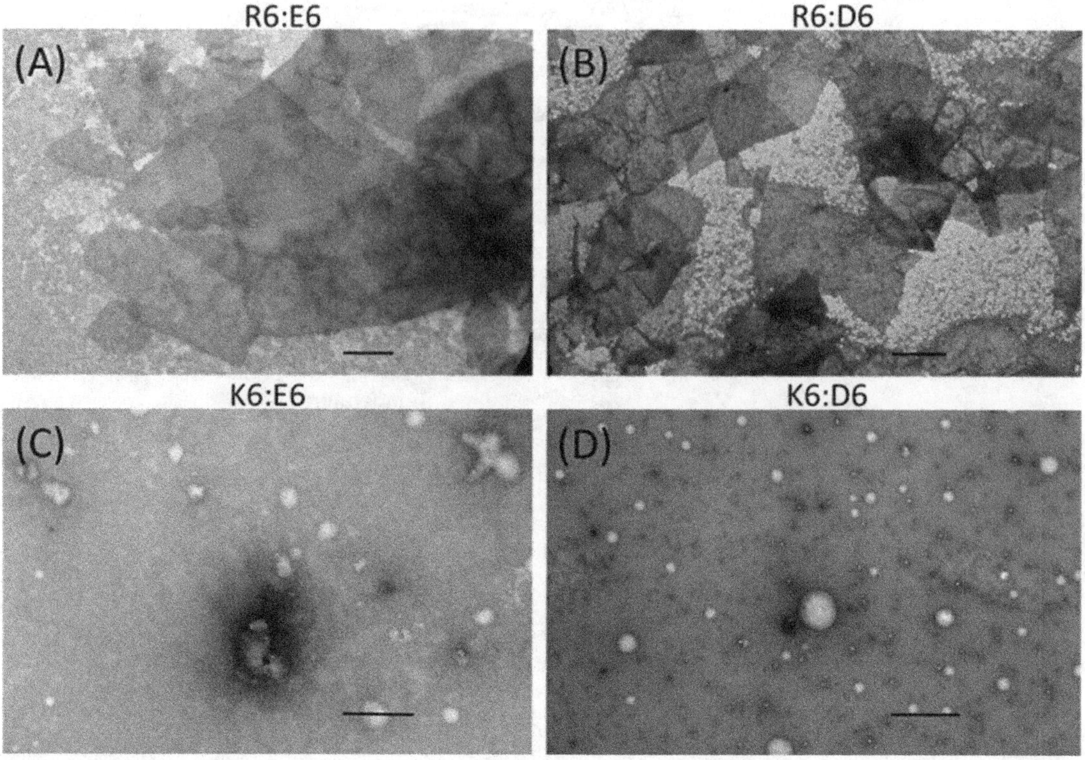

Fig. 2 Transmission electron microscopy of supercharged peptide aggregates. R_6 combined with (**A**) E_6 and (**B**) D_6 form microscale sheets (scale bar = 2 μm). K_6 mixed with (**C**) E_6 or (**D**) D_6 form particulate structures potentially induced by aggregation during drying. (Reprinted (adapted) from Parmar et al. (2016) with permission from the American Chemical Society)

promotes self-assembly into a nanosheet or banded fiber morphology and is statistically overrepresented at interfaces. This is because the polyvalent guanidyl group of arginine is responsible for forming intermolecular as well as intramolecular contacts, in contrast to lysine's side group which prefers intrahelical contacts, promoting a sticky-ended triple helix [22].

Hydrophobicity is another important force responsible for the folding of a triple helix. By varying the hydrophobicity in the different CMP designs, marked changes in morphology are observed ranging from nanofibers to plates. CMPs with varying hydrophobic length show different morphological structures depending on the length of hydrophobic domain ranging from plates, fibres, to nanostars (Fig. 3) [23]. Formed structures show very stable triple-helical characteristics (Fig. 4a, b) and interhelical *d*-spacing consistent with fiber packing observed in natural collagens (Fig. 4c). Hydrophobic residues are buried in the core of the nanostructures (Fig. 5).

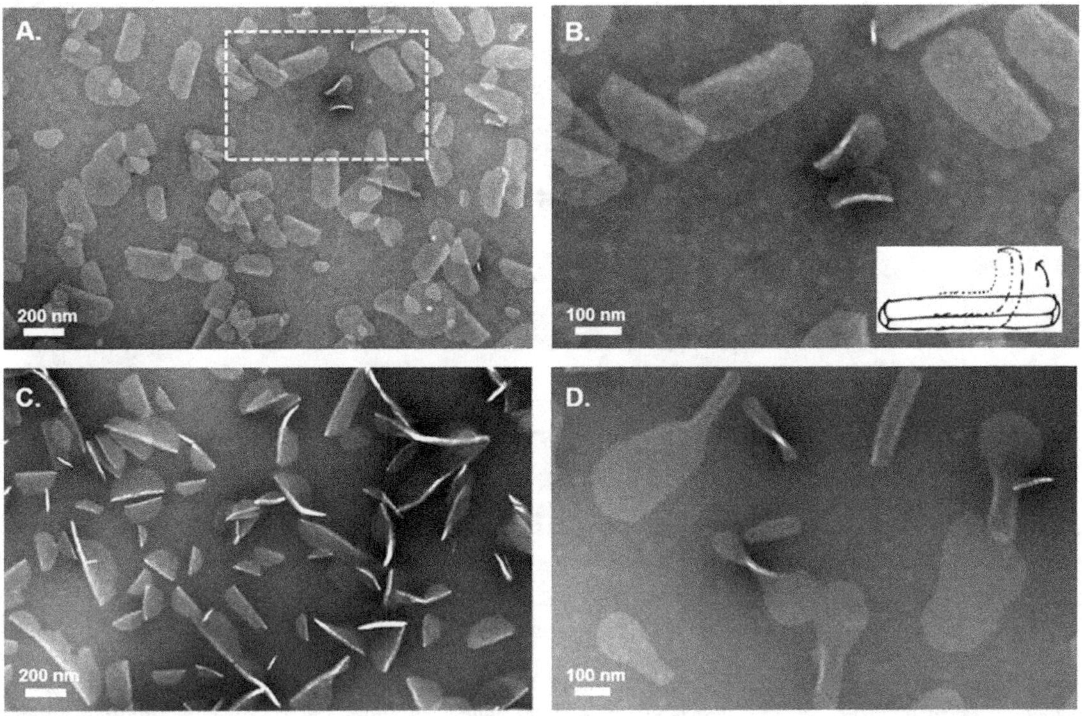

Fig. 3 Transmission electron microscopy of H4 peptide $(POG)_3POGLIGLIG(POG)_4$ morphology. (**A**) This field shows numerous discs alone or overlapping adjacent disks. (**B**) Close-up of two H4 discs in field A curving as shown in the drawing. (**C**) Multiple curved discs and discs perpendicularly aligned to the electron beam. (**D**) Another field highlighting disc flexibility. (Reprinted (adapted) from McGuinness et al. (2014) with permission from the American Chemical Society)

Stereoselectivity is also an important force in guiding the assembly of CMPs. Stereoisomers of $(PPG)_{10}$ clearly show these effects: on mixing both L- and D-forms of $(PPG)_{10}$, i.e., $({}_{L}P_{L}PG)_{10}$ and $({}_{D}P_{D}PG)_{10}$, a supramolecular structure (well-ordered micron-scale sheets) form mainly due to the shape complementarity of the two stereoisomers (Fig. 6). The inter-helical distances between $({}_{L}P_{L}PG)_{10}$ was only found to be larger than of the $({}_{L}P_{L}PG)_{10}$:$({}_{D}P_{D}PG)_{10}$ mixture, consistent with the computationally predicted ones. However, this general rule of helix–helix association seems to best apply in cases where complete mirror symmetry exists [24]. X-ray scattering measurements of interhelical spacing in these sheets support a tight ridges-in-grooves packing of left- and right-handed triple helices (Fig. 7).

In short, this chapter focuses on describing the different characterization methods employed to study the role of bimolecular interactions in the formation of diverse morphology self-assembled higher order structures of CMPs.

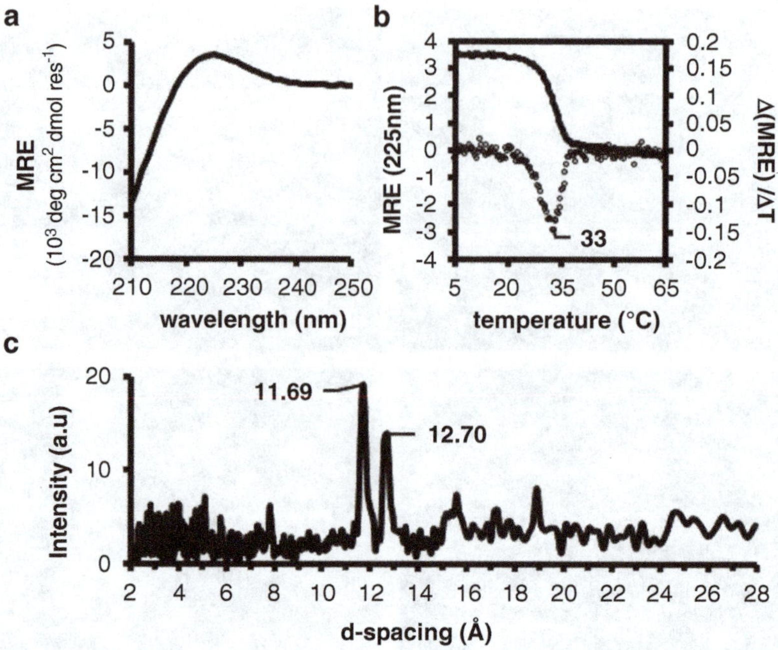

Fig. 4 H4 peptide ($(POG)_3POGLIGLIG(POG)_4$) forms triple helices. (**a**) CD spectrum shows positive mean residue ellipticity (MRE) band at 225 nm, consistent with triple-helical structure. (**b**) Thermal denaturation profile shows cooperative unfolding of H4 with a T_m of 33 °C. (**c**) WAXS of H4 shows interhelical *d*-spacing consistent with fiber packing observed in natural collagens. (Reprinted (adapted) from McGuinness et al. (2014) with permission from the American Chemical Society)

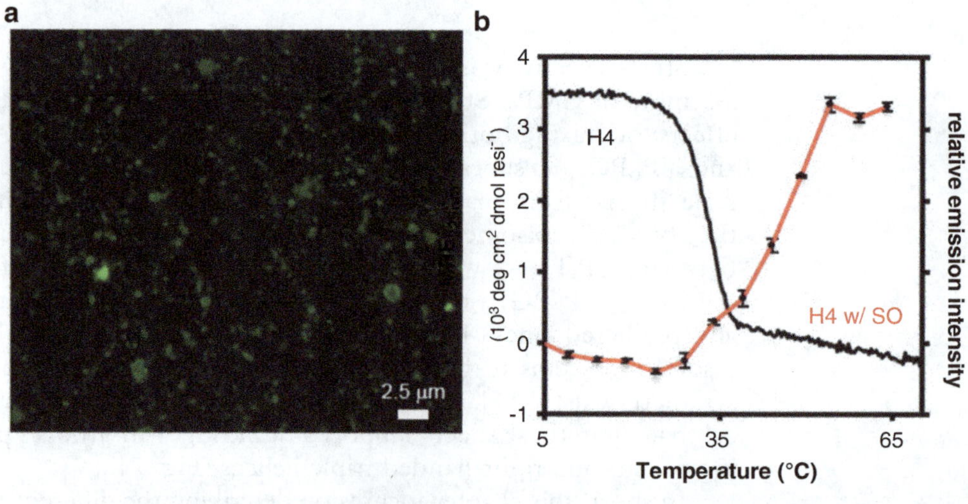

Fig. 5 (**a**) Confocal fluorescence microscopy of solution of Sypro Orange with H4 peptide ($(POG)_3POGLIGLIG(POG)_4$). (**b**) Differential scanning fluorimetry of H4 with Sypro Orange (red) and circular dichroism thermal denaturation melt of H4 (black) (Reprinted (adapted) from McGuinness et al. (2014) with permission from the American Chemical Society)

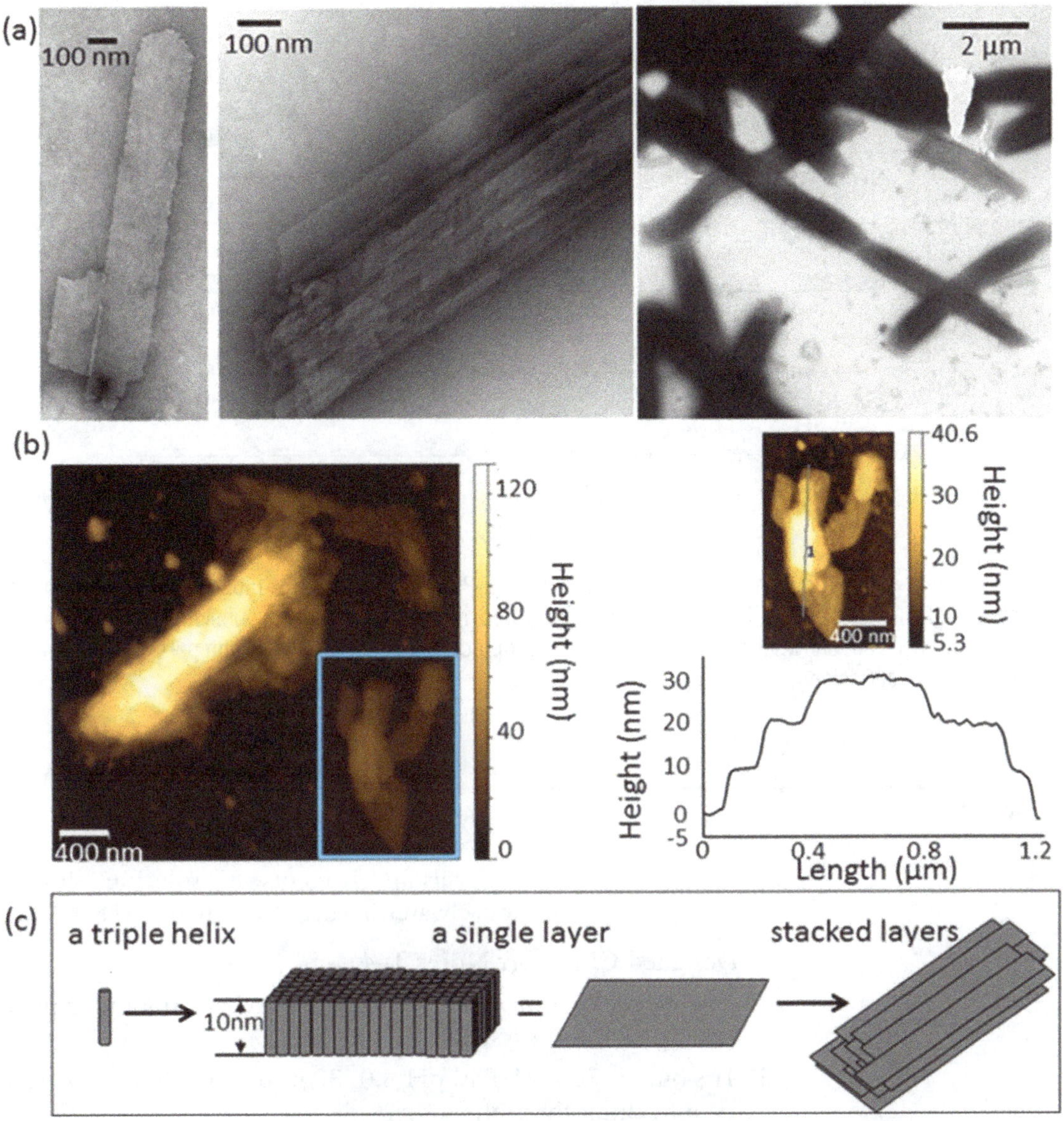

Fig. 6 Sheet morphology. (**a**) Transmission electron microscopy and (**b**) atomic force microscopy of a 1:1 L to D mixture. Contour heights varied by integer multiples of 10 nm. (**c**) A proposed model of assembly where triple helices align into a 10 nm thick layer, which further stack. Samples were incubated at 4 °C for ~4 weeks prior to imaging. (Reprinted (adapted) from Xu et al. (2013) with permission from the American Chemical Society)

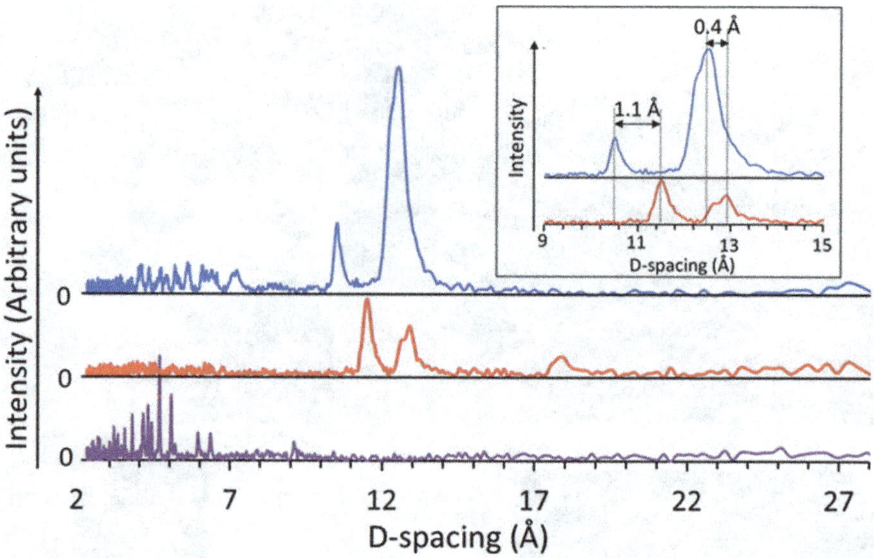

Fig. 7 *d*-spacing measured with WAXS of mixture 1 mM $(_LP_LPG)_{10}$:1 mM $(_DP_DPG)_{10}$ (upper) compared to oversaturated $(_LP_LPG)_{10}$ suspension (middle) and oversaturated mutant I-peptide suspension (bottom). The *d*-spacing differences are shown in the inset. (Reprinted (adapted) from Xu et al. (2013) with permission from the American Chemical Society)

2 Materials

Use analytical grade chemicals and reagents. All solutions should be prepared using ultrapure water having a resistivity of 18.2 MΩ.

1. Designed CMPs (*see* **Notes 1–4**).
2. Deionized filtered water using Millipore's Ultrapure type I water purification system.
3. Tris buffer: 50 mM Tris, pH 8.0, 30 mM NaCl, 2.5% glycerol, 10 mM dithiothreitol.
4. Phosphate Buffer: 10 mM sodium phosphate, pH 7.0.
5. TEM copper grids.
6. Petri dish.
7. Filter paper.
8. 1% phosphotungstic acid in ultrapure water.
9. 0.2 μm syringe filters.
10. Syringes: 2 mL, 5 mL, 10 mL.
11. TEM grid forceps.
12. Glass/quartz cuvette (*see* **Note 5**).
13. Silicon wafers.
14. Mica.

15. Sypro Orange 50× stock solution.
16. Micro quartz cells.
17. Circular dichroism spectrophotometer (*see* **Note 6**) with Peltier temperature controller.
18. Light scattering and zeta potential instrument (*see* **Note 7**).
19. Transmission electron microscope (*see* **Note 8**).
20. Atomic force microscope (*see* **Note 9**).
21. X-ray diffractometer for wide angle scattering measurements (*see* **Note 10**).
22. Differential scanning fluorimeter with micro quartz cell (*see* **Note 11**).
23. Microscope for confocal microscopy measurements (*see* **Note 12**).
24. UV-Vis spectrophotometer.
25. Mass spectrometer (electrospray ionization).
26. 1 mm diameter glass capillaries.
27. Ethanol.

3 Methods

Carry out all procedures at room temperature, unless otherwise specified.

3.1 Peptide Preparation

Filter all buffer and peptide solutions using 0.2 μm syringe filters.

1. Verify the mass of the peptide molecules using a mass spectrometer.
2. Prepare 2 mM peptide stock solution in phosphate buffer to maintain pH = 7.0 (*see* **Note 13**).
3. Determine the peptide concentrations in solution by measuring the absorbance at 214 nm by UV-Vis spectrophotometer using $\varepsilon_{214} = 2200/\text{M/cm}$.
4. Allow the peptides to fold at 4 °C at least for 48 h (*see* **Note 14**).

3.2 Circular Dichroism Spectroscopy (CD)

For CD spectra maintain the total peptide concentration at 0.2 mM in pH 7, 10 mM phosphate buffer (*see* **Note 15**).

3.2.1 Isothermal Spectra

1. For wavelength spectra, perform wavelength scans at 4 °C with an averaging time of 10 s at each wavelength ranging from 190 to 260 nm with an increment of 0.5 nm (Fig. 4a).
2. Convert observed ellipticity to mean residual ellipticity (MRE) by dividing measured values by the peptide concentration, number of residues, and cell path length:

$$\text{MRE} = (\text{CD signal} \times \text{molecular weight}) / (\text{path length} \times \text{peptide bonds} \times \text{concentration})$$

3.2.2 Thermal Denaturation

1. Obtain a spectrum of buffer in the cuvette by averaging three scans. This is the blank to be subtracted from the sample signal.
2. Obtain a melt of the sample in the same cuvette for a temperature range of 4–80 °C at 0.33 °C/step and 2 min equilibration time for better resolution and to reduce the signal to noise ratio (Figs. 1a and 4b).
3. Convert CD signal to MRE:

$$\text{MRE} = (\text{CD signal} \times \text{molecular weight}) / (\text{path length} \times \text{peptide bonds} \times \text{concentration})$$

4. Determine the low temperature (folded) and high temperature (unfolded) linear parts of the curve. Fit a linear curve to determine the CD signal of folded state called CDf(t) and CD signal of the unfolded state called CDu(t).
5. Calculate the melting temperature (T_m) and convert the CD signal to fraction folded:

$$\text{FF}(t) = (\text{CD}(t) - \text{CDu}(t)) / (\text{CDf}(t) - \text{CDu}(t))$$

Melting temperature (T_m) is the point where the FF(t) value is 0.5.

3.3 Static and Dynamic Light Scattering (SLS & DLS)

1. Prepare a 1 mM peptide solution in pH 7, 10 mM phosphate buffer such that no bubbles are formed (*see* **Notes 15** and **16**).
2. Determine the autocorrelation functions with an acquisition time of 2 min per correlation function for better resolution and to reduce the signal to noise ratio (*see* **Notes 17–21**).
3. Determine the hydrodynamic radius of the particle using the Stokes-Einstein equation (Fig. 1c).

3.4 Zeta Potential

1. Gently transfer the sample (2 mM peptide solution in pH 7, 10 mM phosphate buffer) into the cuvette using a syringe to avoid the formation of bubbles (*see* **Notes 15** and **16**).
2. Perform five measurements for each sample. The final result should be the average of these measurements (Fig. 1b).

3.5 Transmission Electron Microscopy (TEM)

1. Centrifuge the aggregated sample (2 mM peptide solution in pH 7, 10 mM phosphate buffer) at 10,000 × *g* for 10 min.
2. Discard the supernatant and resuspend in an identical buffer to homogenously disperse the aggregates.

3. Mix the sample thoroughly and then take 10 μL and drop cast it onto the TEM grid placed on a petri dish (*see* **Note 22**).
4. Leave the sample on the grid for 5 min and absorb the excess sample using clean filter paper.
5. Stain the grid negatively with 10 μL of 1% phosphotungstic acid and leave it overnight in an incubator set at 37 °C after covering the petri dish with parafilm with small holes poked through for ventilation.
6. Analyze the sample using TEM to get the morphology of higher order structures (Figs. 2 and 3).

3.6 Wide-Angle X-Ray Scattering (WAXS)

WAXS is employed to investigate the interhelical *d*-spacing between the triple-helical molecules of the formed nanostructures (Figs. 5c and 7)

1. Prepare supersaturated solutions by placing peptide powder in a glass capillary (1 mm diameter), and add 10 mM phosphate buffer, pH 7 on top of the powder (*see* **Note 15**).
2. Allow the powder to soak in the solvent for at least 2 weeks at 4 °C to ensure sufficient hydration.
3. Mount the capillary in an X-ray diffractometer with a sealed Cu target X-ray tube and mirror focusing monochromator (Cu Kα; $\lambda = 1.5418$ Å).
4. Set the nominal sample to detector distance at 2.2 cm.
5. Collect the data by scanning along 2θ from 3° to 40° (step size 0.04°) with a 60 s count per step (Figs. 4 and 7).

3.7 Confocal Fluorescence Microscopy

Confocal microscopy is employed to investigate the binding of Sypro Orange with CMP nanodiscs. Images of H4 (sequence $(POG)_3POGLIGLIG(POG)_4$, where O is hydroxyproline) with Sypro Orange show effective nanodisc sequestration of the dye (Fig. 5).

1. Take a 10–20 μL aliquot of the aggregated solution (2 mM peptide in pH 7, 10 mM phosphate buffer) and place on microscope slides covered with glass coverslips.
2. Incubate the samples on ice for 20 min to ensure settling.
3. Image the sample using a confocal microscope with a red filter.

3.8 Differential Scanning Fluorimetry (DSF)

DSF can be employed to study the binding of H4 nanodiscs to the solvatochromatic dye Sypro Orange which has an affinity for hydrophobic residues (*see* **Note 23**) (Fig. 5).

1. Dilute the 50× stock solution of Sypro Orange with either H4 disc (2 mM peptide in 10 mM phosphate buffer, pH 7) or phosphate buffer to 5× (*see* **Note 24**).

2. Monitor the samples using a spectrofluorimeter with λ_{ex} 470 nm and λ_{em} 610 nm wavelengths, with 2 nm and 6 nm slit-widths, respectively.
3. Correct the samples with Sypro Orange for Sypro Orange fluorescence.
4. Run three measurements for each temperature step (4–64 °C in 5 °C increments, at the same rate as the CD experiments), and compute an average and standard deviation.
5. Use standard deviations as error bars.

3.9 Atomic Force Microscopy (AFM)

1. Clean the silicon wafers with ethanol
2. Dry the silicon wafers with nitrogen gas and place them on a glass slide with adhesive. Cool the wafers before sample deposition.
3. Deposit the supersaturated solution (2 mM peptide in 10 mM phosphate buffer, pH 7) and then remove the excess sample using deionized water and dry it for 20 min at room temperature.
4. Rinse the samples with deionized water and dry with nitrogen gas.
5. Load the sample on the stage and image the sample using contact mode in air on samples that are prepared on freshly cleaved mica.
6. Analyze the data using a combination of software packages (*see* **Note 25**).

4 Notes

1. Peptides can be synthesized in-house or through a third party facility via standard FMOC chemistry. Peptide design is crucial. Thoroughly study all associated literature and procure some initial computer generated modeling and simulation data before the peptide synthesis. The N-terminal of the synthesized peptides must be acetylated and the C-terminal must be amidated.
2. *Peptide design for studying electrostatics:* In order to understand how subtle differences between arginine and lysine (both possessing the same charge) can affect the stability and higher order assembly, the following peptide sequences can be used. Bold residues represent the supercharged part of the peptide. The single-letter abbreviation O represents hydroxyproline.

   First set of peptides: It is used to investigate the role of Arginine and Lysine residues on the folding and self-assembly of CMPs.

   cpB-RE: PEGEOGROGROGPRGPRGPRGROGROGPRG,

   cpC-RE: PRGROGROGROGPRGEOGPEGPEGEOGEOG,

cpB-KD: PDGDOGKOGKOGPKGPKGPKGKOGKOGPKG,

cpc-KD: PKGKOGKOGKOGPKGDOGPDGPDGDOGDOG

Second set of peptides: It is used to investigate the effect of charge crowding on the folding and self-assembly of higher-order structures.

R6: (**RRG**)$_3$(POG)$_7$, K6: (**KKG**)$_3$(POG)$_7$, E6: (**EEG**)$_3$(POG)$_7$, D6: (**DDG**)$_3$(POG)$_7$

Third set of peptides: It is used to investigate the effect of position of charge residues on the folding and self-assembly of higher-order structures.

PRG3: (**PRG**)$_3$(POG)$_7$, ROG3: (**ROG**)$_3$(POG)$_7$, PKG3: (**PKG**)$_3$(POG)$_7$, KOG3: (**KOG**)$_3$(POG)$_7$, PEG3: (**PEG**)$_3$(POG)$_7$, EOG3: (**EOG**)$_3$(POG)$_7$, PDG3: (**PDG**)$_3$(POG)$_7$, DOG3: (**DOG**)$_3$(POG)$_7$

3. *Peptide design for studying the role of hydrophobicity:* CMPs should be designed incorporating hydrophobic residues at the X and Y positions to understand the role of length and position of hydrophobic residues. Bold residues represent the hydrophobic residues.

   H2: (POG)$_3$POGPOG**LIG**(POG)$_4$, H3: (POG)$_3$POGP**IGL**
   IG(POG)$_4$,

   H4: (POG)$_3$POG**LIGLIG**(POG)$_4$, QH4: (POG)$_3$QQG**LIG**
   LIG(POG)$_4$,

   H5: (POG)$_3$P**IGLIGLIG**(POG)$_4$, H6: (POG)$_3$**LIGLIGLI**
   G(POG)$_4$

4. *Peptide design for studying the role of stereoselectivity:* A ridges-in-grooves model using right and left handed CMPs for forming supramolecular structures can be evaluated for a system where shape is the main feature determining intermolecular packing. The following two peptides can be used: $({}_{L}P_{L}PG)_{10}$ and $({}_{D}P_{D}PG)_{10}$.
5. Use lint-free paper for cleaning all cuvettes before analyzing to prevent lint and fiber contamination which may interfere with the analysis.
6. An Aviv Model 420SF Spectrophotometer was used in our lab for carrying out the experiments.
7. In our lab, a Zetasizer Nano ZS (Malvern Instruments, UK) was used for light scattering and zeta potential measurements
8. We used a Philip 420 Electron Microscope for imaging.
9. We used an Agilent 5500 for tapping mode in air, with a Multi75a probe (Bruker MPP-21120-10). MFP-3D-BioAFM with a Nikon microscope was used for visualization. Measurements were made using contact mode in air with a SNL-10 probe (Bruker).
10. We used a Rigaku Geigerflex X-ray diffractometer.
11. We used a Fluoromax-4 spectrofluorimeter (Jobin Yvon Horiba).

12. A Leica TCS Sp8 G-STED microscope was used in our lab.
13. Be sure to measure the pH of the sample/buffer after uniform mixing since pH plays a very important role in these experiments.
14. Make sure all the proteins and peptides have purity above 95% and store them appropriately at −20 °C to prevent their denaturation and prolong shelf life.
15. Before measurement by DLS, WAXS, CD, etc. use a 0.22 μm syringe filter to remove unwanted aggregates or contaminants which may act as possible nucleation sites or even interfere in the self-assembly process.
16. In light scattering measurements make sure no bubbles are formed during the sample preparation or during measurement.
17. For DLS analysis, parameters like solvent used, the equation model, cell position factor etc. must be properly entered. Use glass/quartz cuvettes for temperature studies.
18. For kinetics measurement, scattered light should be collected continuously using acquisition times for each measurement to have more number of points for a better fit.
19. Measurements should be performed using the right attenuator to avoid saturation of the detector.
20. Count rates should be converted into units of derived count rates, i.e., assuming no attenuation of the laser light source by the attenuator.
21. Use a proper delay time between measurements to allow the machines to reach and maintain the specific conditions required for the analysis.
22. For TEM sample preparation, use small forceps especially designed for TEM for easier handling. Do not press the grid too hard with the forceps.
23. Our result suggests hydrophobic residues are buried in the core of the nanostructure.
24. Subtract (buffer + Sypro Orange) data from (Sample + Sypro Orange) data.
25. Data can be analyzed using a combination of Gwyddion, WSXM, and MATLAB software packages.

Acknowledgments

A.S.P. acknowledges support from Department of Science and Technology (DST-SERB), India—SB/FTP/PS-073/2014 (Start-Up Research Grant for Young Scientist) & SB/S2/RJN-

140/2014 (Ramanujan Fellowship Award). JL acknowledges support from Department of Science and Technology (DST-SERB), India—YSS/2015/000572.

References

1. Merlea C, Perreta S, Lacoura T et al (2002) Hydroxylated human homotrimeric collagen I in Agrobacterium tumefaciens-mediated transient expression and in transgenic tobacco plant. FEBS Lett 515:114–118
2. O'Leary LER, Fallas JA, Bakota EL et al (2011) Multi-hierarchical self-assembly of a collagen mimetic peptide from triple helix to nanofibre and hydrogel. Nat Chem 3:821–828
3. Brodsky B, Persikov AV (2005) Molecular structure of the collagen triple helix. Adv Protein Chem 70:301–339
4. Kar K, Amin P, Bryan MA et al (2006) Self-association of collagen triple helix peptides into higher order structures. J Biol Chem 281:33,283–33,290
5. Persikov AV, Ramshaw JAM, Brodsky B (2000) Collagen model peptides: sequence dependence of triple-helix stability. Pept Sci 55:436–450
6. Persikov AV, Brodsky B (2002) Unstable molecules form stable tissues. PNAS 99:1101–1103
7. Lee CH, Singlaa A, Lee Y (2001) Biomedical applications of collagen. Int J Pharm 221:1–22
8. Hartgerink JD, Beniash E, Stupp SI (2001) Self-assembly and mineralization of peptide-amphiphile nanofibers. Science 294:1684–1688
9. Jiang T, Xu C, Liu Y et al (2014) Structurally defined nanoscale sheets from self-assembly of collagen-mimetic peptides. J Am Chem Soc 136:4300–4308
10. Rele S, Song Y, Apkarian RP et al (2007) D-periodic collagen-mimetic microfibers. J Am Chem Soc 129:14,780–14,787
11. Pires MM, Lee J, Ernenwein D, Chmielewski J (2012) Controlling the morphology of metal-promoted higher ordered assemblies of collagen peptides with varied core lengths. Langmuir 28:1993–1997
12. Przybyla DE, Rubert Perez CM, Gleaton J et al (2013) Hierarchical assembly of collagen peptide triple helices into curved disks and metal ion promoted hollow spheres. J Am Chem Soc 135:3418–3422
13. Cui H, Webber MJ, Stupp SI (2010) Self-assembly of peptide amphiphiles: from molecules to nanostructures to biomaterials. Biopolymers 94:1–18
14. Xu F, Li F, Jain V et al (2013) Compositional control of higher order assembly using synthetic collagen peptides. J Am Chem Soc 134:47–50
15. Persikov AV, Ramshaw JAM, Brodsky B (2005) Prediction of collagen stability from amino acid sequence. J Biol Chem 280:19,343–19,349
16. Rainey JK, Goh MC (2004) An interactive triple-helical collagen builder. Bioinformatics 20:2458–2459
17. Xu F, Zahid S, Silva T (2011) Computational design of a collagen A:B:C-type heterotrimer. J Am Chem Soc 133:15,260–15,263
18. Xu F, Zhang L, Koder RL (2010) De novo self-assembling collagen heterotrimers using explicit positive and negative design. Biochemistry 49:2307–2316
19. Litowski JR, Hodges RS (2002) Designing heterodimeric two-stranded α-helical coiled-coils. J Biol Chem 277:37,272–37,279
20. Vizcarra CL, Mayo SL (2005) Electrostatics in computational protein design. Curr Opin Chem Biol 9:622–626
21. Egelman EH, Xu C, Maio FD et al (2015) Structural plasticity of helical nanotubes based on coiled-coil assemblies. Structure 23:280–289
22. Parmar AS, James JK, Grisham DR et al (2016) Dissecting electrostatic contributions to folding and self-assembly using designed multicomponent peptide systems. J Am Chem Soc 138:4362–4367
23. McGuinness K, Khan J, Nanda V (2014) Morphological diversity and polymorphism of self-assembling collagen peptides controlled by length of hydrophobic domains. ACS Nano 8:12,514–12,523
24. Xu F, Khan J, McGuinness K et al (2013) Self-assembly of left and right-handed molecular screws. J Am Chem Soc 135(50): 18,762–18,765

Chapter 19

Design and Characterization of Fibers and Bionanocomposites Using the Coiled-Coil Domain of Cartilage Oligomeric Matrix Protein

Priya Katyal and Jin Kim Montclare

Abstract

Tremendous effort has been dedicated to the design and assembly of bioinspired protein-based architectures with potential applications in drug delivery, tissue engineering, biosensing, and bioimaging. Here, we describe our strategy to generate fibers and bionanocomposites using the coiled-coil domain of cartilage oligomeric matrix protein (COMPcc). Our construct, Q, engineered by swapping particular regions of COMPcc to optimize surface charge, self-assembles to form nanofibers. The Q protein nanofibers can efficiently bind curcumin to form robust mesofibers that can be potentially used for drug delivery and biomedical applications. In addition, using the same Q protein, we describe the biotemplation of gold nanoparticles (AuNP) in the presence and absence of the hexahistidine tag (His-tag). The Q bearing His-tag·AuNP (Q·AuNP) readily deposits on electrode surfaces, while Q without His-tag·AuNP (Qx·AuNP) stabilizes the soluble protein·gold bionanocomposites for several days without aggregating.

Key words Self-assembly, Fibers, Gold nanoparticles, Coiled-coil protein, Cartilage oligomeric matrix protein, Domain-swapped, Bionanocomposites

1 Introduction

Nature has created a plethora of biomolecular assemblies from nanoscale to mesoscale bearing various shapes and endowed with sophisticated biological functions [1]. Inspired by unique biological self-assembly, biomacromolecules have been increasingly used as building blocks to assemble complex higher order architectures [2, 3]. More specifically, protein molecules that can be engineered to self-assemble into diverse shapes and structures, conferring different functionalities and properties to the complex three-dimensional architectures, have garnered interest in the field of biomaterials design [4]. These complex molecular assemblies have been used for many novel technological and biomedical applications, including their use in drug delivery, tissue engineering, biosensing, and bioimaging [2, 5].

Andrew K. Udit (ed.), *Protein Scaffolds: Design, Synthesis, and Applications*, Methods in Molecular Biology, vol. 1798, https://doi.org/10.1007/978-1-4939-7893-9_19,

Tremendous progress has been made in designing structurally defined assemblies and developing functional architectures with peptides and proteins bearing coiled-coil motifs. These motifs, comprised of 2–7 helical structures, are the most common and versatile frameworks found in self-assembled protein architectures [6–8]. Successful attempts of using coiled-coil motifs include self-assembling fibers (SAFs) [9–11], self-assembling protein nanoparticles (SAPN) [12], spheres [13], self-assembling caGE-like particles (SAGE) [14], and many other complex structures [15].

There has been rising interest in tailoring coiled-coil proteins that can self-assemble to form fibers. The exemplary work by Kojima *et al.* on alpha-helical peptides that self-assemble to form fibers of 5–10 nm in diameter [16], and subsequent studies by Woolfson and colleagues [9, 17] on heterodimeric, head-to-tail sticky-end assembly of fibers [10, 18], represent key landmark designs in engineering SAFs systems. Alternatively, Ogihara *et al.* have redesigned a coiled-coil variant to form a domain swapped three-helix bundle, capable of forming fibers [19]. These studies guide the design and synthesis of various SAFs that offer excellent prospects in developing biomaterials with numerous applications.

Advances in the field of biomaterials include the use of bio-integrated systems, with focus on fabricating functional hybrid materials with better thermal, mechanical and kinetic properties [20, 21]. Of particular interest are bionanocomposites comprised of nanoparticles conjugated to peptide/protein assemblies. Bionanocomposites of protein·gold nanoparticles (AuNPs), obtained through the templation of gold salts by proteins, have the flexibility to assemble/disassemble based on external triggers including changes in pH [22, 23], altering peptide concentrations [23, 24] or through the use of additives [25, 26].

In this chapter we will discuss the design of a recombinant protein based on the coiled-coil domain of cartilage oligomeric matrix protein (COMPcc) for the formation of fibers and bionanocomposites with AuNPs. COMPcc contains six complete heptads, forming a homopentameric structure with a central hydrophobic core [27]. The hydrophobic core bears two cavities capable of binding hydrophobic compounds including vitamin D, all trans-retinol, cyclohexane, and curcumin [28, 29]. In addition, it can also bind saturated and unsaturated fatty acids [30].

Inspired by previous studies that demonstrated fiber formation of domain swapped helical proteins [19], we have designed Q protein, where the last seven residues of COMPcc are eliminated and the N- and C-terminal segments are swapped with respect to glutamine (Q54). Q54 acts as a divider between the two hydrophobic pockets of COMPcc; thus domain swapping with respect to this glutamine residue does not affect the binding pocket. The N-terminal proline residue is displaced to the center of the protein, providing a local break, while maintaining the alpha-helical structure on either side.

Relative to wild type COMPcc, the solvent exposed residues of Q possess a better surface charge distribution allowing lateral assembly of the protein into nanofibers of diameters up to 560 nm at pH 4. We explore the ability of these fibers to bind the hydrophobic small molecule, curcumin, that has anticancer [31], anti-inflammatory [32], antibacterial [33], and wound-healing properties [34]. When bound to curcumin, the fibers further assemble into microfibers with a diameter of 16 μm [35]. The same construct can also be used for gold templation to form bionanocomposites. To address the impact of the hexahistidine tag (His-tag), Q (which bears the His-tag) and Qx (whereby the His-tag is removed) were subjected to AuNP templation and synthesis. In the presence of the tag, large assemblies of Q·AuNP form that precipitate out of solution. By contrast, when the His-tag was cleaved, Qx·AuNP produces bionanocomposites that do not aggregate and remain stable for several days [36].

Overall, the chapter describes the design of Q protein, and its characterization using a combination of techniques including circular dichroism (CD), attenuated total reflectance-fourier transform infrared spectroscopy (ATR-FTIR) and transmission electron microscopy (TEM). In addition, Q assembles into distinct architectures, forming mesofibers upon binding to curcumin and bionanocomposites when crystallized with AuNPs (Fig. 1). Methods for characterizing the curcumin-loaded fibers and the AuNP bionanocomposites are briefly discussed. Upon binding to curcumin, the nanoscale assembly of Q stacks together to form large mesoscale fibers. These fibers have unique opportunities to be used as scaffolds for drug delivery and diagnostic applications. For engineered AuNP bionanocomposites, we present their use in forming a tunable kinetic barrier on an electrode surface. Our results on protein·AuNP bionanocomposites suggest that Q can also assemble into different morphologies with cleaved Qx capable of stabilizing AuNPs and uncleaved Q forming AuNPs embedded protein films. The methods described here serve as examples for design and development of bioinspired assemblies with varied applications.

2 Materials

2.1 Genetic Engineering

1. 2× filtered deionized water (dH_2O): dH_2O filtered twice using 0.2 μm syringe filters.
2. Polymerase Chain Reaction (PCR) reagents: 0.2 mM dNTP mix, DMSO, DNA polymerase enzyme with reaction buffer.
3. 200 ng pQE9-COMPcc template DNA (*see* **Note 1**).
4. Restriction enzymes: BamHI and HindIII.
5. Primers: 10 μM each of Q54 Fwd1/Rev1 and Q54 Fwd2/Rev2 (*see* **Note 2**).

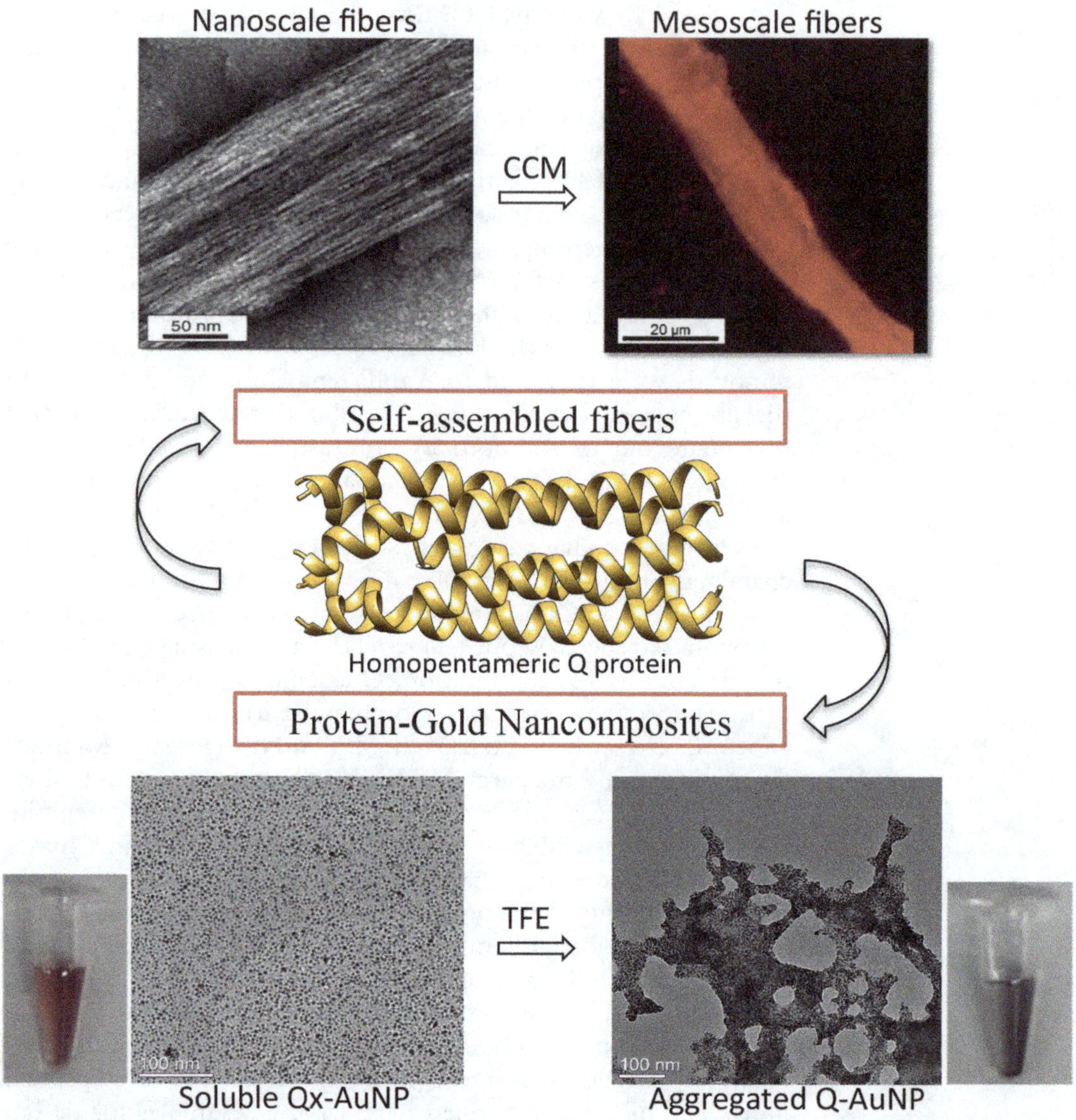

Fig. 1 Assembly of Q protein into distinct architectures. Homopentamers of Q assemble to form protofibrils that bundle together to create fibers ranging from tens to a hundred nanometers in diameter. These nanoscaled fibers are capable of binding curcumin (CCM) that results in mesofibers with diameters on the micrometer scale. The Q protein was also shown to template gold nanoparticles, with morphologies dependent on the secondary structure of protein. In the presence of His-tag, Q templates AuNPs to form large assemblies that cluster together to form aggregates. Upon cleaving the tag, the cleaved protein (Qx) stabilizes the nanoparticles, resulting in dispersed Qx·AuNP. Interestingly, trifluoroethanol (TFE) induces conformational changes in the protein that drives Qx·AuNP to aggregate. Adapted from Hume *et al.* (2014) and Hume *et al.* (2015) with permission from the American Chemical Society

6. Thermocycler.
7. TAE buffer: Prepare a 50× stock using 242 g of Tris base, 57.1 mL of glacial acetic acid, 100 mL of 0.5 M EDTA pH 8.0, make up to 1 L with dH_2O. Dilute 40 mL of 50× TAE buffer in 2 L of dH_2O to get 1× TAE buffer.
8. DNA gel electrophoresis apparatus.
9. 2% agarose gel: 1 g of agarose in 50 mL of 1× TAE buffer.
10. DNA loading dye (6×).
11. DNA gel extraction and purification kit.
12. XL-1-Blue competent cells.
13. MiniPrep Kit.
14. pQE30 vector (*see* **Note 3**).
15. T4 DNA Ligation kit: T4 buffer, T4 ligase.
16. φ ladder.
17. 50% glycerol solution, sterilized by autoclave.
18. 2 mL screw-capped cryo-vial.
19. NanoDrop.

2.2 Protein Expression

1. Ampicillin stock solution: 200 mg/mL ampicillin.
2. Tryptic Soy Agar (TSA) plates: Autoclave 10 g of TSA in 250 mL of dH_2O. Add 250 μL of ampicillin stock solution when cooled. Pour ~23–25 mL of solution per plate.
3. Isopropyl-β-D-thiogalactopyranoside (IPTG), 200 mg/mL.
4. Luria Broth (LB): 10 g of NaCl, 10 g of tryptone, 5 g of yeast extract in 1 L of dH_2O. Autoclave before use.
5. Protein gel electrophoresis apparatus.
6. 12% sodium dodecyl sulfate-polyacrylamide gel electrophoresis (SDS-PAGE) gel.
7. 6× SDS loading dye.

2.3 Protein Purification

All buffer solutions are filter sterilized using 0.22 μm pore filter membranes, and stored at 4 °C.

1. Probe Sonicator.
2. Lysis buffer: 50 mM Tris–HCl, 0.5 M NaCl, 20 mM imidazole, 6 M urea, adjust to pH 8 using HCl/NaOH.
3. Elution buffers: Prepare elution buffers A–D of 100 mM, 200 mM, 500 mM and 1 M imidazole in lysis buffer:
   (a) 100 mM imidazole: add 82 mg of imidazole to 15 mL of lysis buffer.
   (b) 200 mM imidazole: add 184 mg of imidazole to 30 mL of lysis buffer.

 (c) 500 mM imidazole: add 490 mg of imidazole to 50 mL of lysis buffer.
 (d) 1 M imidazole: add 1 g of imidazole to 80 mL of lysis buffer.
4. Sodium phosphate buffers (PB): 50 mM sodium phosphate, pH 4, 8, and 10.
5. Dialysis buffers:
 (a) PB, pH 4, 8, 10, each with 3 M urea.
 (b) PB, pH 4, 8, 10, each with 1.5 M urea.
 (c) PB, pH 4, 8, 10, each with 0.75 M urea.
6. Nickel-nitrilotriacetic acid (Ni-NTA) resin.
7. Gravity flow column (diameter = 2.5 cm).
8. Bicinchoninic acid (BCA) kit.
9. Dialysis tubing (3.5 kDa).

2.4 Characterization of Q

2.4.1 Circular Dichroism (CD) Spectroscopy

1. Quartz cuvette (path length 1 mm).
2. Jasco J-815 spectrometer attached with PTC-423S single position Peltier temperature control system.
3. 5 M nitric acid.

2.4.2 Transmission Electron Microscopy (TEM)

1. Copper grids.
2. Blotting paper.
3. 1% w/v uranyl acetate (filtered).
4. Jeol JEM-1400 transmission electron microscope (Jeol, MA, USA).
5. ImageJ software for image analysis.

2.4.3 Attenuated Total Reflectance-Fourier Transform Infrared (ATR-FTIR) Spectroscopy

1. PerkinElmer System 2000 FT-IR with DuraSamplIR II T diamond ATR accessory, equipped with a MCT-A detector.

2.4.4 Confocal Microscopy

1. Glass slides and coverslips.
2. Leica TCS SP2 AOBS confocal microscope system (Leica, IL, USA) equipped with argon ion and HeNe lasers fitted with 63×/1.4 NA oil-immersion objective lens.
3. Software for analysis: ImageJ 64 1.43 along with Amira 5.43, Volren 3D rendering routine.

2.4.5 Curcumin Binding

1. 1 mM curcumin in non-denatured methanol.

2.4.6 Protein Cleavage

1. Factor Xa cleavage kit.

2.4.7 Gold Nanoparticle (AuNP) Bionanocomposites

1. Au salt solution: 0.1 M $HAuCl_3 \cdot 3H_2O$ in dH_2O.
2. 0.1 M sodium borohydride in dH_2O.
3. UV-Vis spectrophotometer.

2.4.8 Cyclic Voltammetry (CV) and Impedance Spectroscopy

1. Fe-cyanide solution: 10 mM potassium hexacyanoferrate (II) trihydrate and 10 mM potassium hexacyanoferrate (III) in PB, pH 8.
2. Platinum wire counter electrode (d = 3.0 mm, l = 7.2 cm).
3. Ag/AgCl reference electrode (d = 4.7 mm).
4. Glassy-carbon disk electrodes (d = 3.0 mm).
5. Potentiostat with computer interface.
6. Electrochemical cell.

3 Methods

3.1 Design of Q

The Q construct is engineered by removal of the last heptad of COMPcc and swapping the domains with respect to Q54. The C-terminal residues following Q54 are swapped to the N-terminus side (Fig. 2). Q is engineered via PCR assembly, in which overlapping DNA fragments are assembled to generate a full gene. DNA fragments Q1 and Q2 are amplified using pQE9-COMPcc [29] as the template (*see* **Note 1**). The generated fragments are then annealed to yield the entire DNA sequence of Q. The resultant DNA encoding for protein Q is inserted into the pQE30 vector to form the final Q DNA, which is used to transform into XL-1 Blue competent cells. An IEGR recognition sequence is added C-terminal to the His-tag, which is recognized and cleaved by Factor Xa, when desired.

3.1.1 PCR to Generate Fragments Q1 and Q2

1. Generate Q1 via PCR with primers Q54 Fwd1/Rev1, and Q2 with Q54 Fwd2/Rev2 (*see* **Note 2**). Set up the reactions using the following concentrations of reagents: 0.7 μL (200ng) pQE9-COMPcc template DNA, 10 μL reaction buffer, 1 μL dNTP mix, 1 μL DMSO, 1 μL forward and reverse primers (10 μM each), and 1 μL DNA polymerase enzyme. Make up the reaction volume to 50 μL using 34.3 μL of 2× filtered dH_2O.
2. Amplify the DNA using the following program:
   (a) Initial denaturation at 98 °C for 2 min
   (b) Repeat for 30 cycles:
      - Denature at 98 °C for 5 s
      - Anneal at 54 °C for 5 s
      - Extend at 72 °C for 20 s
   (c) Incubate at 72 °C for 1 min
   (d) Store at 4 °C

a

	His tag	abcdefg	abcdefg	abcdefg	abcdefg	abcdefg	abcdefg	abcdefg
	15	30	37	44	51	58	65	72
wt	MRGS**HHHHHH**GSDL	APQMLRE	LQETNAA	LQDVREL	LRQQVKE	ITFLKNT	VMESDAS	GKLN
	1	17	20	27	34	41	48	
Q	MRGS**HHHHHH**GSIEGR	VKE	ITFLKNT	APQMLRE	LQETNAA	LQDVREL	LRQQSKL	
		1	4	11	18	25	32	
Qx		VKE	ITFLKNT	APQMLRE	LQETNAA	LQDVREL	LRQQSKL	

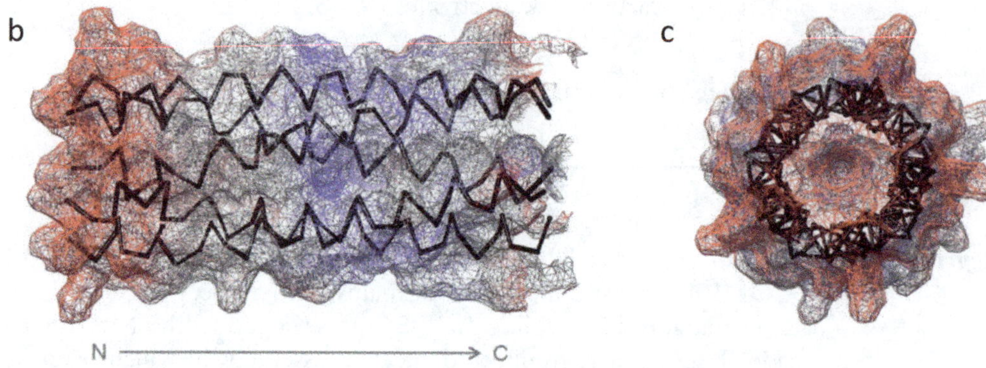

Fig. 2 Protein sequence and surface charge representation of Q. (**a**) Q is generated by removing the last heptad and swapping the N- and C-terminal regions with respect to glutamine at position 54. The sequence of parent COMPcc (wild-type (wt)) is shown for reference. Qx was generated by cleaving the His-tag using Factor Xa, which recognizes the IEGR motif and cleaves after the arginine residue. Surface charge representation of pentameric assembly of Q is shown along (**b**) and down (**c**) the pentameric axis. Adapted from Hume *et al.* (2014) with permission from the American Chemical Society

3. Perform gel electrophoresis of the amplified DNA: Add 5 μL of loading dye to the 50 μL reaction tube and load 15 μL of the sample onto a 2% agarose gel. Run the gel for 30 min at 100 V in 1× TAE buffer.
4. Gel extract and purify the DNA fragments using a commercial kit. The expected base pairs (bp) length of DNA fragments Q1 and Q2 are 64 bp and 101 bp, respectively.
5. Determine the concentrations of purified DNA by measuring the absorbance at 260 nm (*see* **Note 4**).
6. Resuspend the DNA in 2× filtered dH_2O to achieve a concentration of 33 ng/μL.

3.1.2 PCR Assembly of Q1 and Q2 to Generate Q

1. Anneal the DNA fragments Q1 and Q2 (generated in Subheading 3.1.1) via PCR. Set up the reaction using: 3 μL each of Q1 (100 ng/μL) and Q2 (33 ng/μL) template DNA, 10 μL of reaction buffer, 1 μL of dNTPs, 1 μL of DMSO, 29 μL of 2× filtered dH_2O, and 1 μL of DNA polymerase enzyme in a reaction volume of 50 μL.

2. Use the following program for PCR:
   (a) Initial denaturation at 98 °C for 2 min
   (b) Repeat for 10 cycles:
      - Denature at 98 °C for 2 s
      - Anneal at 54 °C for 5 s
      - Extend at 72 °C for 20 s
   (c) Incubate at 72 °C for 1 s
   (d) Store at 4 °C
3. Remove the tube from the thermocycler. Add 1 μL of Q54 Fwd 1 primer and 1 μL of Q54 Rev 2 primer. Amplify using the PCR amplification cycle (described in **step 2**) with 35 cycles.
4. Gel purify the DNA using a commercial kit to yield Q DNA. The expected bp length of Q DNA is 142 bp.
5. Set up separate restriction digestions of Q DNA and pQE30 vector: Use 30 μL of 2× filtered dH_2O, 11 μL of Q DNA, 5 μL of buffer, 2 μL each of BamHI and HindIII in a reaction volume of 50 μL. For the vector, use 41 μL of 2× filtered dH_2O, 2 μL of vector DNA, 5 μL of buffer, 2 μL each of BamHI and HindIII in a reaction volume of 50 μL.
6. Incubate the tubes at 37 °C overnight.
7. Gel purify the restricted DNA and the vector using a commercial kit.
8. Ligate the vector and insert using: 2 μL of vector DNA, 5 μL of insert, 15 μL of dH_2O, 2 μL of T4 buffer, and 1 μL of T4 ligase in a total volume of 20 μL (*see* **Note 5**).
9. Incubate the reaction tube for 2 days and nights at 16 °C (*see* **Note 5**). Confirm ligation by performing gel electrophoresis using a φ ladder as control.
10. Transformed the product into XL-1 Blue competent cells, according to standard manufacturer's protocol.
11. Plate the transformed mixture on a TSA plate with 0.2 mg/mL ampicillin. Incubate the plate overnight at 37 °C.
12. Prepare six culture tubes with 5 mL of LB containing 0.2 mg/mL ampicillin. Inoculate each tube with a single colony. Incubate overnight at 37 °C in a shaking incubator (250 rpm).
13. Prepare glycerol stocks: In a 2 mL screw-capped cryo-vial, add 500 μL of overnight culture and 500 μL of 50% glycerol. Mix gently and freeze the tube at −80 °C.
14. Take the remaining culture and isolate the plasmid DNA using a plasmid MiniPrep Kit.
15. Identify positive clones by setting up a restriction enzyme digestion: 41 μL of 2× filtered dH_2O, 3 μL of plasmid DNA, 5 μL of buffer, and 0.5 μL each of BamHI and HindIII in a

reaction volume of 50 μL. Run the digested product on agarose gel to determine the presence of vector DNA (4512 bp) and target gene DNA (142 bp).

16. Sequence the DNA obtained from positive clones to further confirm successful cloning (*see* **Note 6**).

3.2 Expression of Q

If desired, a small-scale expression can be performed first by scaling back the procedure below to evaluate protein expression.

1. Perform a fresh transformation using DNA showing 100% sequence identity. Plate the transformed mixture on a TSA plate with 0.2 mg/mL ampicillin and incubate the plate overnight at 37 °C.
2. Inoculate 5 mL of LB containing 0.2 mg/mL ampicillin in a culture tube with a single colony from a freshly streaked plate. Incubate this starter culture overnight at 37 °C in a shaking incubator (250 rpm).
3. Inoculate 1 L of LB containing 0.2 mg/mL ampicillin with 1 mL of the starter culture. Incubate at 37 °C, 250 rpm until the optical density at 600 nm >0.8. Aliquot 1 mL of this pre-expression sample for SDS-PAGE analysis.
4. Induce the culture with IPTG at a final concentration of 200 μg/mL, and further incubate for 3 h at 37 °C, 250 rpm.
5. Take out 1 mL of the post-expression sample for SDS-PAGE analysis.
6. Harvest the cells at 4 °C and 4000 rpm (3399 × *g*) for 15 min.
7. Discard the supernatant and store the cell pellet at −80 °C, or use it immediately for purification.
8. Evaluate protein expression using the pre- and post-expression samples:
   (a) Centrifuge the pre- and post-expression samples for 3 min at 13,600 rpm (15,922 × *g*).
   (b) Discard the supernatant and resuspend the cells in 50 μL of lysis buffer.
   (c) Add 5 μL of SDS loading dye and heat the samples at 100 °C for 20 min.
   (d) Add 15 μL of each sample to gel lane.
   (e) Run SDS-PAGE at 85 V for 10 min followed by 200 V for 45 min to evaluate protein expression (*see* **Note 7**).

3.3 Purification of Q

1. Resuspend the cells in 10 mL of lysis buffer and lyse the cells using a probe sonicator for 1 min, with pulse setting of 5 s on and 5 s off at 35% amplitude on ice (*see* **Note 8**).

2. Centrifuge the lysate at 4 °C, 4000 rpm (3399 × *g*) for ~50 min to separate out the insoluble cell debris.
3. Prepare Ni-NTA beads prior to use: Add 5 mL of Ni-NTA slurry for every 40 mL of lysate to a 50-mL conical tube. Spin the beads at 4000 rpm (3399 × *g*) for 10 min. Pour out the supernatant. Wash the beads with ~25 mL of lysis buffer, spin and discard the supernatant.
4. Add the soluble supernatant (from **step 2**) to the prepared Ni-NTA beads. Mix end over end for 3 h at 4 °C.
5. Spin the mixture at 2000 rpm (850 × *g*) for 20 min at 4 °C. Discard the supernatant.
6. Add 10 mL of lysis buffer to resuspend the beads.
7. Set up the gravity flow column. Load the resuspended beads onto the column and collect the flow-through.
8. Wash the beads with 2 × 50 mL of lysis buffer at 4 °C.
9. Elute the protein using a concentration gradient of imidazole ranging from 100 mM to 1 M:
   (a) 5 mL of elution buffer A
   (b) 5 mL of elution buffer B
   (c) 5 mL of elution buffer B
   (d) 5 mL of elution buffer C
   (e) 5 mL of elution buffer C
   (f) 5 mL of elution buffer C
   (g) 10 mL of elution buffer D
   (h) 10 mL of elution buffer D
   (i) 10 mL of elution buffer D
10. The purity of the fractions is checked by SDS-PAGE analysis (*see* **Note 9**).
11. Dialyze the pure fractions against 2 L of the buffers below using three different buckets for each of pH 4, 8, and 10 at 4 °C:
    (a) dialysis buffer A (2 h)
    (b) dialysis buffer B (2 h)
    (c) dialysis buffer C (2 h)
    (d) PB (3 h)
    (e) PB (3 h)
    (f) PB (3 h)
12. Estimate the protein concentration using a BCA kit.

3.4 Characterization of Q

3.4.1 Circular Dichroism (CD) Spectroscopy

CD analysis reveals the secondary structure characteristics of Q, which can be studied under different pH conditions (Fig. 3) (*see* **Note 10**). The thermodynamic parameters of Q at pH 4, 8 and 10 are listed in Table 1 (*see* **Note 11**).

1. Clean the quartz CD cuvette with 5 M nitric acid and rinse thoroughly with dH_2O prior to use.
2. Prepare a 10 μM (6.3×10^{-2} mg/mL) protein sample.
3. Perform wavelength and temperature scans by recording CD spectra using a Jasco J-815 spectrometer attached with PTC-423S single position Peltier temperature control system.
4. Perform a wavelength scan at room temperature over a range from 190 to 250 nm with a step size of 1 nm.
5. Perform a temperature scan at 222 nm ranging from 20 to 85 °C, at a heating rate of 1 °C/min.

a

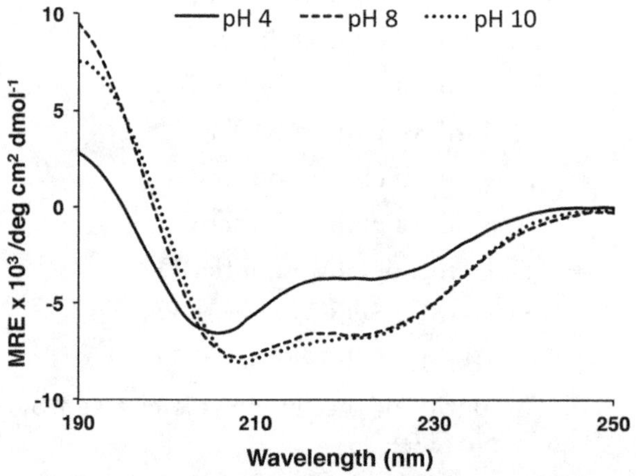

b

		pH 4	pH 8	pH 10
MRE	$-\Theta_{222} \times 10^3$/deg cm^{2}dmol^{-1}	3.71	6.63	6.76
	$-\Theta_{min} \times 10^3$/deg cm^{2}dmol^{-1}	6.54	7.79	8.07
% Composition	α-helix	9	22	25
	β-sheet	36	21	18
	random coil	55	57	57

Fig. 3 Circular dichroism spectra of Q at pH 4, 8, and 10 shown in (**a**). Mean residue ellipticity (MRE) and secondary structure composition based on CD measurements are listed in (**b**). Data is averaged from at least two replicates and was obtained with 10 μM protein concentration. Adapted from Hume *et al.* (2014) with permission from the American Chemical Society

Table 1
Thermodynamic constants obtained from thermal melts of 10 μM Q, measured by CD

pH	Tm (°C)	ΔH° (kcal mol^{-1})[a]	ΔS° (kcal mol^{-1} K^{-1})[b]	ΔG° (kcal mol^{-1})[c]
4	55.3	−32.3	−97.1	−3.3
8	63.5	−32.6	−96.8	−3.8
10	46.4	−46.4	−145.2	−3.1

Adapted from Hume *et al.* (2014) with permission from the American Chemical Society
[a]van't Hoff enthalpy calculated as described in Subheading 4, **Note 11**
[b]At equilibrium, ΔG° = 0. Hence, the change in entropy ΔS° = ΔH°/Tm
[c]Free energy of folding at 25 °C calculated according to the expression ΔG° = ΔH° − TΔS°

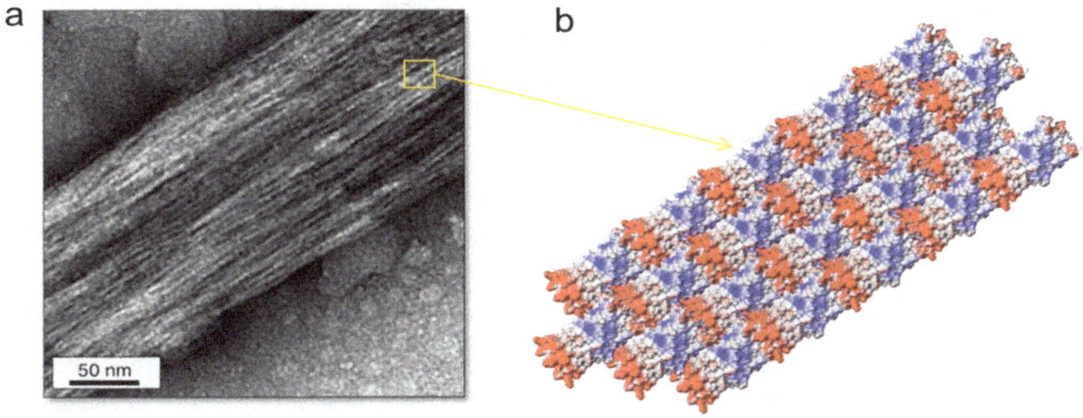

Fig. 4 Microscopy and modeling of protein fibers. (**a**) Transmission electron micrograph of Q fiber, 10 μM, pH 4. (**b**) Schematic representation of Q fiber assembly with staggered positive and negative regions of the pentamer. Scale bar in (**a**) represent 50 nm. Adapted from Hume *et al.* (2014) with permission from the American Chemical Society

3.4.2 Transmission Electron Microscopy (TEM)

TEM is employed to study if Q self-assembles into fibers (Fig. 4a, b) (*see* **Note 12**).

1. Prepare 10 μM protein samples in PB, pH 4, 8 and 10.
2. Place 3 μL of the sample on copper grids.
3. After 1 min, blot the excess solution using filter paper. Wash the grids with 3–4 drops of dH_2O.
4. Negatively stain the grids with 3 μL of 1% w/v uranyl acetate (filtered) for 1 min.
5. Blot the excess liquid and dry at room temperature for 10–15 min (*see* **Note 13**).
6. Acquire electron micrographs of protein fibers using Jeol JEM-1400 transmission electron microscope.
7. Analyze the dimensions of fibers using ImageJ software.

3.4.3 Attenuated Total Reflectance Fourier Transform Infrared Spectroscopy (ATR-FTIR)

To further determine the secondary structure characteristics of Q in its self-assembled fibers state, ATR-FTIR is used, which enables investigation of the protein's conformation in its solid/insoluble state (Table 2) (*see* **Note 14**).

1. Prepare 10 μM protein samples in PB, pH 4, 8, and 10. Add 5 μL of protein sample onto the diamond ATR surface.
2. Collect spectra (128 scans) at room temperature over a range of 4000–400/cm with 0.5/cm resolution using PerkinElmer System 2000 FT-IR with DuraSamplIR II T diamond ATR accessory, equipped with a MCT-A detector.
3. Perform a second derivative zero baseline correction of the amide I region (1700–1600/cm), and deconvolute the peaks with a Gaussian function.

3.5 Mesofibers Formation on Curcumin Binding

SAFs based on Q protein have the potential to leverage the therapeutic effects of curcumin, which suffers from low aqueous solubility and poor bioavailability [37]. Here, we describe the effect of curcumin on secondary structure, fiber formation, and assembly of Q (*see* **Notes 15** and **16**).

Table 2
Secondary structure conformation based on ATR-FTIR data for 10 μM Q in 50 mM PB

	Conformation		
Sample	**α-helix**	**β-sheet**	**Random coil**
pH 4	62	38	
pH 8	63	37	
pH 10	68	32	
Curcumin	79	10	12
Q	65	35	
Q+AuNP	29.9	42.9	27.2
Q+TFE	70.7	29.3	
Q+AuNP+TFE	36.8	36.3	26.9
Qx	44.6	26	29.4
Qx+AuNP	48.2	32.1	19.6
Qx+TFE	51.2	25	23.8

Percent composition was determined from relative areas of peaks fit to spectra (data not shown)

The percent composition is shown for Q at pH 4, 8, and 10, in the presence of curcumin at a 5:1 molar ratio with Q at pH 4. In addition, values are also shown for Q before and after templation with AuNP and in the presence of 25 v/v% TFE. Adapted from Hume et al. (2014) and Hume et al. (2015) with permission from the American Chemical Society

3.5.1 Curcumin Binding

1. Sonicate the 1 mM curcumin stock for ~15 min to remove any aggregates.
2. Incubate Q with varying concentrations (0–80 μM) of curcumin in PB at pH 4.
3. Perform wavelength scan (described in Subheading 3.4.1) of Q with varying molar ratios of curcumin.
4. Perform ATR-FTIR (described in Subheading 3.4.3) for 10 μM Q in PB at pH 4, in the presence of curcumin at a 5:1 molar ratio.

3.5.2 Confocal Microscopy

The fluorescent properties of curcumin can be exploited to visualize and measure the fibers dimensions under confocal microscope (Fig. 5a) (*see* **Note 17**). The 3D reconstructed image of Q fibers (Fig. 5b) shows curcumin being uniformly distributed throughout the fiber.

1. Set up the Leica TCS SP2 AOBS confocal microscope system (Leica, IL, USA) equipped with argon ion and HeNe lasers for excitation at 458 nm and detection window set between 465 and 560 nm
2. Scan the sample using a 63×/1.4 NA oil-immersion objective lens.
3. Reconstruct the images using ImageJ 64 1.43 along with Amira 5.43, using the Volren 3D rendering routine.

3.6 Q·AuNP Bionanocomposites

Conjugates of protein·AuNPs have a broad range of applications in biomedicine and drug delivery [38]. Below we describe templation of AuNP onto protein with (Q) and without (Qx) the hexahistidine

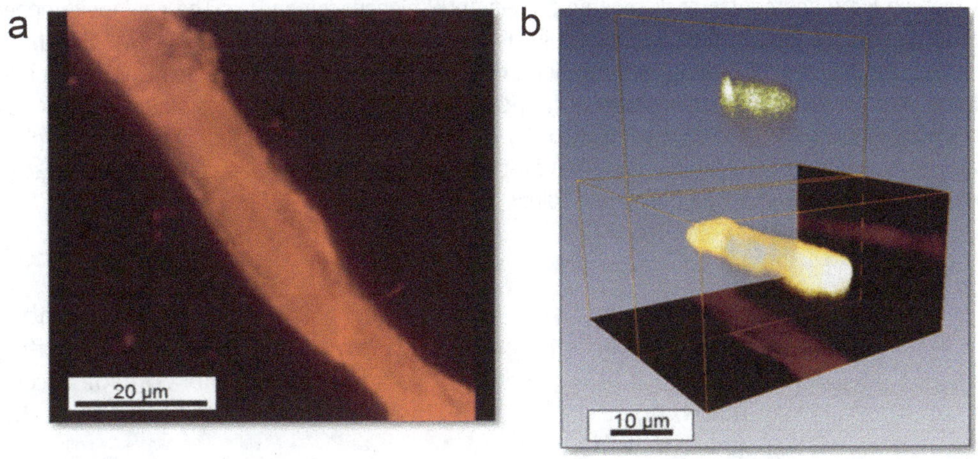

Fig. 5 (**a**) Reconstruction of 3D confocal XYZ data of 10 μM Q protein in the presence of 50 μM curcumin (5:1 molar ratio of curcumin: protein). (**b**) 3D representation of the same Q fiber showing XZ and YZ orthogonal views and an oblique slice cross-section above the 3D bounding box. Scale bars in (**a**), and (**b**) represent 20 μm, and 10 μm, respectively. Adapted from Hume *et al.* (2014) with permission from the American Chemical Society

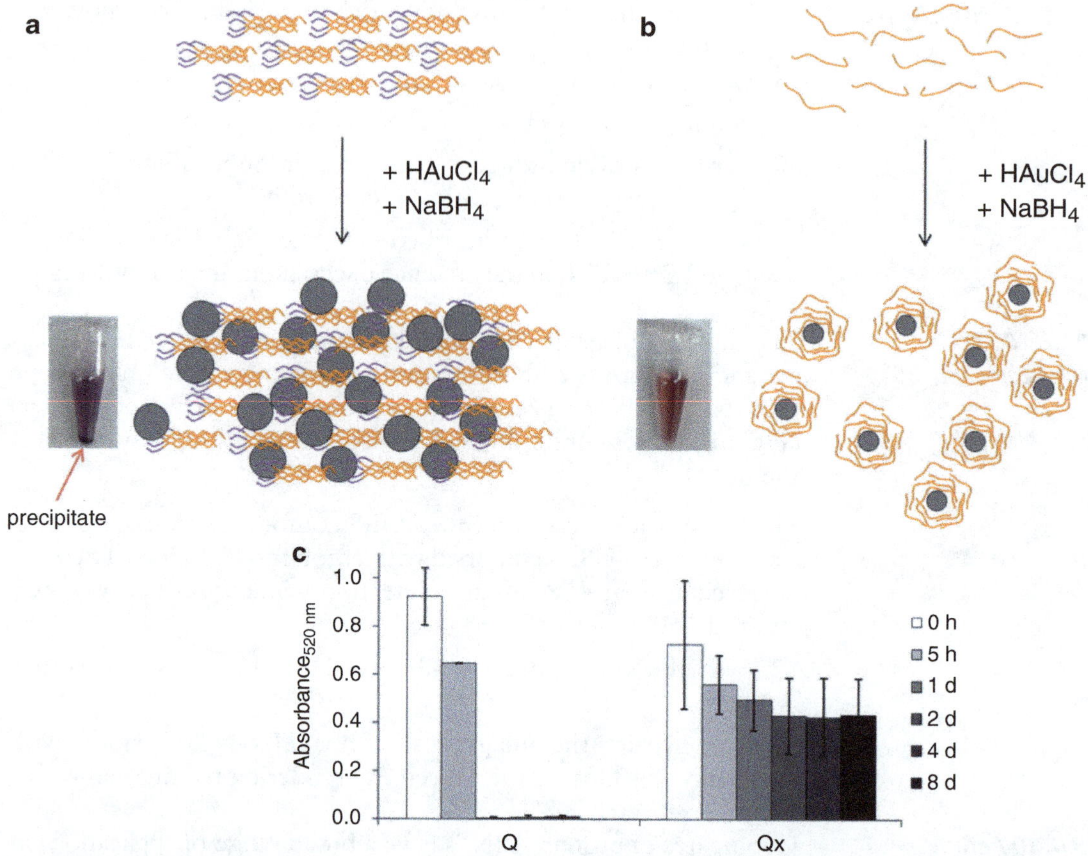

Fig. 6 Schematic illustration of protein and nanoparticle assembly upon gold templation. Protein units are illustrated in orange, His-tags in purple, and AuNPs as gray spheres. (**a**) α-helical protein Q with His-tag forms ordered pentamer assemblies, and upon gold templation exhibit a decrease in helical content and template AuNPs forming large aggregates of disordered protein·AuNP bionanocomposites. The photo with eppendorf tube demonstrates the formation of precipitates in Q·AuNP 24 h post templation. (**b**) Cleaved Q with His-tag removed is unstructured and upon gold templation exhibit an increase in helical content, surrounding AuNPs and stabilizing them in solution. The photo with eppendorf tube demonstrates characteristic pink color of protein stabilized Qx·AuNP 24 h post-templation. (**c**) Normalized absorbance of protein·AuNP complexes at 520 nm. Error bars represent standard deviation of triplicate measurements. Adapted from Hume *et al.* (2015) with permission from the American Chemical Society

tag (His-tag) to modify the macromolecular assembly of protein·AuNPs (Fig. 6). We also describe the effect of trifluoroethanol (TFE), which provides additional control over the macromolecular assembly by inducing and stabilizing alpha-helical structure of Q protein.

3.6.1 Cleavage of Protein Q to Yield Qx

To prepare Qx, the His-tag is cleaved off using Factor Xa. The IEGR motif engineered into the primary sequence is recognized by Factor Xa, which allows cleavage after the arginine residue.

1. Use the Factor Xa cleavage kit to cleave the His-tag off of the protein according to the manufacturer's protocol. Use 10 μM protein in a total volume of 204 μL and a protein–enzyme molar ratio of 50:1; the protein to 10× cleavage buffer volume ratio is maintained at 9:1.
2. Incubate the reaction mixture at room temperature for 4 h.
3. Prepare Ni-NTA beads as described in Subheading 3.3 **step 3**.
4. Incubate the reaction mixture with the prepared Ni-NTA beads and purify the cleaved protein (Qx). The Qx protein elutes out as flow-through while the His-tag remains bound to the beads.
5. Confirm the cleavage by SDS-PAGE (*see* **Note 18**).
6. Dialyze the isolated cleaved protein (Qx) against 2 L of PB, pH 8 at 4 °C for 4 h.
7. Determine the concentration of Qx using the BCA kit.

3.6.2 Generation of Protein·AuNP Bionanocomposites

1. Incubate the proteins (Q and Qx, respectively) with Au salt solution at 1:40 molar ratio for 15 min in a reaction volume of 1 mL. A final concentration of 10 μM protein (in PB, pH 8) is used with 400 μM Au salt solution.
2. Reduce the metal using sodium borohydride at a final concentration of 1 mM (*see* **Notes 19** and **20**).
3. Measure the absorbance of 10 μM protein with and without 400 μM Au salt solution, over a range of 400–700 nm with peak absorption at 520 nm. Take measurements immediately after reduction (**step 2**), after 5 h, and 1, 2, 4, and 8 days (*see* **Note 21**).
4. Perform CD (Subheading 3.4.1), TEM (Subheading 3.4.2), and ATR-FTIR (Subheading 3.4.3) analysis of Q and Qx in the presence and absence of AuNPs (*see* **Notes 22** and **23**). Perform additional experiments with TFE (*see* **Note 24**) (Fig. 7).

3.6.3 Characterization of Bionanocomposites

Described below are cyclic voltammetry (CV) and electrochemical impedance spectroscopy to evaluate the conductivity of the protein·AuNP bionanocomposites (*see* **Note 25**). The AuNP is templated with Q and the resultant bionanocomposites are deposited on an electrode surface.

1. Use a three-electrode system to study the electrochemical properties of protein·AuNP bionanocomposites.
   (a) Platinum wire as the counter electrode
   (b) Ag/AgCl reference electrode
   (c) Glassy-carbon disk electrode (d = 3.0 mm) as the working electrode

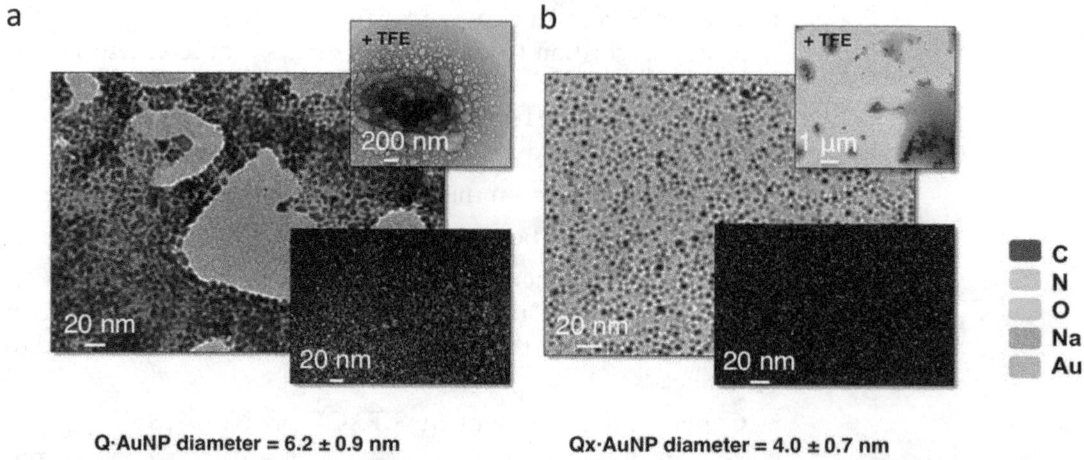

Fig. 7 Transmission electron micrographs and elemental maps for Q (**a**), and Qx (**b**) templated AuNP obtained 8 days after reduction. Proteins in the presence of AuNP in buffer containing 25 v/v % TFE are shown in the top insets. Area analyzed in elemental maps and corresponding TEM image overlap. Maps display location of signals resulting from C (purple), N (pink), oxygen (lime green), Na (light brown), and Au (orange). Scale bars for TFE insets are 200 nm for (**a**) and 1 μm for (**b**). Adapted from Hume *et al.* (2015) with permission from the American Chemical Society

2. Using samples prepared as described in Subheading 3.6.2, expose the disk electrode to varying concentrations of protein (4, 8, 10, 15, 25, and 35 μM) in the presence of AuNPs (*see* **Note 25**).
3. Immediately, record the CV data and Faradaic impedance spectra (Nyquist plots) in 50 mL Fe-cyanide solution at room temperature (*see* **Note 26**).
4. For Impedance measurements, record the data at a bias potential of 210 mV within the frequency range of 10^{-1}–10^{5} Hz (Fig. 8a).
5. For CV measurements, use a concentration of 8 μM protein at a scan rate of 0.1 mV/s with a voltage sweep between −0.8 and 0.9 V (Fig. 8b) (*see* **Notes 27** and **28**).

4 Notes

1. The gene encoding for COMPcc is cloned into the pQE9 vector at BamHI and HindIII sites to generate pQE9-COMPcc [29]. The GenInfo Identifier (i.e., GI#) for COMP is 25304. The DNA sequence for COMPcc is:

   TCA ATT GTG AGC GGA TAA CAA TTT CAC ACA GAA TTC ATT AAA GAG GAG AAA TTA ACT ATG AGA GGA TCG CAT CAC CAT CAC CAT CAC GGA TCC GGT GAC CTG GCG CCG CAG ATG CTG CGT GAA CTG CAG

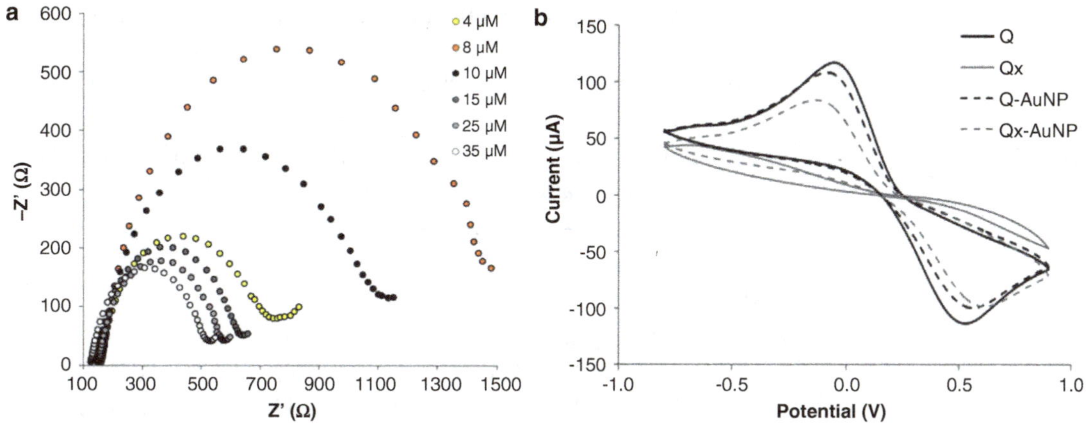

Fig. 8 (**a**) Nyquist plots for varying concentrations (4, 8, 10, 15, 25, and 35 μM) of proteins Q in the presence of AuNPs. Nyquist plots of Q·AuNP shows the electron-transfer resistance to be highest at 902.1 Ω when measured using a concentration of 8 μM. (**b**) Cyclic voltammograms of 8 μM Q, Qx, Q·AuNP, and Qx·AuNP. All electrochemical measurements were carried out in the presence of 10 mM $[Fe(CN)_6]^{3-}$ and 10 mM $[Fe(CN)_6]^{4-}$ in PB, pH 8, on glassy-carbon electrodes. Adapted from Hume *et al.* (2015) with permission from the American Chemical Society

Table 3
Primer sequences for PCR

	Primer sequence (5′→3′)
Q54 Fwd 1	GCATGGGATCCATCGAAGGTCGCGTTAAAGAAATCACCTTC
Q54 Fwd 2	CTGAAAAACACCGCGCCGCAGATGCTGCGT
Q54 Rev1	CATCTGCGGCGCGGTGTTTTTCAGGAAGG
Q54 Rev1	GCATGAAGCTTTGACTGCTGACGCAGCAGTTC

GAA ACC AAC GCG GCG CTG CAG GAC GTT CGT GAA CTG CTG CGT CAG CAG GTT AAA GAA ATC ACC TTC CTG AAA AAC ACC GTT ATG GAA TCT GAC GCG TCT GGT AAG CTT AAT TAG CTG AGC TTG GAC TCC TGT TGA TAG ATC CAG TAA TGA CCT CAG AAC TCC ATC TGG ATT

2. Design the primers carefully so that BamHI and HindIII sites are introduced into the Q DNA. The sequences for the primers used are given in Table 3.

3. The pQE30 plasmid vector used in our lab is a variant of the pQE30 vector containing an *E.coli* phenylalanyl-tRNA synthetase (PheRS**) mutation A294G/T251G [39]. However, we anticipate that commercially available pQE30, without PheRS**, can also be used.

4. The concentration of DNA is measured using the formula:

$$\text{dsDNA concentration} = 50\ \mu g / \text{mL} \times A_{260} \times \text{dilution factor}$$

Lyophilize the DNA for 3 h if the concentrations of DNA fragments Q1 and Q2 are less than 33 ng/μL.

5. The optimal vector to insert molar ratio for ligation varies from 1:1 to 1:10. For optimum ligation of Q, use a temperature of 16 °C.
6. The DNA sequence for the generated Q is: GGATCCATCGAAGGTCGCGTTAAAGAAATCACCT TCCTGAAAAACACCGCGCCGCAGATGCTGCGTGAAC TGCAGGAAACCAACGCGGCGCTGCAGGACGTTCGT GAACTGCTGCGTCAGCAGTCAAAGCTT.
7. A prominent band of 6.3 kD in the post-expression sample indicates expression of Q protein.
8. A freeze/thaw cycle at −80 °C, prior to sonication, facilitates cell lysis. The cell suspension should be maintained on an ice bath or ice water during sonication.
9. Typically, Q is eluted in the range of 100–500 mM imidazole at 4 °C.
10. The Q construct retains the α-helical conformation (Fig. 3a), with the helicity being slightly reduced compared to the COMPcc parent at pH 4, 8, and 10. This can be attributed to the displacement of a proline residue to the center of the protein. The helical content of Q increases from pH 4 to pH 10 (Fig. 3b).
11. To assess the stability of Q at different pH conditions, temperature scans at 222 nm are used to calculate the thermodynamic parameters. The reversibility of Q, confirmed by performing forward and reverse melts in the range of 20–85 °C, allows us to calculate T_m, $\Delta G°$, $\Delta H°$, and $\Delta S°$ using Van't Hoff analysis [40]. Compared to COMPcc, Q exhibited higher T_m at acidic and neutral pH, while showing a similar T_m at pH 10. It could be envisaged that surface charge distribution of Q contributes to its stabilization at different pH conditions. Van't Hoff analysis is performed based on the method described by Greenfield [40]:

    The Gibbs free energy of folding ($\Delta G°$) is given by:

$$\Delta G^{\circ} = -RT \ln K \tag{1}$$

Reworking Eq. 1 we obtain:

$$\ln K = \frac{-\Delta G^{\circ}}{RT} = \frac{-\Delta G^{\circ}}{R}\left(\frac{1}{T}\right)$$

Plot of ln*K* versus T^{-1} (K^{-1}) provides a linear relationship between ln*K* and *T*, allowing us to calculate standard enthalpy change (Δ °) using the following equation:

$$\Delta H^{\circ} = -\text{slope} \times R \tag{2}$$

12. Although Q forms fibers at all the pH conditions tested, numerous fibers are observed at acidic pH (Fig. 4a). Owing to the surface charge distribution of Q (shown in the Fig. 4b), the protofibrils bundle together to form large fibers. TEM reveals protofibrils with diameters of 3.5 ± 0.7 nm (n = 210), while the diameters of the resulting fibers varies from tens to hundreds of nanometers (20–560 nm, n = 14). Hundreds of protofibrils are assembled laterally to form fibers with high aspect ratio.
13. For preparing TEM grids, the samples should be air dried for 10–15 min and carefully blotted by using the filter paper along the edge of the grid.
14. A significant peak in the band region of 1650–1653/cm appears which corresponds to the classical alpha-helical band. Consistent with our CD data, the helical content increases from 62% to 68% with an increase in pH from pH 4 to pH 10.
15. With an increasing concentration of curcumin, CD spectra reveal an increase in the alpha-helical structure of Q, as depicted by an increase in MRE values at 222 nm. Furthermore, a sharp peak at 1653/cm, accompanied with an increase in helical content from 62% to 79% are observed when curcumin bound Q is examined using ATR-FTIR. Overall, curcumin stabilizes the coiled-coil assembly of Q.
16. The following precautions should be taken while working with curcumin:
    (a) A fresh stock solution should be prepared for each use, and preferably use the solution the same day.
    (b) Curcumin is light sensitive, and should be protected from direct exposure to light. Prepare the solution in amber colored vials or cover the vials with aluminum foil.
    (c) Curcumin can permanently stain the work surfaces. Careful housekeeping is necessary while working with curcumin. Neoprene gloves are recommended for handling curcumin.
17. Large fibers averaging a diameter of 16.0 ± 5.6 μm (n = 22) are observed for Q in the presence of curcumin (Fig. 5a). The change from nanoscale fibers to mesoscale fibers is consistent with our ATR-FTIR data, confirming the stabilizing role of curcumin on fiber assembly.

18. The cleaved protein Qx appears at a molecular weight of 4.4 kD.
19. The molar ratio of protein to Au to sodium borohydride is kept constant at a ratio of 1:40:100. Sodium borohydride is very hygroscopic, and must be stored away from moisture. The solutions should be prepared fresh every time.
20. Q forms large aggregates that precipitate out of solution, with the dark purple color of the solution waning after a few hours. On the other hand, Qx solvates AuNPs, stabilizing them in solution, maintaining a pink hue over an extended period of time (Fig. 6a, b). In the absence of protein, AuNPs exhibit a similar hue as the Q protein. The Q·AuNP clusters together to form large aggregates while AuNPs (in buffer alone) remained as discrete particles, as confirmed via TEM.
21. Q exhibits higher absorbance compared to Qx, when measured immediately after the reduction step. After 5 h, Q shows a dramatic decrease in absorbance with values dropping close to zero after 1 day (Fig. 6c). This is attributed to the aggregation of protein·AuNPs forming large assemblies, and consequently causing the solution to lose its color after a few hours. On the contrary, absorbance with Qx is stabilized after 1 day, maintaining 70% of its original value for the remainder of the period studied.
22. Upon templation, the peak deconvolution of ATR-FTIR data indicates the decrease in helical content of Q from 65% to 30% (Table 2). Upon cleavage of the His-tag, Q experiences some loss in secondary structure with Qx exhibiting a decrease in helical content to 45%. Upon templation, Qx demonstrates a modest increase in helicity to 48%.
23. TEM samples with AuNPs are not stained with uranyl acetate. TEM analysis shows Qx as sheet-like structures as opposed to Q forming high aspect ratio fibers. In the presence of AuNP, aggregation is observed with Q·AuNP having a diameter of 6.2 ± 0.9 nm while nanoparticles associated with cleaved protein are randomly dispersed and measured at 4.0 ± 0.7 nm (Fig. 7). Elemental mapping reveals that proteins and AuNPs exist together with proteins enveloping the gold nanoparticles.
24. To study the effect of TFE, experiments are performed by using 25 v/v % TFE and PB, pH 8, with 10 μM protein in the presence and absence of AuNPs. Upon addition of TFE, the helical content of Q and Qx are increased to 70% and 51%, respectively (Table 2). Interestingly, in the presence of AuNP and TFE, Qx assembles to form aggregates as judged by TEM micrographs (Fig. 6). It can be envisaged that TFE induces structural changes that drives Qx to assemble into aggregates, instead of forming soluble dispersed nanoparticles.

25. The Q·AuNP is readily deposited on the electrode surface, while the Qx·AuNP remains soluble.
26. For CV and electrochemical impedance spectroscopy measurements, degas/purge the solutions for 10 min, prior to use.
27. A concentration of 8 μM protein is chosen for CV measurements as it provides peak resistance values in impedance measurements (Fig. 8a).
28. Upon templation with AuNP, a decrease in current equivalent to 9.0 μA is observed for Q, whereas an increase in peak current of 39.9 μA is observed for Qx·AuNP (Fig. 8b). The presence of His-tag allows Q·AuNP to form a film on electrode surface that decreases the electron transfer at the electrode surface, while for the solvated Qx·AuNP, the overall electron transfer is increased.

Acknowledgements

This work was supported by the National Science Foundation under award number NSF DMR-1728858.

References

1. Mann S (1993) Molecular tectonics in biomineralization and biomimetic materials chemistry. Nature 365(6446):499–505
2. Busseron E, Ruff Y, Moulin E et al (2013) Supramolecular self-assemblies as functional nanomaterials. Nanoscale 5(16):7098–7140. https://doi.org/10.1039/C3NR02176A
3. Stephanopoulos N, Ortony JH, Stupp SI (2013) Self-assembly for the synthesis of functional biomaterials. Acta Mater 61(3):912–930. https://doi.org/10.1016/j.actamat.2012.10.046
4. Woolfson DN, Ryadnov MG (2006) Peptide-based fibrous biomaterials: Some things old, new and borrowed. Curr Opin Chem Biol 10(6):559–567. https://doi.org/10.1016/j.cbpa.2006.09.019
5. Maham A, Tang Z, Wu H et al (2009) Protein-based nanomedicine platforms for drug delivery. Small 5(15):1706–1721. https://doi.org/10.1002/smll.200801602
6. Yu YB (2002) Coiled-coils: stability, specificity, and drug delivery potential. Adv Drug Deliv Rev 54(8):1113–1129
7. Zuccola HJ, Rozzelle JE, Lemon SM et al (1998) Structural basis of the oligomerization of hepatitis delta antigen. Structure 6(7):821–830
8. Gunasekar SK, Haghpanah JS, Montclare JK (2008) Assembly of bioinspired helical protein fibers. Polym Adv Technol 19(6):454–468. https://doi.org/10.1002/pat.1136
9. Bromley EH, Channon KJ, King PJ et al (2010) Assembly pathway of a designed alpha-helical protein fiber. Biophys J 98(8):1668–1676. https://doi.org/10.1016/j.bpj.2009.12.4309
10. Pandya MJ, Spooner GM, Sunde M et al (2000) Sticky-end assembly of a designed peptide fiber provides insight into protein fibrillogenesis. Biochemistry 39(30):8728–8734
11. Smith AM, Acquah SF, Bone N et al (2004) Polar assembly in a designed protein fiber. Angew Chem Int Ed Engl 44(2):325–328. https://doi.org/10.1002/anie.200461599
12. Indelicato G, Wahome N, Ringler P et al (2016) Principles governing the self-assembly of coiled-coil protein nanoparticles. Biophys J 110(3):646–660. https://doi.org/10.1016/j.bpj.2015.10.057
13. Pamies P (2013) Coiled-coil balls. Nat Mater 12(6):472–472. https://doi.org/10.1038/nmat3680
14. Fletcher JM, Harniman RL, Barnes FR et al (2013) Self-assembling cages from coiled-coil peptide modules. Science 340(6132):595–599. https://doi.org/10.1126/science.1233936

15. Gradišar H, Božič S, Doles T et al (2013) Design of a single-chain polypeptide tetrahedron assembled from coiled-coil segments. Nat Chem Biol 9(6):362–366. https://doi.org/10.1038/nchembio.1248
16. Kojima S, Kuriki Y, Yoshida T et al (1997) Fibril formation by an amphipathic α-helix-forming polypeptide produced by gene engineering. Proc Jpn Acad Ser B 73(1):7–11. https://doi.org/10.2183/pjab.73.7
17. Papapostolou D, Bromley EH, Bano C et al (2008) Electrostatic control of thickness and stiffness in a designed protein fiber. J Am Chem Soc 130(15):5124–5130. https://doi.org/10.1021/ja0778444
18. Ryadnov MG, Bella A, Timson S et al (2009) Modular design of peptide fibrillar nano- to microstructures. J Am Chem Soc 131(37):13,240–13,241. https://doi.org/10.1021/ja905539h
19. Ogihara NL, Ghirlanda G, Bryson JW et al (2001) Design of three-dimensional domain-swapped dimers and fibrous oligomers. Proc Natl Acad Sci U S A 98(4):1404–1409. https://doi.org/10.1073/pnas.98.4.1404
20. Saha K, Agasti SS, Kim C et al (2012) Gold nanoparticles in chemical and biological sensing. Chem Rev 112(5):2739–2779. https://doi.org/10.1021/cr2001178
21. Daniel MC, Astruc D (2004) Gold nanoparticles: assembly, supramolecular chemistry, quantum-size-related properties, and applications toward biology, catalysis, and nanotechnology. Chem Rev 104(1):293–346. https://doi.org/10.1021/cr030698+
22. Stevens MM, Flynn NT, Wang C et al (2004) Coiled-coil peptide-based assembly of gold nanoparticles. Adv Mater 16(11):915–918. https://doi.org/10.1002/adma.200306430
23. Bayraktar H, Srivastava S, You C-C et al (2008) Controlled nanoparticle assembly through protein conformational changes. Soft Matter 4(4):751–756. https://doi.org/10.1039/B716386J
24. Tullman JA, Finney WF, Lin Y-J et al (2007) Tunable assembly of peptide-coated gold nanoparticles. Plasmonics 2(3):119–127. https://doi.org/10.1007/s11468-007-9033-z
25. Aili D, Enander K, Rydberg J et al (2008) Folding induced assembly of polypeptide decorated gold nanoparticles. J Am Chem Soc 130(17):5780–5788. https://doi.org/10.1021/ja711330f
26. Gunasekar SK, Anjia L, Matsui H et al (2012) Effects of divalent metals on nanoscopic fiber formation and small molecule recognition of helical proteins. Adv Funct Mater 22(10):2154–2159. https://doi.org/10.1002/adfm.201101627
27. Malashkevich VN, Kammerer RA, Efimov VP et al (1996) The crystal structure of a five-stranded coiled coil in COMP: a prototype ion channel? Science 274(5288):761–765
28. Guo Y, Bozic D, Malashkevich VN et al (1998) All-trans retinol, vitamin D and other hydrophobic compounds bind in the axial pore of the five-stranded coiled-coil domain of cartilage oligomeric matrix protein. EMBO J 17(18):5265–5272. https://doi.org/10.1093/emboj/17.18.5265
29. Gunasekar SK, Asnani M, Limbad C et al (2009) N-terminal aliphatic residues dictate the structure, stability, assembly, and small molecule binding of the coiled-coil region of cartilage oligomeric matrix protein. Biochemistry 48(36):8559–8567. https://doi.org/10.1021/bi900534r
30. McFarlane A, Orriss G, Okun N et al (2012) The pentameric channel of COMPcc in complex with different fatty acids. PLoS One 7(11):e48130. https://doi.org/10.1371/journal.pone.0048130
31. Wilken R, Veena MS, Wang MB et al (2011) Curcumin: a review of anti-cancer properties and therapeutic activity in head and neck squamous cell carcinoma. Mol Cancer 10:12. https://doi.org/10.1186/1476-4598-10-12
32. Zsila F, Bikadi Z, Simonyi M (2004) Induced circular dichroism spectra reveal binding of the antiinflammatory curcumin to human alpha1-acid glycoprotein. Bioorg Med Chem 12(12):3239–3245. https://doi.org/10.1016/j.bmc.2004.03.074
33. Rai D, Singh JK, Roy N et al (2008) Curcumin inhibits FtsZ assembly: an attractive mechanism for its antibacterial activity. Biochem J 410(1):147–155. https://doi.org/10.1042/BJ20070891
34. Panchatcharam M, Miriyala S, Gayathri VS et al (2006) Curcumin improves wound healing by modulating collagen and decreasing reactive oxygen species. Mol Cell Biochem 290(1–2):87–96. https://doi.org/10.1007/s11010-006-9170-2
35. Hume J, Sun J, Jacquet R et al (2014) Engineered coiled-coil protein microfibers. Biomacromolecules 15(10):3503–3510. https://doi.org/10.1021/bm5004948
36. Hume J, Chen R, Jacquet R et al (2015) Tunable conformation-dependent engineered protein·gold nanoparticle nanocomposites. Biomacromolecules 16(6):1706–1713. https://doi.org/10.1021/acs.biomac.5b00098

37. Prasad S, Tyagi AK, Aggarwal BB (2014) Recent developments in delivery, bioavailability, absorption and metabolism of curcumin: the golden pigment from golden spice. Cancer Res Treat 46(1):2–18. https://doi.org/10.4143/crt.2014.46.1.2

38. Arvizo R, Bhattacharya R, Mukherjee P (2010) Gold nanoparticles: opportunities and challenges in nanomedicine. Expert Opin Drug Deliv 7(6):753–763. https://doi.org/10.1517/17425241003777010

39. Datta D, Wang P, Carrico IS et al (2002) A designed phenylalanyl-trna synthetase variant allows efficient in vivo incorporation of aryl ketone functionality into proteins. J Am Chem Soc 124(20):5652–5653. https://doi.org/10.1021/ja0177096

40. Greenfield NJ (2006) Using circular dichroism collected as a function of temperature to determine the thermodynamics of protein unfolding and binding interactions. Nat Protoc 1(6):2527–2535. https://doi.org/10.1038/nprot.2006.204

Chapter 20

Preparing Genes for Repetitive Elastin-Like Polypeptides Using Gibson Assembly

James K. Deyling, Edward M. Turk, and Nolan B. Holland

Abstract

Elastin-like polypeptides (ELPs) are stimuli-responsive polymers consisting of peptide repeat units from the elastin protein. Most commonly these materials are biosynthesized in *E. coli* using genes that have been built by standard molecular biology techniques. The inherent repetitive nature of these polypeptides makes the construction of genes that encode them more challenging than for typical recombinant proteins. Described here is a robust technique to produce genes encoding long chains of ELP through successive rounds of Gibson assembly that nearly double the ELP length with each cloning iteration. Using Gibson assembly to produce the genes for high molecular weight ELPs requires fewer steps and significantly less time than current techniques. This approach is modular and allows for the flexible assembly of ELPs with various compositions. The procedure can be adapted for preparing genes encoding other repetitive proteins.

Key words Gibson assembly, Elastin-like polypeptides, Molecular biology, Ligation, High molecular weight proteins, Repetitive sequence proteins

1 Introduction

Elastin-like polypeptides (ELPs) are stimuli-responsive synthetic biopolymers that have applications as hydrogel scaffolds for tissue engineering, as drug-delivery systems, and as purification tags for recombinant proteins [1, 2]. A prominent feature of ELPs is that they exhibit an inverse transition temperature (T_t). Below the T_t the polypeptide is completely soluble, while above the T_t the system separates into two phases: a polymer rich coacervate and a polymer poor aqueous phase. This transition can be affected by the properties of the polymer (the amino acid sequence, length, molecular architecture) and the solution conditions (pH, ionic strength, concentration), resulting in their responsive behavior.

The fundamental structure of ELPs is the repeat of the pentapeptide sequence Glycine–Valine–Glycine–Valine–Proline (GVGVP). Variations of this repeat unit, through substitution of

Andrew K. Udit (ed.), *Protein Scaffolds: Design, Synthesis, and Applications*, Methods in Molecular Biology, vol. 1798, https://doi.org/10.1007/978-1-4939-7893-9_20,

either of the valine residues, can retain the general characteristic properties of the material, but change the conditions for which the polymer is soluble [3]. The relative hydrophobicity of the guest residues and the number of repeats impact the T_t value for a particular ELP, making them a versatile polymer.

Recombinant DNA technology has made many advances over the last 20 years and has significantly impacted the study of proteins. Traditionally, DNA sequences possessing matching sticky or blunt ends made by enzymatic restriction were the only building blocks for a new piece of synthetic DNA [4]. Currently, standard commercial gene synthesis combines overlapping oligonucleotides to assemble genes. However, this requires unique sequences which preclude the construction of repetitive genes such as those encoding ELPs. Different approaches have been utilized to synthesize repetitive genes, including recursive directional ligation [5] and golden-gate cloning [6]. However, these approaches can require cleanup steps to remove undesired fragments to self-ligation, multiple modifications to anterior and posterior regions of the sequences, and/or polymerase chain reaction (PCR) for gene amplification. With the Gibson assembly system, no cleanup steps are required at any stage, a single modification to a gene is used to prepare the modular construct, and no time-consuming PCR steps are used during cloning [7].

In this protocol, we demonstrate the preparation of repetitive ELP gene structures that can effectively be doubled in length through Gibson assembly and can be modified to have desired terminal modifications. This is accomplished by successive rounds of cloning, the first of which primes the sequence with new homologous sites by the addition of a double stranded oligonucleotide. The second and subsequent rounds of cloning use only the product from the first round of cloning, by excising the entire sequence from one aliquot of plasmid by double digestion and opening another aliquot by single digestion, and ligating the excised piece into the opened piece. This process may be repeated indefinitely to continuously produce repetitive sequences essentially doubled in length [8].

2 Materials

Prepare all solutions with nuclease-free water and molecular biology grade reagents. All reagents should be prepared and stored at room temperature, unless otherwise specified. The methods described work well for our group but are not necessarily the only way to prepare these materials.

```
A.            i.                    v.
                          M  G  H  G  V  G  V  P  G  V  G  V  P  G  V  G  V  P
5'-TTTAAGAAGGAGATATACATATGGGCCACGGCGTGGGTGTTCCGGGCGTAGGTGTCCCAGGTGTGGGCGTACCG...
3'-AAATTCTTCCTCTATATGTATACCCGGTGCCGCACCCACAAGGCCCGCATCCACAGGGTCCACACCCGCATGGC...
                      NdeI      PflMI

                  iii.
   G  V  G  V  P  G  V  G  V  P  G  W  P  *  *
...GGCGTTGGTGTTCCTGGTGTGGGCGTGCCGGGCTGGCCGTGATAATTCGAGCTCCGTCGACAAGCTTGCGGCCGCACT-3'
...CCGCAACCACAAGGACCACACCCGCACGGCCCGACCGGCACTATTAAGCTCGAGGCAGCTGTTCGAACGCCGGCGTGA-5'
                                 BglI

B.         i.             ii.             iii.            iv.             v.
                     M  G  C  N  H  G  V  G  V  P  G  E  G  V  P  G  V  G  V  P
5'-AGAAGGAGATATACATATGGGTTGTAACCACGGTGTGGGCGTGCCGGGCGAAGGTGTTCCTGGCGTGGGTGTTCCG-3'
3'-TCTTCCTCTATATGTATACCCAACATTGGTGCCACACCCGCACGGCCCGCTTCCACAAGGACCGCACCCACAAGGC-5'
                 NdeI            PflMI
```

Fig. 1 The DNA sequences utilized for the gene assembly. The original ELP gene (**A**) and the designed insert (**B**) corresponding to I and II of Fig. 2, respectively. The restriction enzyme recognition sites are underlined and overlaps for Gibson assembly are shaded

1. *E. coli* containing plasmid with elastin-like gene to be repeated: The bacteria are ideally from individual colonies grown on agar plates, but can be obtained from a frozen stock. The gene utilized in this example is based on a previously reported elastin-like polypeptide sequence (Fig. 1) utilized for directional ligation methods [9].
2. Double-stranded DNA (dsDNA): Fragments of dsDNA with the appropriate sequences (Subheading 3.1) can be directly purchased from any number of vendors. If the DNA fragments are short, complimentary single stranded oligonucleotides can be purchased and annealed to obtain the desired dsDNA.
3. Buffer: 1 M Tris–HCl, pH 8.0. Add 10 mL of water to a glass beaker. Weigh 12.1 g of Tris–HCl and add it to the beaker. Add water to a volume of 90 mL. Mix and adjust to pH 8.0 with HCl. Make up the volume to 100 mL with water. Refrigerate at 4 °C.
4. Salt solution: 50 mM $MgCl_2$. Add 10 mL of water to a glass beaker. Weigh 476 mg of $MgCl_2$ and add it to the beaker. Make up the volume to 100 mL with water. Refrigerate at 4 °C.
5. Lysogeny broth (LB): Weigh 5 g of yeast extract, 5 g of NaCl, and 10 g of peptone and add to a sterile 2 L Erlenmeyer flask. Mark with autoclave tape between the fluid and air layers on the flask. Cover loosely with aluminum foil. Add 1 L of deionized (DI) water and autoclave at an appropriate level until autoclave tape indicates temperature has been reached. Let cool to 50 °C, and then add 100 mg of ampicillin sodium salt. Store at 4 °C.
6. Selection plates: Make LB medium as above, with the addition of 15 g of agar to the Erlenmeyer flask before autoclaving.

After adding the ampicillin to the autoclaved media, pour approximately 15 mL of fluid media into sterile 10 cm diameter plates (*see* **Note 1**).

7. Glycerol solution: Fill a flask with 65 mL of 100% glycerol. Weigh out 1.2 g of $MgSO_4$ and 0.3 g of Tris–HCl and add to the flask. Fill the flask to 100 mL with DI water. Autoclave to sterilize.
8. 50 mL culture tubes.
9. Centrifuge.
10. Qiagen Spin Miniprep kit.
11. Petri dishes, 10 cm diameter.
12. Qubit fluorometer and assay kit.
13. 2 mL screw-cap conical tubes.
14. Gel electrophoresis system with loading and running buffers and precast 2% agarose gels, and imager.
15. DNA ladder.
16. PCR reagents: Q5 High-fidelity 2× Master Mix (New England Biolabs), T7 Promoter and T7 Terminator primers at 10 μM.
17. 5-alpha competent *E. coli* (High Efficiency) (New England Biolabs).
18. BL21(DE3) competent *E. coli* (New England Biolabs).
19. NEBuilder HiFi Master Mix 2× (New England Biolabs).
20. PflMI restriction enzyme.
21. NdeI restriction enzyme.
22. 10× NEBuffer 3.1 (New England Biolabs).

3 Methods

Carry out all reactions at room temperature, unless otherwise specified.

3.1 Gene Design

1. Obtain the sequence of the gene in a plasmid that encodes the protein that is to be repeated (*see* **Note 2**).
2. Define the region that is to be repeated, which needs to be flanked by restriction enzyme sites. For this example, we start with a pET 20b plasmid with five repeats of GVGVP flanked by the restriction sites PflMI and BglI (Fig. 1A) [9].
3. Determine the oligonucleotide sequence (Fig. 1B) to be utilized in the first round of cloning, which introduces the DNA homology necessary for Gibson assembly in subsequent rounds of protein length doubling (Fig. 2, Structure II):
   (a) Take 15 to 25 bases prior to the first restriction enzyme cut site of the plasmid in **step 1**. These will make up region i (Fig. 2) that overlaps with the receiving plasmid.

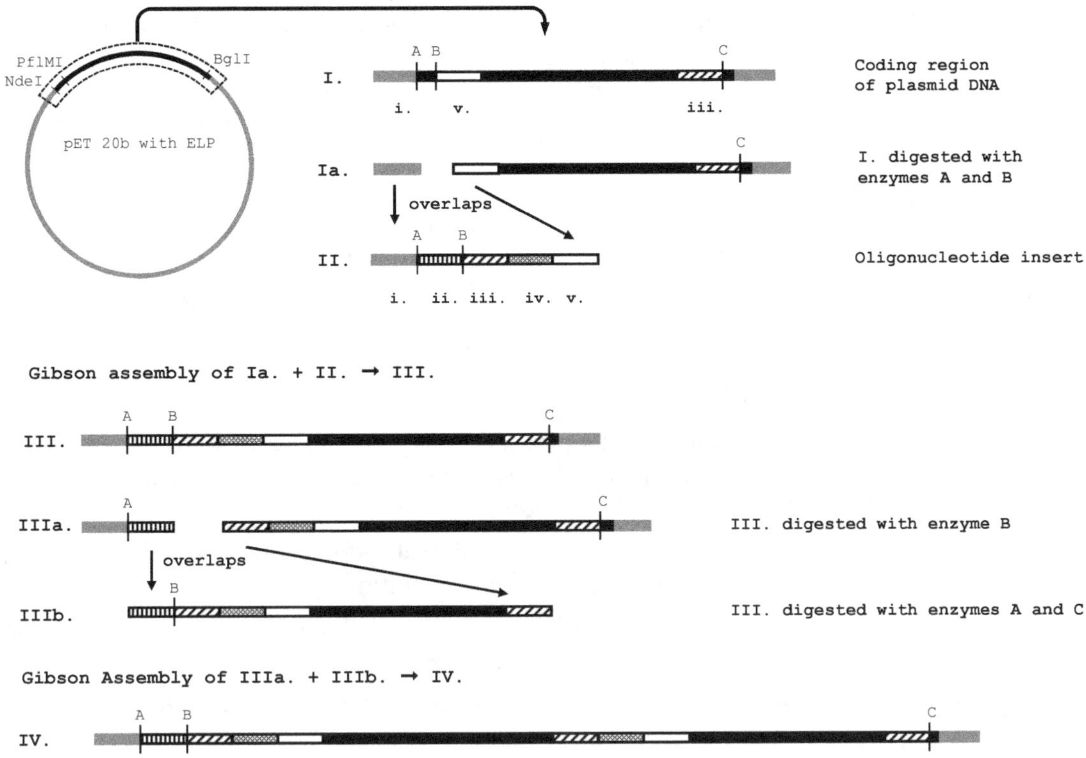

Fig. 2 Schematic steps of assembly of elastin-like polypeptide genes. The original plasmid encoding the elastin-like polypeptide (I) is based on reference [9]. The shaded regions in the designed insert (II) represent DNA sequences that can be used as overlaps for Gibson assembly or for adding DNA to the gene, including overlap for the pET 20b sequence before the NdeI cut site (i), region of new DNA that produces a new overlap sequence (ii), overlap sequence that matches the final pentapeptide of the ELP gene (iii), optional region containing any new DNA (iv), and overlap sequence that matches the initial pentapeptide of the ELP gene (v). Restriction enzyme recognition sequences for NdeI, PflMI, and BglI are denoted by A, B, and C, respectively. The final product (IV) can be used as the fragment III to again double the gene

(b) Incorporate two restriction enzyme cut sites (A and B) separated by 15 to 25 bases (region ii). We have utilized cut-sites for NdeI and PflMI enzymes, which are compatible with previously reported directional ligation methods [9]. The bases in this region will remain to code for N-terminal residues of interest (in our example this includes a cysteine group).

(c) Add the 15 to 25 bases prior to the terminal restriction enzyme cut site of the plasmid in **step 1** (region iii).

(d) Add any number of additional bases after the second restriction enzyme (region iv). These bases end up between cut sites B and C in construct III (Fig. 2) and will be included in the repeated sequence (*see* **Note 3**).

(e) Add the 15 to 25 bases that are found after the third restriction enzyme cut site of the plasmid in **step 1** (region v).

4. Verify that the oligonucleotide sequence obtained has a start codon, that all of the desired codons are in the proper reading frame, and that the DNA translates to the proper amino acid sequence. This gene fragment can be ordered as double stranded.

3.2 Overnight Culture and Plasmid Preparation

1. Place 10 mL of LB media into a culture tube and inoculate with bacteria containing plasmid from a single colony selected from an agar plate (*see* **Note 4**). Place into an incubator shaker for 16 h at 37 °C and 275 rpm.
2. Pellet the bacteria from the media at 8500 × *g* for 3 min at 15 °C.
3. Miniprep the plasmid using the Qiagen spin miniprep kit according to the supplied directions, with the following deviations (*see* **Note 5**):
   (a) For the alcohol removal step, centrifuge for a total of 15 min to dry the spin column (*see* **Note 6**).
   (b) For the elution step, elute with 34 μL of elution buffer (*see* **Note 7**). Store refrigerated.
4. Use a Qubit fluorometer and assay kit to quantify the plasmid yield (*see* **Note 8**).

3.3 First Round Cloning

This section describes the assembly of construct III in Fig. 2.

1. Obtain a PCR tube and add to it 1 μL of PflMI restriction enzyme, 1 μL of NdeI restriction enzyme, 1 μL of 10× NEBuffer 3.1, and a volume of plasmid isolated in Subheading 3.2 and nuclease-free water, bringing the tube to a final volume of 10 μL and a plasmid concentration of 20 ng/μL (*see* **Note 9**).
2. Mix the reaction by pipetting up and down and incubate overnight at 37 °C.
3. Heat inactivate the enzymes at 65 °C for 1 h. Refrigerate until use.
4. To another PCR tube add 10 μL of NEBuilder HiFi Master Mix 2×, 0.5 μL of a 10 μM stock of dsDNA insert, and a volume of digested plasmid and nuclease-free water bringing the tube to a volume of 20 μL and a plasmid concentration of 2 ng/μL. Pipette to mix the contents of the tube, incubate at 50 °C for 15 min, and then refrigerate until ready for transformation.
5. Transform the DNA by following the NEB 5-alpha competent *E. coli* (High Efficiency) transformation protocol as provided by the manufacturer, using 2 μL of the HiFi assembly product and spreading 100 μL of cells onto a selection plate.

6. Allow the plates to grow overnight in an incubator at 37 °C (*see* **Note 10**).
7. Refrigerate plates with colonies.

3.4 PCR Screening and Sequencing

1. Obtain a new selection plate. Make and number circles 1–9 on the glass face exterior of the plate.
2. Obtain microcentrifuge tubes equivalent to the number of plates and label them sequentially, filling each with 25 μL of DI water.
3. Select colonies from a selection plate (prepared in Subheading 3.3) by stabbing each colony with a fresh micropipette tip, then stab the respective circle on the newly labeled plate, and then finally inoculate the water with the similarly numbered tube. Place the new plate into a 37 °C incubator overnight.
4. Bring 250 mL of water to a boil on a hot plate, and drop the numbered tubes into the water, boiling them for 2 min.
5. Remove the tubes from the boiling water and centrifuge them at 15000 × *g* for 2 min. Take 2 μL of each supernatant and add to separate PCR tubes.
6. Prepare a standard PCR reaction within each tube, adding 12.5 μL of Q5 High-fidelity 2× Master Mix, 1.25 μL of 10 μM T7 Promoter primer, 1.25 μL of 10 μM T7 Terminator primer, and 8 μL nuclease-free water. Prepare one additional control tube using 2 μL of plasmid from the original miniprep.
7. Conduct a PCR with the parameters listed in Table 1 to run 35 cycles of amplification.
8. Take 5 μL of each PCR reaction and add to separate tubes loaded with 15 μL of loading buffer and vortex. Take 5 μL of DNA ladder and add to another tube with 15 μL of loading buffer and vortex. Pipette the ladder mix into the marker well of a 2% agarose precast gel and the samples and control into wells 1–10.

Table 1
Parameters for the PCR

Step	Temperature (°C)	Time (s)	
Initial denaturation	98	30	
Denaturation	98	8	Repeat 34×
Annealing	51	30	
Extension	72	120	
Hold	4	Infinite	

9. Run the gel until the leading dye front travels three-quarters of the way through the gel.
10. Compare samples with the ladder. Verify that the control is at an appropriate location compared to the ladder, and determine which samples are larger than the control (*see* **Note 11**).
11. Take samples that present the proper level of DNA chain size change and prepare a plasmid preparation from an overnight culture (Subheading 3.2) of the respective colony that was grown on the plate stabbed at the beginning of the PCR colony screening.
12. Mix either a T7 promoter or T7 terminator primer at 2.5 μM and 4 μL with 8 μL with the plasmid and send the plasmid for sequencing to verify the construct sequence.

3.5 Preparing the Frozen Stock

1. For each sample that returned good sequences from colony screening, prepare a 5 mL culture by inoculating LB media from the stab made during the colony screening (Subheading 3.4). Place into an incubator shaker for 4 h at 37 °C and 275 rpm.
2. For each culture, take a 2 mL screw-cap conical tube and add 750 μL of glycerol solution. Add 750 μL of the culture to the conical screw-cap tube. Vortex the tube gently.
3. Prepare a freezing bath in a beaker by combining dry ice and pure ethanol, and suspend the tubes in the bath. Store the tube at −80 °C until further use.

3.6 Second Round and Downstream Cloning

3.6.1 Plasmid Preparation

1. Purify plasmid from one of the bacterial colonies that returned the correct DNA sequence in Subheading 3.5 following the procedures in Subheading 3.2.
2. Refrigerate until needed.

3.6.2 Generating Construct IIIa

1. Obtain a PCR tube and add to it 1 μL of PflMI restriction enzyme, 1 μL of 10× NEBuffer 3.1, and a volume of plasmid from Subheading 3.6.1 and nuclease-free water bringing the tube to a final volume of 10 μL and plasmid concentration of 20 ng/μL (*see* **Note 9**).
2. Mix the reaction by pipetting up and down and incubate overnight at 37 °C.
3. Heat inactivate the enzymes at 65 °C for 1 h. Refrigerate until use.

3.6.3 Generating Construct IIIb

1. Prepare the insert by performing a double digestion reaction consisting of NdeI and BglI. Obtain a PCR tube and add to it 1 μL of NdeI restriction enzyme, 1 μL of BglI restriction enzyme, 1 μL of 10× NEBuffer 3.1, and a volume of plasmid

from Subheading 3.6.1 and nuclease-free water bringing the tube to a final volume of 10 μL and plasmid concentration of 20 ng/μL (*see* **Note 9**).

2. Mix the reaction by pipetting up and down and incubate overnight at 37 °C.
3. Heat inactivate the enzymes at 65 °C for 1 h. Refrigerate until use.

3.6.4 Generating Construct IV

1. To a PCR tube, add 10 μL NEBuilder HiFi Master Mix 2×, 2 μL of the PflMI digested plasmid (Subheading 3.6.2), 2 μL of the NedI-BglI double-digested plasmid (Subheading 3.6.3), and 6 μL of water (*see* **Notes 12–14**).
2. Transform the DNA by following the NEB 5-alpha competent *E. coli* (High Efficiency) transformation protocol as provided by the manufacturer, using 2 μL of the HiFi assembly product and spreading 100 μL of cells onto a selection plate. Allow the plates to grow overnight in an incubator at 37 °C (*see* **Note 10**).
3. Refrigerate plates with colonies.
4. Screen the colonies following the procedures in Subheading 3.4 and prepare frozen stocks as in Subheading 3.5 for samples returning positive results.
5. Repeat the procedures in this Subheading 3.6 with bacteria from **step 4** in order to generate genes containing additional repetitions of the protein sequences.

3.7 Preparation of Expression System

1. Transform plasmid DNA obtained during the screening process to an expression strain of *E. coli* by following the NEB BL21(DE3) competent *E. coli* transformation protocol as provided by the manufacturer.
2. Allow selection plates to grow overnight in an incubator at 37 °C.
3. Screen the colonies as in Subheading 3.4 to ensure plasmid contains the correct sequence and prepare frozen stocks for successful transformations. The gene is ready to be expressed.

4 Notes

1. This media becomes much more viscous as it cools, so work quickly to ensure the plates are poured before the media ceases to flow.
2. If an existing gene is not already available, the initial gene fragment with the appropriate structure (i.e., Fig. 2 III) can be directly ordered and the first round cloning step (Subheading 3.3) can be skipped.

3. As a result, the longer this sequence is, the fewer the steps needed to reach the desired length. We have added one additional pentapeptide with a glutamic acid residue (i.e., GEGVP).
4. Alternatively, using a micropipette tip, scrape a small piece of a frozen bacterial stock stored at −80 °C and add to the culture tube. Take care to keep the frozen stock outside of the freezer for the minimum amount of time possible to prevent harm coming to the bacterial sample.
5. Other commercially available miniprep kits may also be used.
6. This is **step 9** in the Qiagen manual. Spinning for an extended period of time increases the likelihood of drying the column of alcohol, thus increasing eluted plasmid yield.
7. This is **step 10** in the Qiagen manual. The elution volume is reduced from the standard protocol to increase the plasmid concentration.
8. If the plasmid concentration is found to be below 10 ng/μL, attempt another overnight culture/DNA prep with a longer drying spin or a lower plasmid elution volume to increase the yield.
9. Ensure that restriction enzymes are not vortexed or kept above freezing for long periods of time, doing so will result in their denaturing and loss of activity. An excess of restriction enzyme has shown higher success rates in ensuring that digestion occurs over standard reaction conditions.
10. Ideally, following transformation one should see between 1 and 100 colonies on the plate and they should be distinguishable from one another. If no growth occurs, one can attempt to incubate the cells in S.O.C. media for 1 h, spin down the tube for 30 s, pipette out 100 μL of the highly concentrated media at the bottom of the tube and plate again overnight. If overgrowth occurs and there are no distinguishable colonies, take 200 μL of S.O.C. media in a tube, scrape one line of cells from the selection plate and inoculate this tube, vortex, and plate 100 μL of this mix onto another plate and grow overnight. Positive and negative controls should be incorporated into this step, consisting of uncut plasmid from the initial DNA prep being transformed to ensure that the frozen stock is still usable, and competent cells with no plasmid to ensure that the ampicillin is properly selecting for plasmid.
11. For the Cys-VEV oligonucleotide insert, plasmids should be 36 nucleotides longer (going from MGHGVGVP to MGCNHGVGVPGEGVPGVGVP), which is easier to visualize on the higher concentration 2% gel as it provides greater separation for DNA of shorter lengths. For subsequent doubling reactions, changes in DNA length will

be more visible, increasing hundreds of nucleotides in length from the control.

12. For the assembly of the recombinant DNA, oligonucleotides are not used beyond the first round of cloning.
13. The single digested and double-digested plasmids should both be at a concentration of 2 ng/μL. If this cannot be met by plasmid yielded from the miniprep/digestion, extra volume of either plasmid can be added at the cost of an equal volume of water.
14. The opened plasmid cut with PflMI will anneal to the homologous ends of the excised double-digested fragment (Fig. 2).

Acknowledgments

This chapter was prepared with support from the Multi-college Interdisciplinary Research Program of Cleveland State University.

References

1. Cole JT (2016) The synthesis and characterization of multifunctional nanoparticles of elastin-like polypeptides for theranostic applications. Doctoral dissertation, Cleveland State University
2. Kowalczyk T, Hnatuszko-Konka K, Gerszberg A, Kononowicz AK (2014) Elastin-like polypeptides as a promising family of genetically-engineered protein based polymers. World J Microbiol Biotechnol 30:2141–2152. https://doi.org/10.1007/s11274-014-1649-5
3. Urry DW (1997) Physical chemistry of biological free energy transduction as demonstrated by elastic protein-based polymers. J Phys Chem B 101:11,007–11,028. https://doi.org/10.1021/jp972167t
4. Olivera BM, Lehman IR (1967) Linkage of polynucleotides through phosphodiester bonds by an enzyme from Escherichia coli. Proc Natl Acad Sci U.S.A 57(5):1426–1433
5. McDaniel JR, MacKay JA, Quiroz FG, Chilkoti A (2010) Recursive directional ligation by plasmid reconstruction allows rapid and seamless cloning of oligomeric genes. Biomacromolecules 11:944–952. https://doi.org/10.1021/bm901387t
6. Ott W, Nicolaus T, Gaub HE, Nash MA (2016) Sequence-independent cloning and post-translational modification of repetitive protein polymers through sortase and Sfp-mediated enzymatic ligation. Biomacromolecules 17:1330–1338. https://doi.org/10.1021/acs.biomac.5b01726
7. Gibson DG, Young L, Chuang R, Venter JC, Hutchinson CA, Smith HO (2000) Enzymatic assembly of DNA molecules up to several hundred kilobases. Nature 6:343
8. Deyling JK (2016) Design and production of a hydrogel forming polypeptide: engaging high school students in protein design. Master's thesis. Cleveland State University
9. Meyer DE, Chilkoti A (2002) Genetically encoded synthesis of protein-based polymers with precisely specified molecular weight and sequence by recursive directional ligation: examples from the elastin-like polypeptide system. Biomacromolecules 3:357–367. https://doi.org/10.1021/bm015630n

Chapter 21

Peptide Amphiphile Micelles for Vaccine Delivery

John C. Barrett and Matthew V. Tirrell

Abstract

Peptide subunit vaccines enable the specific activation of an immune response without the shortcomings of killed or attenuated pathogens. However, peptide subunit vaccines tend to be less immunogenic than those based on whole organisms. To improve peptide immunogenicity, biomaterials-based platforms have been developed. One such platform, the peptide amphiphile micelle platform, has displayed a unique ability to dramatically improve observed immune responses. Here we describe the design, synthesis, characterization, and application of peptide amphiphile micelles to elicit a robust immune response.

Key words Vaccine, Peptide amphiphile micelle, Self-assembly, Peptide synthesis, Modular

1 Introduction

Peptides fulfill many roles in immunology, yet none is more important than their role as immunogenic epitopes driving the adaptive immune response. This makes peptides attractive vaccine candidates since they can be precisely designed to contain the minimal epitopes necessary to stimulate an immune response, avoiding common problems associated with killed or attenuated pathogen vaccines such as autoimmunity, risk of infection, and allergic reaction [1, 2]. Unfortunately, the largest challenge facing peptide-based vaccines is the fact that peptides alone are weak immunogens [3]. In response to this, various biomaterial platforms have been developed to improve the stability of peptides [4]. Among these platforms are peptide amphiphiles (PAs). PAs are a class of peptide-based biomaterial consisting of bioactive peptide head groups conjugated to hydrophobic alkyl tails, which self-assemble in aqueous solution into micellar structures [5]. Here we describe how PA micelles can be used as a vaccine platform to improve peptide immunogenicity [6–8]. Modular PA micelle features, including micelle size, shape, composition, and targeting abilities, are significant contributors to the platform's ability to raise an immune response [1, 7–12]. For exam-

Andrew K. Udit (ed.), *Protein Scaffolds: Design, Synthesis, and Applications*, Methods in Molecular Biology, vol. 1798, https://doi.org/10.1007/978-1-4939-7893-9_21,

ple, PA micelles can be engineered to be cylindrical or spherical, contain precise amounts of components such as antigens and amphiphilic adjuvants, and efficiently traffic to the lymph node [5, 8, 13, 14]. Using these features, PA micelles can be synthesized and deployed to enhance an immune response and induce strong antigen-specific antibody responses. When delivered subcutaneously to mice, heterogeneous micelles induced a stronger peptide antigen-specific IgG1 antibody response than seen with a conventional gold standard vaccine formulation [7, 8].

We have engineered a modular peptide amphiphile micelle subunit vaccine platform that can enhance lymph node codelivery of antigens and adjuvants leading to a dramatically improved antigen-specific antibody response. The results of previous experiments and the modular nature of the platform combine to provide considerable evidence that PA micelles may be used to achieve desirable immune responses in a wide range of applications. In this protocol paper, we explain the synthesis, purification, and characterization methods of PAs and PA micelles for vaccine applications. Furthermore, we explain the administration of PA micelle vaccines in vivo, as well as methodology used to examine the efficacy of PA micelle vaccines.

2 Materials

2.1 Peptide Amphiphile Synthesis

1. Fmoc-amino acids with appropriate side chain protection groups (*see* **Note 1**).
2. Fmoc-Rink amide 4-methylbenzhydrylamine (MBHA) resin (*see* **Note 2**).
3. Shaker arm apparatus.
4. Peptide synthesis vessel (solid phase, with vacuum and medium frit).
5. 1-Hydroxybenzotriazole (HOBt).
6. *N*-[(1*H*-benzotriazol-1-yl)(dimethylamino)methylene]-*N*-methylmethanaminium hexafluorophosphate *N*-oxide (HBTU).
7. *N*,*N*-dimethylformamide (DMF).
8. *N*,*N*-diisopropylethylamine (DIPEA).
9. *N*-methylpyrrolidone (NMP).
10. Fluorophores: 5(6)-carboxy-rhodamine, 5(6)-carboxyfluorescein (optional).
11. Acetic anhydride.

12. Hydrophobic tail: The dialkyl ester [$(C_{16})_2$-Glu-C_2, designated diC_{16}] chain is used throughout the experiments described below, unless otherwise stated. diC_{16} is prepared in-house (*see* **Note 3**).
13. Peptides prepared on a peptide synthesizer utilizing well-established 9-fluorenylmethoxycarbonyl (Fmoc) solid-phase methodology [15–19].
14. Piperidine.
15. Hydroxylamine hydrochloride.
16. Imidazole.
17. Reagent A: Dissolve 16.5 mg of KCN in 25 mL of distilled water. Dilute 1 mL of the above solution with 49 mL of pyridine (freshly distilled from ninhydrin). Store in a small reagent bottle.
18. Reagent B: Dissolve 1 g of ninhydrin in 20 mL of *n*-butanol. Store in a small reagent bottle.
19. Reagent C: Dissolve 40 g of phenol in 20 mL of *n*-butanol. Store in a small reagent bottle.
20. Silicone Oil bath.
21. Cleavage solution: H_2O–triisopropylsilane (TIS)–trifluoroacetic acid (TFA) (2.5:2.5:95) [20, 21].
22. Diethyl ether.
23. Dichloromethane (DCM).
24. Methanol.
25. 50 mL plastic tubes.
26. 20 mL scintillation vials.
27. Nitrogen gas.

2.2 Purification, Preparation, and Characterization of PA

1. High pressure liquid chromatograph with reversed-phase C8 column coupled to time-of-flight mass spectrometry, utilizing photodiode array detection (PDA) controlled fraction collection (HPLC-MS).
2. Acetonitrile, HPLC-grade.
3. 0.1% formic acid.
4. Hydrochloric acid.
5. Sodium hydroxide.
6. Milli-Q water.
7. PBS (phosphate buffered saline): To create 10× PBS, dissolve 80 g of NaCl, 2 g of KCl, 27.2 g of $Na_2HPO_4{\cdot}7H_2O$ (dibasic heptahydrate), and 2.4 g of KH_2PO_4 (monobasic anhydrous) in 800 mL of distilled H_2O. Adjust pH to 7.4 with HCl, then

add H_2O to 1 L. Dilute to 1× as needed. All PBS in this protocol follows this recipe.

8. Chloroform.
9. Methanol.
10. Adjuvants: Amphiphilic adjuvants such as Toll-Like Receptor 4 (TLR4) agonist Monophosphoryl Lipid A (MPLA, Sigma-Aldrich) or TLR2 agonist $Pam_2Cys\text{-}SK_4$ ($P_2C\text{-}SK_4$, Invivogen). Positive control adjuvants such as Incomplete Freund's Adjuvant (IFA, Sigma-Aldrich).
11. 70 °C water bath.
12. Lyophilizer.
13. 1,6-diphenyl-1,3,5-hexatriene (DPH).
14. 1% w/w aqueous phosphotungstic acid.
15. Fluorimeter plate reader with appropriate well plates.
16. Transmission electron microscope (TEM).
17. Formvar-coated copper grid.
18. Circular dichroism (CD) spectrometer.

2.3 Murine Vaccination and Antibody Response Characterization

1. Female BALB/c mice 6–8 weeks old. The mouse strain/breed can be substituted for the experimental conditions of the specific study.
2. 23 gauge needles.
3. 25 gauge needles.
4. BD Slip Tip 1 mL syringes.
5. Vaseline petroleum jelly.
6. Microcap 100 μL capillary tubes.
7. Gauze.
8. Sodium bicarbonate coating buffer: 7.13 g sodium bicarbonate, 1.59 g sodium carbonate, 1 L Milli-Q water, pH to 9.5 with NaOH or HCl.
9. PBS-T wash buffer: add 500 μL of Tween 20 to 1 L of 1× PBS.
10. Assay diluent: 10% fetal bovine serum (FBS) in 1× PBS.
11. Flat-bottom 96-well EIA microtiter plates.
12. Detection antibody (e.g., IgM, IgG, IgA, IgG1, IgG2a, IgG3, or IgG4).
13. Ultra TMB-ELISA substrate solution (Pierce).
14. 1 M H_3PO_4.

3 Methods

3.1 Peptide Amphiphile Synthesis

Determine a synthetic scheme prior to synthesis, drafting out the location of all chemical additions/modifications (tail, fluorophore, etc.) and corresponding orthogonal protection groups necessary to facilitate synthesis (*see* **Note 4**). For practical examples of chemical synthesis schemes, please see these previous publications [22–24]. Here, we follow an example synthesis scheme for a PA antigen vaccine commonly performed in our laboratory (*see* **Note 5**). However, the steps themselves are general and can be reordered based on the designed synthesis scheme. Figure 1 illustrates a PA molecule synthesized with the following protocol. Specifically, the below scheme describes the synthesis of a N-terminally acylated peptide (J8) with a diC_{16} tail coupled on resin to the side chain amine of the C-terminal lysine.

1. Incorporate each individual amino acid by Fmoc solid-phase methodology on a peptide synthesizer.
2. Transfer beads from the synthesizer vessel to a peptide synthesis vessel.
3. Swell the beads in 10–15 mL of DCM while shaking slowly on a shaker arm for 20–30 min.
4. Drain the swelling solution and wash the beads 3 times (3×) with DMF (10–15 mL, 5 min each) (*see* **Note 6**).
5. For Fmoc deprotection, add 5 mL of 20% Piperidine in DMF. Shake for 5 min and repeat for a total of four times, draining and replacing the piperidine in DMF each time.
6. Drain the deprotection solution and wash the beads 3× with DMF (10–15 mL, 5 min each).
7. To test for the presence of free amines, perform the ninhydrin (i.e., Kaiser) test. Collect approximately 20 beads from the reaction solution with a glass pipette and put them into a small glass test tube (*see* **Note 7**). Add 2–3 mL of methanol into the glass tube with the beads, and use the glass pipette to mix the beads in the methanol as a wash. Aspirate off the methanol, and add one drop each of Reagent A, Reagent B, and Reagent C. Heat in a 110 °C silicone oil bath for 3–5 min (*see* **Note 8**) [25, 26].
8. To acylate the peptide N-terminus on resin (*see* **Notes 9** and **10**), add a tenfold molar excess of acetic anhydride to the Fmoc deprotected peptide, with a 3.8-fold molar excess of HBTU in DMF, and shake for 15 min. Initiate the reaction by the addition of an eightfold molar excess of DIPEA and proceed to shake for 3 h, or until a ninhydrin test indicates the reaction is complete [27].

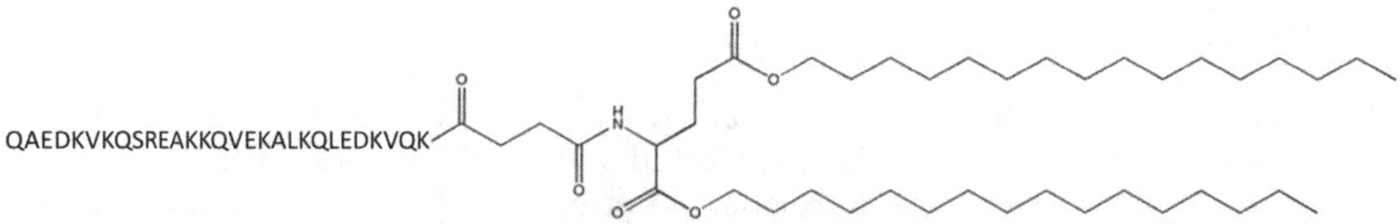

Fig. 1 Chemical structure of a PA molecule containing the alkyl tail off the C-terminal lysine of the J8 epitope (QAEDKVKQSREAKKQVEKALKQLEDKVQK)

9. Drain the coupling solution and wash the beads 3× with DMF (10–15 mL, 5 min each).
10. To deprotect any lysines with Dde groups, add 1 mL of DCM to the beads, then add a solution of 1.25 g of hydroxylamine hydrochloride and 0.918 g of imidazole in 5 mL of NMP. Allow the deprotection solution to shake for 3 h, then wash 3× with DMF (10–15 mL, 5 min each).
11. Deprotect any lysines with Mtt groups with a solution of TFA-TIS-DCM (1:2:97), shake for 1 h. Follow each deprotection with three DMF washes (10–15 mL, 5 min each).
12. Prepare the hydrophobic tail coupling solution in a 20 mL scintillation vial. First measure dry components: fourfold molar excess of diC_{16} and 3.8-fold molar excess of HBTU. Add 10 mL of 1:1 DCM:DMF (5 mL DCM, 5 mL DMF) to the scintillation vial and shake vigorously by hand until the tail is dissolved. Add this solution to the beads, let shake on the shaker arm apparatus for 1 h.
13. Add eightfold molar excess of DIPEA and let shake for an additional 4 h on the shaker arm apparatus (*see* **Note 11**) [28].
14. Drain the coupling solution and follow with a series of washes: 2× in 1:1 DMF:DCM (the base solvent the tail was dissolved in), 2× in DMF, 2× in DCM, 2× in methanol, all with 10–15 mL for 5 min each. Dry under vacuum thoroughly, for at least 10 min (*see* **Note 12**).
15. Cleave the lipidated peptide and deprotect the side-chains by treating the resin for 1.5–3 h with cleavage solution (*see* **Notes 13** and **14**) [20, 21].
16. Collect the cleavage solution by draining into a 50 mL plastic tube (*see* **Note 15**).
17. Evaporate off most of the TFA via nitrogen gas purging (*see* **Note 16**).
18. Add approximately 30 mL of cold diethyl ether (−80 °C) to PA oil, then shake/vortex/sonicate until precipitates are visible (*see* **Note 17**).
19. Centrifuge ether/precipitates at −10 °C and 1700 × g for 5 min.

20. Decant the supernatant, dry the collected crude PA pellet with nitrogen gas purging until the ether has visibly evaporated, then dry the ether supernatant (resulting solids may be additional product). The sample can be dried further under vacuum. If storing, place in −20 °C freezer.

3.2 Purification and Preparation

3.2.1 PA Purification

1. After solubilizing the pellet (*see* **Note 18**), purify the PA product(s) with preparative high pressure liquid chromatography coupled to time-of-flight mass spectrometry, utilizing photodiode array detection (PDA) controlled fraction collection (HPLC-MS). Make use of a reversed-phase C8 column at 50 °C and a method that calls for the following: flow rate of 10 mL/min with Milli-Q water containing 0.1% formic acid (A) and acetonitrile (B), and detection at 220 nm. Begin the solvent gradient at 97% A/3% B and move to 100% B over 30 min. Stay at 100% B for 5 min before ramping back to 97% A/3% B over the course of 3 min.
2. Collect different peaks individually and evaporate the acetonitrile under air or nitrogen overnight (*see* **Note 19**).
3. During purification, determine PA composition by the mass spectrometer attached to the HPLC (*see* **Note 20**).
4. Freeze, lyophilize and store all samples at −20 °C until used.

3.2.2 PA Micelle Preparation

1. Dissolve 12 nmol of PA (J8-diC$_{16}$) in a mixture of chloroform–methanol (3:1) in a 1.5 mL microcentrifuge tube. If adjuvants (MPLA and/or P$_2$C-SK$_4$) are to be included, also add them at 0.1 mol equivalent (1.33 nmol) before dissolving. Use enough solvent to completely dissolve the product (~200 μL) (*see* **Note 21**).
2. Film cast the solution on the walls of the centrifuge tube by evaporating off the organic phase using nitrogen as a drying gas (*see* **Note 22**).
3. Hydrate the resulting thin lipid film with 100 μL or to desired concentration with a PBS or water aqueous phase. Extensively vortex to induce spontaneous micelle formation.
4. Heat the micelle dispersion in a water bath at 70 °C for 60 min and allow to equilibrate overnight at room temperature.

3.3 PA Micelle Characterization

The size, structure, and stability of PA micelles can be evaluated by various methods: critical micelle concentration (CMC), transmission electron microscopy (TEM), CD spectroscopy, and fluorescence spectroscopy. These procedures have been performed previously for studying PA micelles and other particles [6, 29–31]. Below, each method is described. Figure 2 provides an example TEM image and CD spectra that resulted from the following methods.

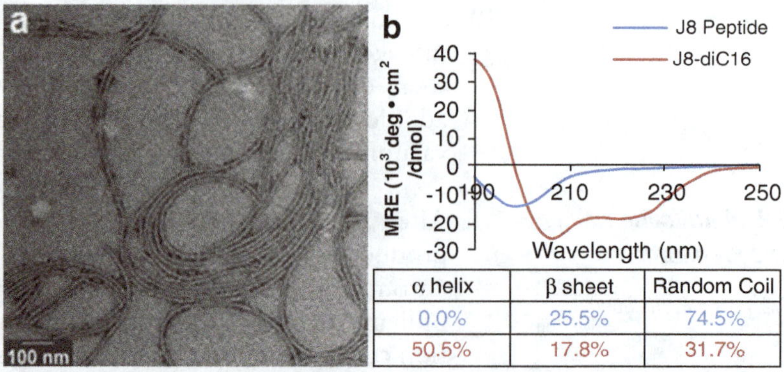

α helix	β sheet	Random Coil
0.0%	25.5%	74.5%
50.5%	17.8%	31.7%

Fig. 2 (**a**) Negative stain TEM image of J8-diC_{16} after film casting and heat annealing. (**b**) CD revealed that the micelle-based J8 peptide has higher α-helicity than J8 peptide in solution. Reprinted from Trent et al. [7]; Copyright (2014) with permission from Springer

3.3.1 Critical Micelle Concentration

CMC is measured by fluorescent sequestration. Varying concentrations of J8-diC_{16} are exposed to 1 mM 1,6-diphenyl-1,3,5-hexatriene (DPH) which greatly increases in fluorescence intensity when trapped within the micelle core.

1. Make a 1 μM DPH in water stock solution (*see* **Note 23**).
2. Make a 1000 μM stock solution of PA in water.
3. Choose a starting concentration, then make the first dilution by taking an appropriate amount of PA and the stock DPH solution. Make serial dilutions (*see* **Note 24**). Vortex after adding the PA and stock solutions.
4. Allow solutions to equilibrate for 1 h prior to fluorescent measurement.
5. Pipet solutions into a well plate for fluorescence.
6. Measure fluorescence utilizing a plate reader (ex. 350 nm, em. 428 nm).
7. Graph fluorescence intensity vs. concentration, and the inflection point will indicate the CMC.

3.3.2 Transmission Electron Microscopy

Negative stain TEM is a useful technique to investigate micelle morphology.

1. To prepare a grid, incubate a PA micelle solution (1 μL of 200 μM) on a Formvar-coated copper grid for 1 min.
2. After 1 min, wick away excess liquid with filter paper or Kimwipes.

3. Incubate Milli-Q water on the grid for 1 min, and then wick away with a Kimwipes.
4. Incubate with 1% w/w aqueous phosphotungstic acid for 1 min, and then wick solution away.
5. Air-dry samples for 5 min on the lab bench.
6. Image on a TEM using an accelerating voltage of 120 kV.

3.3.3 Circular Dichroism

Assess peptide secondary structure using CD.

1. Prepare 30 μM solutions of PAs in water.
2. Measure CD spectra of solutions at 25 °C for a total of five times.
3. The data can be averaged and a curve from 190 to 250 nm can be fit using a linear combination of polylysine basis spectra [32] to determine approximate α-helix, β-sheet, and random coil peptide secondary structure.

3.3.4 Spectrofluorometry

Fluorescence can be used to investigate whether multiple components of a self-assembled heterogeneous micelle are residing in the same micelle versus creating segregated populations of components.

1. Fluorescently labeled components are needed for this experiment (*see* **Note 10**).
2. For Förster resonance energy transfer (FRET) experiments, make three different micelle formulations each with a final formulation concentration of 120 μM (*see* **Note 25**).
3. Excite micelles in spectrofluorometer at 490 nm and record emission between 490 and 700 nm.

3.4 Murine Vaccination and Antibody Response Characterization

The experimental conditions of the study will dictate which mouse strain/breed is required, as well as the vaccination schedule. Below, female BALB/c mice 6–8 weeks old are immunized to investigate the capacity for micelle formulations to induce an antibody-mediated immune response.

3.4.1 Vaccination

1. Use Incomplete Freund's Adjuvant (IFA) delivered with soluble peptide for a positive control group.
2. Use a 23 gauge needle attached to a 1 mL syringe to vaccinate mice subcutaneously with 100 μL of vaccine at the nape of the neck on days 0 (prime), 21 (boost 1), 28 (boost 2), and 35 (boost 3).
3. Before vaccination on days 21, 28, and 35, as well as on day 42, collect whole blood to analyze for peptide-specific antibodies induced by the previous round of immunization.

Collect blood from saphenous veins by first shaving hair from the leg, covering the vein with a thin layer of Vaseline and then pricking the vein with a 25 gauge needle. The blood will pool on top of the Vaseline at which time it can be collected by the capillary tube. Squeeze the blood into a microcentrifuge tube (*see* **Note 26**).

4. Centrifuge the blood at 9600 × *g* for 10 min to separate out red blood cells. Harvest the serum supernatant and store at −20 °C until analysis.

3.4.2 Antibody Response Characterization

Utilize an enzyme-linked immunosorbent assay (ELISA) to determine peptide-specific antibody titers. Figure 3 serves as an example antibody titer that resulted from the following protocol.

1. Coat flat-bottom 96-well EIA microtiter plates overnight at 4 °C with 100 μL of 10 μg/mL peptide in sodium bicarbonate coating buffer in each well.
2. The following day, add 200 μL of PBS-T and incubate for 30–60 s. Vigorously shake the solution out of the plate into the sink and tap inverted plates on paper towels. Repeat two more times.
3. Add 200 μL of assay diluent to each well, cover and incubate for 1 h to block the plates.
4. Remove the assay diluent solution and add 200 μL of 1:1000 sera samples (collected from Subheading 3.4.1) diluted in assay diluent to the top row. Serially dilute the top row into assay diluent with twofold steps down the plate. Cover and incubate for 2 h.
5. Add 200 μL of Wash Buffer and incubate for 30–60 s. Vigorously shake the Wash Buffer out of the plate into the sink and tap inverted plates on paper towels. Repeat two more times.
6. Add 100 μL of 1:3000 diluted detection antibody to each well, cover and incubate for 1 h.
7. Wash wells 3× with 200 μL PBS-T.
8. Add 100 μL of Ultra TMB-ELISA substrate solution to each well.
9. Develop/incubate plates for 15 min in darkness (*see* **Note 27**).
10. Measure the optical density for each well at 650 nm using a plate reader.
11. End point antibody titers are defined as the greatest serum dilution where OD is at least twice that of normal mouse serum at the same dilution (*see* **Note 28**).

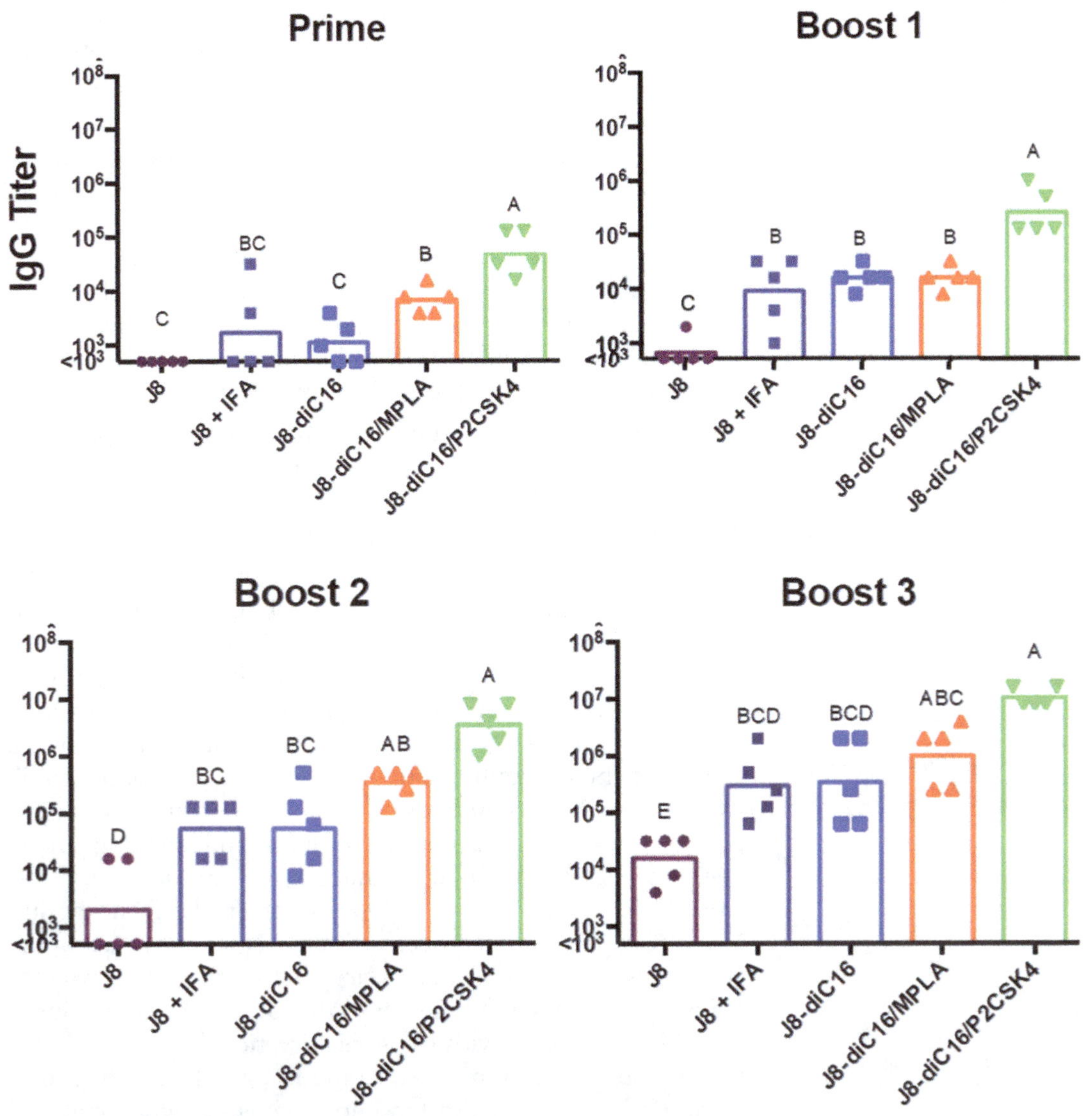

Fig. 3 Mixed peptide amphiphile micelles induced strong antibody isotype responses in vivo. J8-specific antibody titers induced by J8 + IFA and J8-diC$_{16}$ micelles were compared to titers stimulated by mixed micelles containing 10% of either MPLA or P_2C-SK_4. Each secondary signal incorporated into the micelles resulted in an enhanced antibody response. While IgA titers were assessed, no mouse produced above background levels. Each data point represents one mouse ($N = 5$); bars represent the geometric mean. Within a graph, groups that possess different letters have statistically significant differences in mean ($p \leq 0.05$) whereas those that possess the same letter are similar ($p > 0.05$). Reprinted with permission from Barrett et al. [8]; Copyright (2017) American Chemical

4 Notes

1. Orthogonally protected lysines can be useful in a synthesis scheme to have access to another primary amine for further side group modification. Groups such as 1-(4,4-dimethyl-2,6-dioxocyclohex-1-ylidene)ethyl (Dde) or methyltrityl (Mtt) are commonly used to protect lysine in PA synthesis.

2. The use of lower substitution resins can give improved yields for longer sequences, or for those incorporating difficult couplings.
3. Briefly, synthesis first involves the acid-catalyzed condensation of glutamic acid (Glu) with the appropriate fatty acid alcohol to form the dialkyl ester of Glu. The free amino group of Glu is then treated with succinic anhydride and triethylamine to create a free carboxylic acid. The dialkyl ester tail may then be used directly for solid-phase synthesis or converted to a *p*-nitrophenyl ester with *p*-nitrophenol and dicyclohexylcarbodiimide and then coupled [16]. Other tails can be used, based upon the specific application, including other dialkyl chains and monoalkyl chains of various lengths.
4. By utilizing the array of readily available orthogonally protected Lys derivatives, including, but not limited to, Fmoc-Lys(Dde), Dde-Lys(Fmoc), Fmoc-Lys(Fmoc), Fmoc-Lys(Mtt), a variety of chemical designs/modifications can be created. Refer to Novabiochem and Sigma-Aldrich websites for lists of derivatives.
5. The peptide used in this protocol is the J8 epitope (QAEDKVKQSREAKKQVEKALKQLEDKVQK). J8 is derived from the conserved C terminal region of the alpha helical M1 surface protein from Group A *Streptococcus.* An ideal peptide for a PA is a polar peptide that is soluble in water and can offset the hydrophobicity of the lipid tail. Beyond solubility, the most important consideration for choosing the vaccine peptide is the identification of immunodominant domains of epitopes that are capable of inducing an immune response [33]. In its natural conformation, M1 protrudes from the surface of Group A Streptococcus with the N terminus facing out (C terminus attached to the surface). This is the reason why the tail is added to the C terminus: so that immune cells can interact with the peptide with the same directionality that it would see in nature. Additionally, since the J8 peptide is only a small section of the M1 protein (and not completely at a terminus), the N terminus is acylated to "cover" the N terminal charge and more accurately represent the peptide as it is in nature (as part of the larger protein with no end charges). Therefore, acylation and tail location can be included/excluded/modified based on the properties of the chosen peptide.
6. A vacuum can be used to accelerate the draining of solvent. Vacuum for a longer time if switching between solvents (a minute or two). A couple of seconds is fine for iterations of the same solvent. To save time, reaction solutions can be made during DMF washes.
7. Fewer than 10 beads may not give an accurate representation of the status of the reaction. More than 20 beads will be a waste of product since one cannot reuse these beads.
8. Free primary amines result in a deep blue/purple color change. Absence of free primary amines results in no color

change or light brown color after heating. This test can be used during any deprotection or conjugation step to indicate whether the reaction is complete. While the bead washing method described does generally work, it is mostly used to save time. However, it is known that solvent can interfere with the ninhydrin reaction. Therefore, if it is suspected that the ninhydrin test is not producing a reliable readout of whether there is a free amine, wash and dry down the bead for a couple of minutes before testing.

9. Any available deprotected amines from previous steps will also be acylated by this step, so be sure to reference the synthesis scheme and perform acylation only when desired.
10. Fluorophores such as rhodamine (Rho) can also be covalently coupled by amidation reactions with HBTU and DIPEA on the deprotected amine group (either on the side chain from orthogonally deprotected lysine or on the N-terminus of the peptide, depending on the synthesis scheme). Fluorescent dye coupled peptides should be protected from light as best as possible at all stages of synthesis, purification, and application.
11. The reaction can be left to shake overnight to maximize the yield. All shaking should be vigorous enough to disrupt the meniscus, get the beads off the bottom, and distribute the beads throughout the solvent, but not too violent that the beads get stuck high on the walls or on the inside of the cap. If beads stick to the side of the glass vessel, shearing them off with DMF is easier than DCM. Before long reactions (such as conjugation or cleavage), add approximately 70% of reaction solution to the reaction vessel and use this time to optimize shaking. If beads accidentally get stuck to the walls during optimization, use the rest of the reaction solution to rinse the beads off the walls as you continue to optimize shaking.
12. Before cleavage, it is important to wash and dry the beads. If cleavage is to be done the following day, or exceptional drying is desired (though 10 min of vacuum drying is usually enough), remove the plastic drain and cap, cover the opening with a Kimwipes and rubber band, and place in a vacuum oven or desiccator at room temperature overnight.
13. If performing cleavage in a reaction vessel, cap the vessel with aluminum foil rather than the plastic cap as this will get degraded by TFA. If a cap is used, never close the cap too tightly during cleavage, since the reaction can produce gas. MAKE SURE THE DRAIN IS CLOSED before adding approximately 70% of cleavage solution to the dried beads using a glass pipette. Do not shake vigorously as the beads in TFA get stuck to the sides very easily. If they do get stuck, rinse them down with the remaining cleavage solution.
14. If a fluorophore is included in the synthesis scheme, perform resin cleavage and side chain deprotection using a solution of

TIS–phenol–H_2O–TFA (2.5:5:5:87.5). The added phenol protects the fluorophore.

15. The beads and the reaction vessel walls can be rinsed with DCM after TFA cleavage to make sure to collect all of the cleaved product.
16. Let evaporate until only a few milliliters are left, which should look oily. If the solution dries out, add more TFA and attempt to dissolve by vortexing. Use metal tubing for TFA evaporation to prevent corrosion of plastic. For nitrogen gas, give enough pressure to make ripples in TFA in tube, but not enough to displace much liquid.
17. This washes the product to remove cleaved protecting groups and impurities from the solution.
18. After cleavage and ether wash, when dissolving the peptide pellet in water after cleavage, it helps to have a charged amino acid in the sequence. However, regardless of the sequence, begin dissolving the peptide by adding several milliliters of Milli-Q water to the PA pellet, vortex and sonicate until it clearly dissolves. If not adequately dissolved, attempt the following steps in order and stop when solubilized: perform pH exchange of up to 8 units, add acetonitrile, add methanol, add DMSO. Alternatively, GenScript has a useful peptide solubility guideline to follow that can be found on their website, genscript.com.
19. The amount of acetonitrile to evaporate off is dependent on the time and gradient at which the HPLC fraction was collected. For example, if the fraction was collected while the gradient was at 30% acetonitrile and 70% Milli-Q with 0.1% formic acid, evaporate off approximately 30% of the collected solution.
20. Alternative methods such as amino acid analysis and/or MALDI-TOF MS can be used to confirm product composition [34].
21. All concentrations are subject to optimization based on the specific application. To create mixed micelles with multiple components, amphiphilic adjuvants such as Toll-Like Receptor 4 (TLR4) agonist monophosphoryl Lipid A (MPLA) or TLR2 agonist Pam_2Cys-SK_4 (P_2C-SK_4) can be added to the solution to create adjuvant supplemented micelles. Because of the amphiphilic structure, amphiphilic adjuvants will coassemble into a micelle with the synthesized PAs. Coassembling adjuvants into micelles has many advantages compared to covalently linking the adjuvant to a peptide including the advantage that multiple agonists and multiple PA antigens can be incorporated into micelles in a modular fashion.
22. Vaccines can easily be made in 1.5 mL micro centrifuge tubes. To evaporate off the organic phase, make a homemade nitrogen drying apparatus. Find a short tube that can attach to the nitrogen spigot on one side, and attach a 200 μL pipette tip to the other end with Parafilm. Set the nitrogen gas to a manageable level, hold the centrifuge tube in one hand at a 30–45°

angle to expand the meniscus and continually rotate the tube while carefully blowing off organic phase to create a film.

23. To do this, start by making a 100 mM DPH solution in THF solution. Take 10 μL of this solution and add it to 1 mL of water. Sonication may be required to mix. Next, take 20 μL of the previous solution and add it to 20 mL of water to make the final working solution. Vortex to mix. Keep any vial or tube with DPH in it covered with foil because DPH is light sensitive. Each 1 mL vial can be set in a tube holder and the holder covered with foil.
24. In order to make concentrations for a log plot, choose a starting concentration and then divide that by $10^{1/4}$ to make the next concentration, and so on. Dilute each subsequent concentration by using the DPH stock solution and vortex after adding the PA. For example: 316 μM, 178 μM, 100 μM, etc., down to approximately 0.01 μM.
25. Rhodamine only micelles contain unlabeled J8-diC_{16} with 10% rhodamine labeled J8-diC_{16} and 10% unlabeled P_2C-SK_4. Fluorescein only micelles contain unlabeled J8-diC_{16} and 10% fluorescein labeled P_2C-SK_4. Fluorescein and rhodamine micelles contain unlabeled J8-diC_{16} with 10% rhodamine labeled J8-diC_{16} and 10% fluorescein labeled P_2C-SK_4.
26. Other blood collection techniques may be used, including, but not limited to maxillary facial vein route or retro-orbital sinus route depending on experimental conditions and approved protocols.
27. If more time is needed to perform measurement, the reaction can be stopped with 100 μL of 1 M H_3PO_4 and measured within 30 min of stopping reaction.
28. If end-point titers are not reached with one plate, then additional titrations should be utilized until ODs are diluted to background.

References

1. Black M, Trent A, Tirrell M, Olive C (2010) Advances in the design and delivery of peptide subunit vaccines with a focus on Toll-like receptor agonists. Expert Rev Vaccines 9:157–173. https://doi.org/10.1586/erv.09.160
2. Li W, Joshi M, Singhania S et al (2014) Peptide vaccine: progress and challenges. Vaccine 2:515–536. https://doi.org/10.3390/vaccines2030515
3. Flower DR (2013) Designing immunogenic peptides. Nat Chem Biol 9:749–753. https://doi.org/10.1038/nchembio.1383
4. Skwarczynski M, Toth I (2016) Peptide-based synthetic vaccines. Chem Sci 7:842–854. https://doi.org/10.1039/C5SC03892H
5. Trent A, Marullo R, Lin B et al (2011) Structural properties of soluble peptide amphiphile micelles. Soft Matter 7:9572–9582. https://doi.org/10.1039/c1sm05862b
6. Black M, Trent A, Kostenko Y et al (2012) Self-assembled peptide amphiphile micelles containing a cytotoxic T-cell epitope promote a protective immune response in vivo. Adv Mater 24:3845–3849. https://doi.org/10.1002/adma.201200209
7. Trent A, Ulery BD, Black MJ et al (2014) Peptide amphiphile micelles self-adjuvant group A streptococcal vaccination. AAPS J 17:380–388. https://doi.org/10.1208/s12248-014-9707-3
8. Barrett JC, Ulery BD, Trent A et al (2016) Modular peptide amphiphile micelles improving an antibody-mediated immune response to group A streptococcus. ACS Biomater Sci Eng 3:144–152. https://doi.org/10.1021/acsbiomaterials.6b00422

9. Mora-Solano C, Collier JH (2014) Engaging adaptive immunity with biomaterials. J Mater Chem B 2:2409–2421. https://doi.org/10.1039/c3tb21549k
10. Liu H, Irvine DJ (2015) Guiding principles in the design of molecular bioconjugates for vaccine applications. Bioconjug Chem 26:791–801. https://doi.org/10.1021/acs.bioconjchem.5b00103
11. Irvine DJ, Hanson MC, Rakhra K, Tokatlian T (2015) Synthetic nanoparticles for vaccines and immunotherapy. Chem Rev 115:11109–11146. https://doi.org/10.1021/acs.chemrev.5b00109
12. Liu H, Moynihan KD, Zheng Y et al (2014) Structure-based programming of lymph-node targeting in molecular vaccines. Nature 507:519–522. https://doi.org/10.1038/nature12978
13. Wen Y, Collier JH (2015) Supramolecular peptide vaccines: tuning adaptive immunity. Curr Opin Immunol 35:73–79. https://doi.org/10.1016/j.coi.2015.06.007
14. Acar H, Srivastava S, Chung EJ et al (2016) Self-assembling peptide-based building blocks in medical applications. Adv Drug Deliv Rev 110–111:1–57. https://doi.org/10.1016/j.addr.2016.08.006
15. Yu YC, Berndt P, Tirrell MV, Fields GB (1996) Self-assembling amphiphiles for construction of protein molecular architecture. J Am Chem Soc 118:12515–12520. https://doi.org/10.1021/ja9627656
16. Berndt P, Fields GB, Tirrell M (1995) Synthetic lipidation of peptides and amino acids: monolayer structure and properties. J Am Chem Soc 117:9515–9522
17. Yu YC, Tirrell MV, Fields GB (1998) Minimal lipidation stabilizes protein-like molecular architecture. J Am Chem Soc 120:9979–9987. https://doi.org/10.1021/ja981654z
18. Missirlis D, Khant H, Tirrell M (2009) Mechanisms of peptide amphiphile internalization by SJSA-1 cells in vitro. Biochemistry 48:3304–3314. https://doi.org/10.1021/bi802356k
19. Knorr R, Trzeciak A, Bannwarth W, Gillessen D (1989) New coupling reagents in peptide chemistry. Tetrahedron Lett 30:1927–1930
20. King DS, Fields CG, Fields GB (1990) A cleavage method which minimizes side reactions following Fmoc solid-phase peptide-synthesis. Int J Pept Protein Res 36:255–266
21. Fields CG, Fields GB (1993) Minimization of tryptophan alkylation following 9-fluorenylmethoxycarbonyl solid-phase peptide-synthesis. Tetrahedron Lett 34:6661–6664.https://doi.org/10.1016/S0040-4039(00)61669-6
22. Lauer-Fields JL, Minond D, Brew K, Fields GB (2007) Application of topologically constrained mini-proteins as ligands, substrates, and inhibitors. In: Fields GB (ed) Methods in molecular biology: peptide characterization and application protocols. Humana Press, Totowa, NJ, pp 125–166
23. Rezler EM, Khan DR, Tu R et al (2007) Peptide-mediated targeting of liposomes to tumor cells. In: Fields GB (ed) Methods in molecular biology: peptide characterization and application protocols. Humana Press, Totowa, NJ, pp 269–298
24. Mata A, Palmer L, Tejeda-Montes E, Stupp SI (2011) Design of biomolecules for nanoengineered biomaterials for regenerative medicine. In: Navarro M, Planell JA (eds) Nanotechnology in regenerative medicine: methods and protocols. Humana Press, Totowa, NJ, pp 39–49
25. Sarin VK, Kent SBH, Tam JP, Merrifield RB (1981) Quantitative monitoring of solid-phase peptide synthesis by the ninhydrin reaction. Anal Biochem 117:147–157
26. Wellings DA, Atherton E (1997) Methods in enzymology. Solid-phase peptide synthesis. Academic Press, San Diego
27. Fields CG, Lloyd DH, Macdonald RL et al (1991) HBTU activation for automated Fmoc solid-phase peptide synthesis. Pept Res 4:95–101
28. El-Faham A, Albericio F (2011) Peptide coupling reagents, more than a letter soup. Chem Rev 111:6557–6602. https://doi.org/10.1021/cr100048w
29. Mlinar LB, Chung EJ, Wonder EA, Tirrell M (2014) Active targeting of early and mid-stage atherosclerotic plaques using self-assembled peptide amphiphile micelles. Biomaterials 35:8678–8686. https://doi.org/10.1016/j.biomaterials.2014.06.054
30. Kastantin M, Ananthanarayanan B, Karmali P et al (2009) Effect of the lipid chain melting transition on the stability of DSPE-PEG(2000) micelles. Langmuir 25:7279–7286. https://doi.org/10.1021/la900310k
31. Lin BF, Marullo RS, Robb MJ et al (2011) De novo design of bioactive protein-resembling nanospheres via dendrimer-templated peptide amphiphile assembly. Nano Lett 11:3946–3950. https://doi.org/10.1021/nl202220q
32. Greenfield NJ, Fasman GD (1969) Computed circular dichroism spectra for the evaluation of protein conformation. Biochemistry 8:4108–4116
33. Demento SL, Siefert AL, Bandyopadhyay A et al (2011) Pathogen-associated molecular patterns on biomaterials: a paradigm for engineering new vaccines. Trends Biotechnol 29:294–306. https://doi.org/10.1016/j.tibtech.2011.02.004
34. Lauer-Fields JL, Nagase H, Fields GB (2000) Use of Edman degradation sequence analysis and matrix-assisted laser desorption/ionization mass spectrometry in designing substrates for matrix metalloproteinases. J Chromatogr A 890:117–125. https://doi.org/10.1016/S0021-9673(00)00396-4

Chapter 22

Controlled Assembly of the Filamentous Chaperone Gamma-Prefoldin into Defined Nanostructures

Douglas S. Clark and Dominic J. Glover

Abstract

Self-assembling protein templates have enormous potential for the fabrication of multifunctional nanostructures that require precise positioning of individual molecules, such as enzymes and inorganic moieties, in regular patterns. A recently described approach uses ultrastable filaments composed of the gamma-prefoldin (γPFD) protein and engineered connector proteins to construct novel architectures useful for basic research and practical applications in nanobiotechnology. Here we describe the production of the γPFD and connector proteins from *E. coli*, and the assembly of γPFD with connector proteins into macromolecular structures with defined shapes.

Key words Filament, Protein, Self-assembly, Biomaterials, Prefoldin, Template, Nanomaterials

1 Introduction

Self-assembling protein structures are ideal templates upon which to position and align functional materials for the creation of nanoscale devices. Naturally occurring structural protein complexes have been exploited as nanoscale scaffolds, including viral capsids, S-layer coat proteins, and chaperonin capsules [1–3]. While slight adjustments in the dimensions of these assemblies are possible, in most cases the dimensions are limited to those of the natural molecules, generating structures that do not meet the specifications desired for a specific application. In addition, many native proteins have specific environmental limitations, such as a narrow range of stability, which make them impractical for the large-scale fabrication of molecular devices.

An alternative strategy seeks to build complex protein architectures from simple subunits that are assembled with precision over large surface areas [4, 5]. A recently described approach uses a filamentous protein (Fig. 1a) called gamma-prefoldin (γPFD) from a deep-sea archaeon [6] as a building block to create geometrically ordered shapes of controllable

Andrew K. Udit (ed.), *Protein Scaffolds: Design, Synthesis, and Applications*, Methods in Molecular Biology, vol. 1798, https://doi.org/10.1007/978-1-4939-7893-9_22,

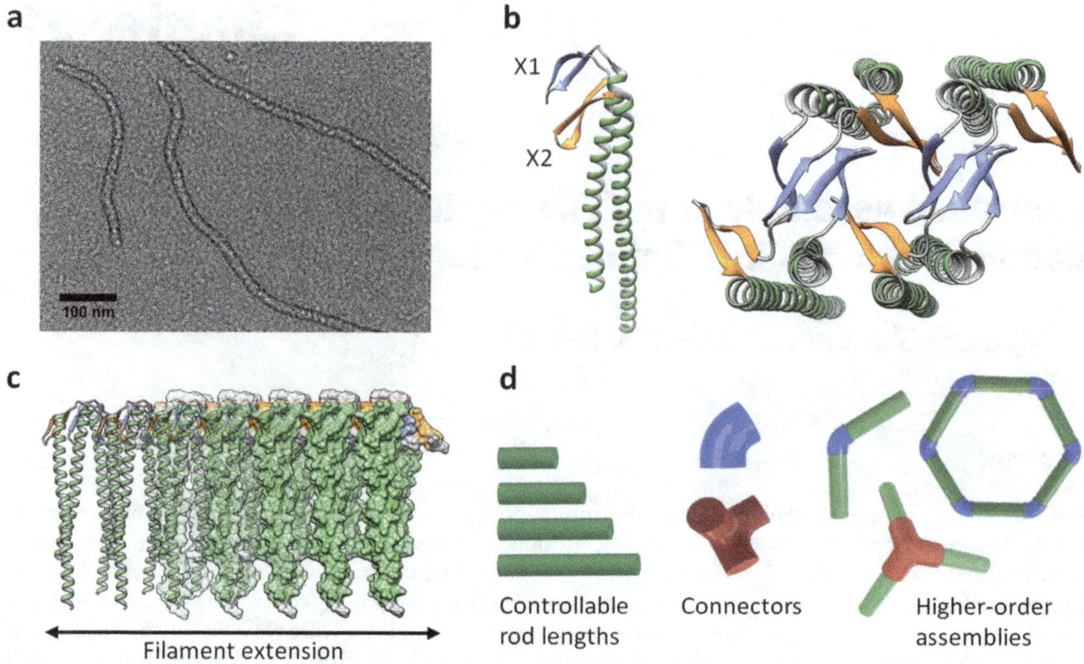

Fig. 1 Predicted structure and assembly of filaments into higher-order structures. (**a**) TEM image of γPFD filaments (scale bar = 100 nm). (**b**) The γPFD subunit is composed of two β-sheets, designated X1 and X2, which are flanked by helical regions that form a coiled-coil. (**c**) The β-sheet domains are responsible for the assembly of the subunits into long filaments. (**d**) Schematics for the design and assembly of filament rods and connectors into higher-order structures. Reproduced from Glover et al. (2016) with permission from Nature Publishing Group

dimensions [7]. The γPFD subunit consists of two β-sheet regions, designated X1 and X2, which are flanked by long α-helical domains that associate together as a coiled-coil (Fig. 1b). The β-sheets are responsible for the oligomerization of the subunits into highly stable filaments that can exceed several microns in length (Fig. 1c) [8]. The γPFD filament functions as a molecular chaperone with its coiled-coils serving to arrest protein misfolding and aggregation [9].

Initial shape engineering of γPFD was achieved by the creation of a capping protein called TERM (thermophilic extension resistant mutant) to control filament length [10]. The TERM contains mutations in the X2 β-sheet region of γPFD (Fig. 1b) that prevent filament elongation but still enable γPFD binding through the functional X1, thereby allowing TERM to attach to the termini of nascent filaments and cap their extension. Although the γPFD originates from an archaeon, the protein and variants such as TERM can be efficiently expressed in and purified from *E. coli*.

The TERM protein can be used to control the length distribution of γPFD filaments [9]; however, these capped-filaments are unable to assemble into complex shapes. To overcome this limitation, connector proteins were developed (Fig. 1d) that used the strong and specific interaction of an engineered heterodimer coiled-coil to link capped-filaments together into complex shapes [7]. The coiled-coil consists of two oppositely charged helical domains (E-coil and K-coil), which bind each other but not themselves [11]. The E-coil and K-coil were substituted into the X2 domain of TERM to create the bait–prey proteins, TERM-Ecoil and TERM-Kcoil [7]. When refolded in the presence of γPFD (Fig. 2a), these proteins function as a TERM to cap filament extension in a concentration-dependent manner (Fig. 2b). The resulting capped filaments have a narrow distribution of filament lengths that can be further reduced by size-exclusion chromatography (SEC) (Fig. 2c). Furthermore, as these capped filaments contain either TERM-Ecoil or TERM-Kcoil subunits at the termini, the filaments can bind to their respective binding partner.

To link multiple capped filaments together, a 3-way connector protein was created by fusing a trimerization domain called foldon [12] to the TERM-Kcoil. The resulting protein, TERM-Kcoil-foldon, forms a trimeric assembly that joins three TERM-Ecoil capped filaments into "pinwheel" structures with relatively uniform arms (Fig. 2d) [7]. Here we describe the production of the γPFD and connector proteins from an *E. coli* system, the assembly and capping of filaments with the TERM-Ecoil, and their linkage into complex shapes using the TERM-Kcoil-foldon 3-way connector protein (Fig. 2e). Subsequently, the resulting protein structures are routinely visualized and imaged by transmission electron microscopy (TEM).

2 Materials

All solutions are prepared using ultrapure water and analytical grade reagents. Adjust the pH of buffers containing NaH_2PO_4 with 2 M NaOH. Diligently follow all waste disposal regulations when disposing waste material. Special consideration must be given to waste contaminated by uranyl acetate.

2.1 Protein Expression

1. Lysogeny broth (LB) media: Weigh 10 g of tryptone, 5 g of yeast extract, 10 g of NaCl, and add water to a volume of 1 L. Autoclave.
2. Ampicillin: Prepare a 100 mg/mL solution in water, filter-sterilize through a 0.22 μm filter into a sterile tube, and store at −20 °C.
3. LB-agar ampicillin plates: Weigh 5 g of tryptone, 2.5 g of yeast extract, 5 g of NaCl, 7.5 g of agar, and add water to a volume of 500 mL in a 1 L bottle. Autoclave and cool LB-agar to 60 °C. Add 0.5 mL of ampicillin stock to achieve a final concentration of 100 μg/mL and mix by gentle swirling of the bottle. Pour

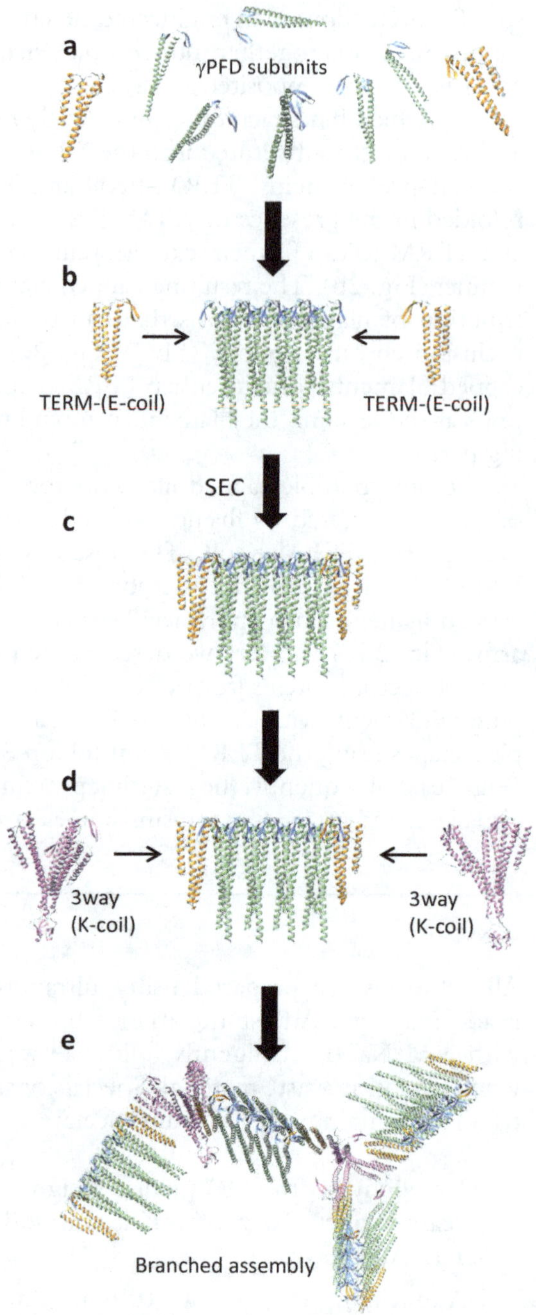

Fig. 2 Strategy for the assembly of γPFD filaments with engineered connector proteins. (**a**) Denatured γPFD and TERM-Ecoil subunits are mixed together at specific molar ratios. (**b**) The proteins are refolded and the TERM-Ecoil incorporates at both termini of the nascent filaments. (**c**) The TERM-Ecoil capped filaments are separated by size-exclusion chromatography (SEC) to produce filaments of a defined length. (**d**) The capped filaments are assembled with the TERM-Kcoil-foldon connector (**e**), which links multiple filaments together into branched structures

LB-agar ampicillin into petri dishes under sterile technique, let plates cool for 30 min and store plates inverted at 4 °C.

4. Isopropyl β-D-1-thiogalactopyranoside (IPTG): Prepare a 1 M solution in water, filter-sterilize through a 0.22 μm filter into a sterile tube, and store at −20 °C.
5. *E. coli* competent cells for protein expression compatible with the T7 promoter, such as BL21(DE3).
6. Phosphate buffered saline (PBS): 20 mM NaH_2PO_4, pH 8.0, 150 mM NaCl in water. Store at 4 °C.
7. Shaker incubator and 2 L baffled shake flasks.
8. UV-Vis spectrophotometer and cuvettes.
9. French pressure cell press or cell homogenizer.
10. Centrifuge with appropriate rotors and centrifugation tubes.

2.2 Purification of the γPFD Protein

1. Phosphate buffered saline (PBS): 20 mM NaH_2PO_4, pH 8.0, 150 mM NaCl in water. Store at 4 °C.
2. High salt buffer: 50 mM NaH_2PO_4, pH 8.0, 1 M NaCl in water. Store at 4 °C.
3. 0.45 μm syringe filters (PVDF or cellulose acetate membrane).
4. Chromatography resins and columns: 40 mL Q Sepharose fast flow column (strong anion ion-exchange resin) and 40 mL Toyopearl Butyl-650C column (hydrophobic interaction resin).
5. Fast Protein Liquid Chromatography (FPLC) system with UV-Vis detector.
6. Reagents for SDS-PAGE gels: 10–20% polyacrylamide tricine protein gels, prestained denaturing protein ladder, SDS-PAGE running buffer, and a sample loading buffer.
7. Dialysis membrane with a molecular weight cutoff of 10 kDa or less.
8. Freeze-dryer/Lyophilizer.
9. 15 mL centrifuge filter columns with a 10 kDa molecular mass cutoff.

2.3 Purification of the Connector Proteins

1. Strep-Tactin wash buffer: 50 mM NaH_2PO_4, pH 8.0, 300 mM NaCl in water. Store at 4 °C.
2. Strep-Tactin elution buffer: 5 mM desthiobiotin, 50 mM NaH_2PO_4, pH 8.0, 300 mM NaCl in water. Store at 4 °C.
3. 5 mL prepacked Strep-Tactin sepharose columns.
4. Reagents for SDS-PAGE gels: 10–20% polyacrylamide tricine protein gels, pre-stained denaturing protein ladder, SDS-PAGE running buffer, and a sample loading buffer.
5. Dialysis membrane with a molecular weight cutoff of 10 kDa or less.

6. 15 mL centrifuge filter columns with a 10 kDa molecular mass cutoff.

2.4 Filament and Connector Assembly

1. Protein solubilization buffer: 8 M guanidinium-HCl, 10 mM NaH_2PO_4, pH 8.0. Store at room temperature (*see* **Note 1**).
2. Phosphate buffered saline (PBS): 20 mM NaH_2PO_4, pH 8.0, 150 mM NaCl in water. Store at 4 °C.
3. Bradford reagent and bovine serum albumin (BSA) protein concentration standard.
4. Strep-Tactin sepharose resin.
5. Strep-Tactin elution buffer: 5 mM desthiobiotin, 50 mM NaH_2PO_4, pH 8.0, 300 mM NaCl in water. Store at 4 °C.
6. Size-exclusion chromatography (SEC) column and resin: Sephacryl S-400 HR resin in a column with a bed height of 90 cm and diameter of 1.6 cm.
7. SEC buffer: 20 mM NaH_2PO_4, pH 8.0, 300 mM NaCl in water. Store at 4 °C.

2.5 Imaging of Filament Assemblies

1. Transmission electron microscope (TEM).
2. 400-mesh carbon-formvar coated copper grids that have been freshly glow discharged/plasma cleaned.
3. Phosphate buffer saline (PBS): 20 mM NaH_2PO_4, pH 8.0, 150 mM NaCl in water. Store at 4 °C.
4. Molecular biology grade DEPC-treated water.
5. 2% uranyl acetate in water that has been filtered through a 0.22 μm syringe filter. Store at room temperature in dark.
6. Small strips of Whatman filter paper (approximately 1 cm × 2 cm in size).

3 Methods

Due to the high thermal stability of the γPFD and connector proteins, these procedures can be carried out at room temperature. Long-term storage of proteins however, should be at 4 °C to prevent contamination. Biological waste including growth media should be disposed of using appropriate procedures.

3.1 Protein Expression

1. Transform plasmids encoding ampicillin resistance and either the wild-type γPFD, TERM-Ecoil, or TERM-Kcoil-foldon into a T7 promoter compatible *E. coli* strain such as BL21(DE3). Plate the transformed bacteria onto LB-agar ampicillin plates, and incubate overnight at 37 °C (*see* **Note 2**).

2. Pick colonies on the plates the next day and inoculate 5 mL LB starter cultures containing 100 μg/mL ampicillin. Incubate starter culture overnight on a shaker at 37 °C.
3. Inoculate flasks containing 1 L of LB and 100 μg/mL ampicillin with 1 mL of overnight starter culture.
4. Grow cultures on a shaker at 37 °C to an optical density of 0.6 at A_{600}, and induce protein expression for 15 h at 25 °C with the addition of 100 μL of IPTG to the 1 L culture (to achieve a final concentration of 0.1 mM IPTG).
5. Centrifuge the expression cultures at 6000 × *g* for 10 min to pellet the bacterial cells, and discard the supernatant.
6. Suspend the cells in 30 mL of PBS per liter of culture and lyse using either a French pressure cell press or homogenizer (*see* **Note 3**).
7. Clarify the cell lysates by centrifugation at 22,000 × *g* for 40 min, and keep the supernatant that contains the soluble proteins.

3.2 Purification of the γPFD Protein

1. Heat the cell lysate containing the wild-type γPFD at 80 °C for 1 h. The cell lysate will turn into an opaque white color (*see* **Note 4**).
2. Centrifuge the heat-treated cell lysate at 22,000 × *g* for 20 min to remove precipitated proteins. Keep the supernatant that contains the soluble native γPFD.
3. Filter the supernatant through a 0.45 μm syringe filter immediately before loading onto a Q sepharose fast flow column (*see* **Note 5**).
4. Wash the resin with at least 5 column volumes of PBS or until the FPLC UV trace has returned to the baseline.
5. Elute the γPFD using a NaCl gradient from 150 mM in PBS to the 1 M NaCl in the high salt buffer over 8 column volumes while collecting individual 1 mL fractions.
6. Run the eluted fractions on an SDS-PAGE gel to visualize proteins, and pool fractions containing γPFD (Fig. 3).
7. Add an equal volume of high salt buffer to the pooled fractions to ensure that the sample contains at least 500 mM NaCl for effective binding to the hydrophobic interaction column.
8. Load pooled fractions containing γPFD onto a Toyopearl Butyl-650C column.
9. Wash the resin with at least 5 column volumes of high salt buffer or until the UV trace has returned to the baseline.
10. Elute the γPFD using a NaCl gradient from 1 M NaCl (high salt buffer) to 150 mM NaCl in PBS over 8 column volumes while collecting individual 1 mL fractions.
11. Run the eluted fractions on an SDS-PAGE gel to visualize proteins, and pool fractions containing γPFD.

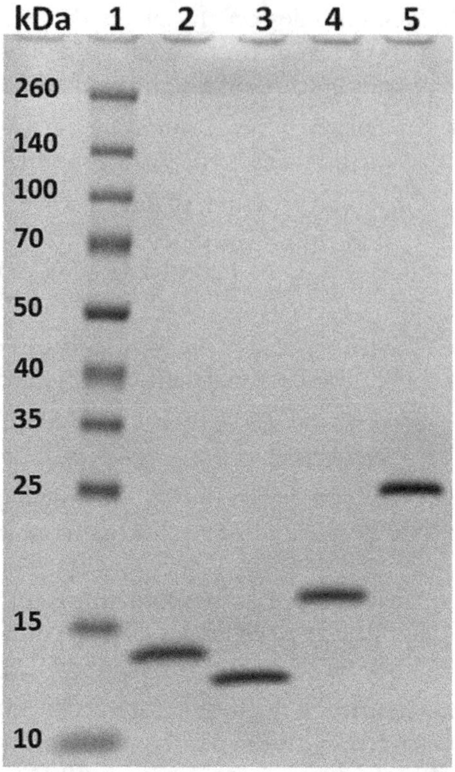

Fig. 3 SDS-polyacrylamide gel electrophoresis of bacterially expressed and purified proteins. Lane 1: Spectra BR protein ladder; lane 2: the γPFD protein purified by ion-exchange chromatography followed by hydrophobic interaction chromatography; lane 3: hexa-histidine-tagged TERM protein purified using nickel-NTA affinity chromatography; lane 4 and 5: strep-tagged 3-way and 2-way connector proteins, respectively, purified by Strep-Tactin affinity chromatography. Reproduced from Glover et al. (2016) with permission from Nature Publishing Group

12. Dialyze the pooled fractions of γPFD against PBS for 4 h at 4 °C. Change the dialysis buffer and dialyze for an additional 4 h at 4 °C.
13. Concentrate the γPFD protein to 100 μL using 10 kDa molecular mass cutoff centrifuge filter columns (*see* **Note 6**).
14. Transfer the concentrated γPFD protein to 1.5 mL tubes and freeze at −80 °C.
15. Lyophilize the γPFD protein for 24 h and store at −20 °C.

3.3 Purification of the Connector protein

1. Load the bacterial lysates containing either connector protein (TERM-Ecoil or TERM-Kcoil-foldon) onto Strep-Tactin columns (*see* **Note 7**).
2. Wash the Strep-Tactin resin using Strep-Tactin wash buffer with at least 5 column volumes or until the UV trace has returned to the baseline.
3. Elute connector proteins using the Strep-Tactin elution buffer until the UV trace has returned to baseline.

4. Run the eluted connector proteins on an SDS-PAGE gel to visualize proteins and confirm their purity (Fig. 3) (*see* **Note 8**).
5. Dialyze the eluted connector proteins against PBS for 4 h at 4 °C. Change the dialysis buffer and dialyze for an additional 4 h at 4 °C.
6. Concentrate the dialyzed protein samples to 100 μL using 10 kDa molecular mass cutoff centrifuge filter columns.
7. Transfer the concentrated proteins to 1.5 mL tubes and freeze at −80 °C.
8. Lyophilize the protein samples for 24 h and store at −20 °C.

3.4 Filament and Connector Assembly

1. Suspend lyophilized protein stocks (γPFD, TERM-Ecoil, and TERM-Kcoil-foldon) in solubilization buffer. Add 100 μL of the buffer per 2 mg of powered protein (*see* **Note 9**).
2. Incubate for at least 3 h, then pipette the solution up and down to ensure that the protein is fully dissolved in the solubilization buffer (*see* **Note 10**).
3. Centrifuge the dissolved proteins at 10,000 × *g* for 2 min to pellet insoluble salts and buffers, and transfer supernatants to fresh tubes.
4. Determine the concentration of the dissolved proteins via a Bradford assay against a standard curve of BSA (*see* **Note 11**).
5. Combine 3.5 mg of denatured γPFD with 0.44 mg of denatured TERM-Ecoil to a final volume of 2 mL using solubilization buffer in a 50 mL tube (*see* **Note 12**).
6. Add 46 mL of PBS to the protein mixture and immediately mix by gently inverting the tube several times (*see* **Note 13**).
7. Incubate for 18 h at 40 °C to refold the proteins and assemble filaments.
8. Add 2 mL bead-volume of Strep-Tactin resin to the refolded proteins and incubate by gentle inversion for 3 h at room temperature.
9. Centrifuge at 200 × *g* for 5 min to pellet the Strep-Tactin resin. Discard the supernatant.
10. Suspend the Strep-Tactin resin in 48 mL of PBS.
11. Repeat **steps 9** and **10** twice more to wash the resin of any unbound proteins (*see* **Note 14**).
12. Centrifuge the Strep-Tactin once more at 200 × *g* for 5 min, discard the supernatant.
13. Suspend the Strep-Tactin resin in 2 mL of Strep-Tactin Elution buffer. Incubate for 5 min.
14. Centrifuge the resin at 200 × *g* for 5 min. Pipette off and keep the supernatant that contains the TERM-Ecoil capped filaments.

15. Repeat **steps 13** and **14** to ensure that all TERM-Ecoil capped filaments are eluted and pool the two eluted samples (*see* **Note 15**).
16. Equilibrate a Sephacryl S-400 HR SEC column with 5 column volumes of SEC buffer at a flow rate of 1 mL/min.
17. Load the 4 mL of eluted TERM-Ecoil capped filaments into the FPLC injection loop.
18. Inject the TERM-Ecoil capped filaments onto the SEC column with SEC buffer at a flow rate of 0.5 mL/min.
19. Run the SEC column with SEC buffer at a flow rate of 0.5 mL/min and collect 2 mL fractions.
20. Record which fractions contain protein by monitoring the FPLC UV trace (*see* **Note 16**).
21. Determine the concentration of the TERM-Ecoil capped filaments in each fraction by measuring the absorbance at 205 nm. Calculate the protein concentration using the Beer-Lambert law with an extinction coefficient ($\varepsilon_{\text{molar}}$) of 31/M/cm and b as the path length in cm (*see* **Note 17**):

$$\text{Concentration}\left(\text{mg}/\text{mL}\right)=A_{205}/\left(31\times b\right)$$

22. Add between 18.2 and 832 ng of the denatured TERM-Kcoil-foldon to 5 μL of solubilization buffer in a 1.5 mL tube (*see* **Note 18**).
23. Dilute 2 μg of the TERM-Ecoil capped γPFD filaments in PBS to a final volume of 95 μL.
24. Add the diluted TERM-Ecoil capped γPFD filaments to the tube containing the TERM-Kcoil-foldon, mix immediately by pipetting the sample up and down several times, and incubate at 40 °C for 18 h. Subsequently, the assembled protein structures are stored at 4 °C.

3.5 Imaging of Filament Assemblies

1. Glow discharge/plasma clean the formvar-carbon grids (*see* **Note 19**).
2. Prepare dilutions of the γPFD-connector assemblies in PBS to a final concentration of 20 μg/mL (*see* **Note 20**).
3. Deposit 7 μL of a diluted protein sample onto the TEM grid for 2 min at room temperature.
4. Wick off the sample using a strip of filter paper applied to a corner of the TEM grid.
5. Rinse grid with five drops of ultrapure water, and wick off any remaining liquid.
6. Stain protein assemblies on the grid with 2% uranyl acetate for 1 min, then wick off the stain (*see* **Note 21**).

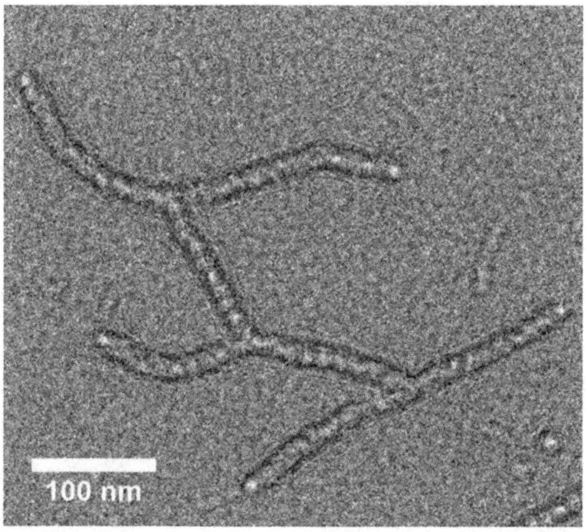

Fig. 4 TEM image of TERM-Ecoil capped γPFD filaments assembled with TERM-Kcoil-foldon connectors into a branched structure. Reproduced from Glover et al. (2016) with permission from Nature Publishing Group

7. The negatively stained filament assemblies on the grid are routinely imaged by TEM at 120 kV accelerating voltage (Fig. 4).

4 Notes

1. The 8 M guanidinium-HCl in the solubilization buffer may precipitate at low temperatures. If crystals are observed in the buffer after storage, heat to 60 °C and stir to redissolve guanidinium-HCl prior to use.
2. The plasmid backbone is the pET-19b vector from Novagen (Merck) that encodes ampicillin resistance, a *lacI* coding sequence, and a T7 promoter upstream of a multiple cloning site. Genes encoding either wild-type γPFD (GenBank accession MJ0648), TERM-Ecoil, or TERM-Kcoil-foldon were cloned into the multiple cloning site of the pET-19b.
3. Nonmechanical cell lysis methods that use detergents and/or lysozyme are generally inefficient in liberating γPFD filaments from *E. coli*.
4. An advantage of working with thermophilic proteins such as the γPFD is the ability to remove most *E. coli* proteins by heating. Contaminating *E. coli* proteins are denatured and precipitate when heated to 80 °C, while the γPFD remains soluble, which simplifies further downstream purification.
5. Additional protein aggregation may occur in the heat-treated lysate if stored for periods greater than a few hours and can rapidly clog the 0.45-μm syringe filter. If this occurs, centrifuge the sample again at 22,000 × *g* for 20 min to remove precipitated proteins.

6. The γPFD tends to form a gel at high concentration at the interface of the filter in the centrifuge column. After concentrating the protein sample to 100 μL, carefully pipette the solution up and down on the filter to suspend and redissolve the protein gel.
7. The connector proteins have also been successfully purified using nickel nitrilotriacetic acid (Ni-NTA) agarose affinity chromatography. Filaments of γPFD, however, naturally have weak affinity for Ni-NTA requiring stringent wash and elution conditions in subsequent steps.
8. The foldon domain in the TERM-Kcoil-foldon protein is resistant to denaturation by SDS, which can result in the protein running as a trimer in the SDS-PAGE gel. Ensure that the sample is fully denatured by addition of the loading buffer and heating at 100 °C for 20 min before loading onto an SDS-PAGE gel.
9. The solution may be cloudy due to trace amounts of salts and buffers being insoluble in the high guanidinium concentration.
10. The solubilization buffer completely denatures the γPFD and connector proteins, which enables these subunits to be subsequently refolded and assembled in a controlled manner.
11. The guanidinium-HCl in the solubilization buffer does not interfere with the Bradford assay if the concentration is kept under 2 M. The stock protein samples can be diluted in PBS to prepare working stocks for the assay.
12. Alternatively, simple branched filament structures can be produced by mixing the denatured γPFD with denatured TERM-foldon and refolding by addition of PBS. However, due to the natural length distribution of the filaments (50 nm–3 μm), the resulting branched structures are nonuniform and random in assembly.
13. Dilution of the guanidinium-HCl by the addition of PBS facilitates the refolding and assembly of the protein subunits into filaments.
14. The γPFD does not contain a Strep-tag and is unable to bind the Strep-tactin resin unless it is capped by the TERM-Ecoil that contains an N-terminal Strep-tag.
15. In addition to removing uncapped filaments, the Strep-Tactin purification also serves to concentrate the filaments into a smaller volume before loading onto the SEC column.
16. The final peak observed in the FPLC UV trace is free TERM-Ecoil that is unbound to any γPFD subunits.
17. The γPFD has a molar extinction coefficient (ε) equal to 5960/M/cm at 280 nm. This low extinction coefficient at 280 nm is due to a lack of tryptophan and cysteine residues in the protein, which makes the measurement of dilute γPFD samples difficult and inaccurate. Therefore, it is recommended to use the Scopes method [13] to measure the concentration of the SEC fractions by measuring absorbance at 205 nm, ideally

in a NanoDrop to minimize the loss of sample. The substantial number of peptide bonds within proteins absorb at 205 nm, resulting in measurements that are more sensitive and display less protein-protein variability than Abs_{280} measurements.

18. The amount of crosslinking between filaments is dependent on the concentration of the TERM-Kcoil-foldon. It is recommended to use a molar ratio of TERM-Kcoil-foldon to capped filaments ranging between 2:3 and 30:1. A greater ratio of the TERM-Kcoil-foldon connector produces a greater amount of branched structures, which may ultimately be used to assemble simple filaments into higher-order structures.
19. This makes the carbon surface hydrophilic, which allows aqueous solutions to spread evenly and improves filament deposition.
20. A higher concentration of protein (0.05–0.1 mg/mL) can be used in preparing TEM grids; however, this may result in individual filament assemblies overlaying each other.
21. Negatively stained proteins are relatively stable on grids, but ideally should be imaged within a few days.

Acknowledgments

We are grateful to Samuel Lim for reviewing the manuscript and making helpful suggestions. This work was supported by the Air Force Office of Scientific Research (FA9550-14-1-0026).

References

1. Oh D, Qi J, Lu Y-C et al (2013) Biologically enhanced cathode design for improved capacity and cycle life for lithium-oxygen batteries. Nat Commun 4:2756
2. Abe S, Ueno T (2015) Design of protein crystals in the development of solid biomaterials. RSC Adv 5(27):21366–21375
3. Bruns N, Pustelny K, Bergeron LM et al (2009) Mechanical nanosensor based on FRET within a thermosome: damage-reporting polymeric materials. Angew Chem Int Ed Engl 48(31):5666–5669
4. Glover DJ, Clark DS (2016) Protein calligraphy: a new concept begins to take shape. ACS Cent Sci 2(7):438–444
5. Luo Q, Hou C, Bai Y et al (2016) Protein assembly: versatile approaches to construct highly ordered nanostructures. Chem Rev 116(22):13571–13632
6. Whitehead TA, Boonyaratanakornkit BB, Höllrigl V et al (2007) A filamentous molecular chaperone of the prefoldin family from the deep-sea hyperthermophile Methanocaldococcus jannaschii. Protein Sci 16(4):626–634
7. Glover DJ, Giger L, Kim SS et al (2016) Geometrical assembly of ultrastable protein templates for nanomaterials. Nat Commun 7:11771
8. Glover DJ, Giger L, Kim JR et al (2013) Engineering protein filaments with enhanced thermostability for nanomaterials. Biotechnol J 8(2):228–236
9. Glover DJ, Clark DS (2015) Oligomeric assembly is required for chaperone activity of the filamentous γ-prefoldin. FEBS J 282(15):2985–2997
10. Whitehead TA, Meadows AL, Clark DS (2008) Controlling the self-assembly of a filamentous hyperthermophilic chaperone by

an engineered capping protein. Small 4(7): 956–960

11. Litowski JR, Hodges RS (2002) Designing heterodimeric two-stranded alpha-helical coiled-coils. Effects of hydrophobicity and alpha-helical propensity on protein folding, stability, and specificity. J Biol Chem 277(40): 37272–37279
12. Bhardwaj A, Walker-Kopp N, Wilkens S et al (2008) Foldon-guided self-assembly of ultra-stable protein fibers. Protein Sci 17(9): 1475–1485
13. Scopes RK (1974) Measurement of protein by spectrophotometry at 205 nm. Anal Biochem 59(1):277–282

Chapter 23

Advances in the Application of Designed Ankyrin Repeat Proteins (DARPins) as Research Tools and Protein Therapeutics

Ykelien L. Boersma

Abstract

Nonimmunoglobulin scaffolds have been developed to overcome the limitations of monoclonal antibodies with regard to stability and size. Of these scaffolds, the class of designed ankyrin repeat proteins (DARPins) has advanced the most in biochemical and biomedical applications. This review focuses on the recent progress in DARPin technology, highlighting the scaffold's potential and possibilities.

Key words Designed ankyrin repeat protein, DARPin, Protein engineering, Protein therapeutic, Research tool, Diagnostics

1 Introduction

For more than a century, antibodies have attracted clinical interest as a means for passive immunization. It was not until the 1970s, when hybridoma technology for the production of monoclonal antibodies (mAbs) was established [1], that mAbs became readily available as a scientific tool and an important pharmaceutical resource. Currently, on average four new therapeutic mAbs are approved for clinical use each year, indicating that by 2020 approximately 70 mAbs will be on the market [2]. This increase in mAbs has been supported by the advent of recombinant DNA technology [3], functional expression of recombinant antibody fragments in *Escherichia coli* [4], and the development of in vitro selection methods such as phage display [5] and ribosome display [6]. Most engineering efforts are directed toward antibody fragments such as scFvs and Fabs. Unfortunately, their biophysical properties are unfavorable as these fragments have a great tendency to aggregate, particularly when used under reducing conditions [7]. This hampers the development of novel molecular formats conjugated to effector functions such as prodrugs or toxins [8–10]. In addition,

Andrew K. Udit (ed.), *Protein Scaffolds: Design, Synthesis, and Applications*, Methods in Molecular Biology, vol. 1798,
https://doi.org/10.1007/978-1-4939-7893-9_23,

for therapeutic purposes scFvs and Fabs are usually converted to the more stable IgG format. Yet this larger molecular size will lead to poorer penetration of the affected tissue [11].

Advances in recombinant DNA technology and selection strategies not only enabled antibody engineering: other binding molecules could be created to overcome the mAb's limitations while retaining its strengths. Hence, these new scaffolds should bind their target antigens with high specificity and affinity. Important considerations for the design of an alternative scaffold are stability and absence of aggregation. Disulfides should be avoided to allow for efficient folding in the cytoplasm of *E. coli*. Combining binding molecules of different specificities in one multimeric format to increase avidity or cross-link epitopes should be easy to produce, preferably maintaining the same production yield as the original monomeric binder. Of course, these assemblies or fusion proteins should not give rise to aggregation [9]. In recent years, a plethora of nonimmunoglobulin-based scaffolds of mostly human origin has been developed (reviewed in [12–14]). These scaffolds meet the described requirements for an alternative scaffold, and share similar characteristics: all possess relatively low molecular weights, containing between 40 and 100 amino acid residues of which 10–20 positions are usually randomized.

Repeat proteins, consisting of a number of repeating structural modules, have attracted much attention as an alternative scaffold. They are more commonly found in eukaryotes, particularly in metazoans, than in prokaryotes [15]. Nature even applies repeat proteins as an adaptive immune system in some jawless invertebrates; as such, repeat proteins could be regarded as "antibodies" [16].The repeats are the result of intragenic duplication and recombination events and are relatively common, occurring in at least 14% of all proteins [17]. Interactions between adjacent repeat modules define and stabilize the shape and curvature of the overall structure [18, 19]. As a result, the elongated structure has a larger surface area to volume ratio than globular proteins, making repeat proteins particularly suited for protein–protein interactions [20]. The structural and sequence similarity between repeating modules permits the design of an ideal module while retaining the characteristics of the family. This principle has been applied to a number of scaffolds, e.g., armadillo repeat proteins [21], alpha-helicoidal repeat proteins [22], and tetratricopeptide repeats [23]. The class of designed ankyrin repeat proteins (DARPins), however, has advanced the most in biochemical and biomedical applications [24]; therefore, this review is focused on the recent progress in DARPin technology and applications.

2 DARPin Architecture

In Nature, ankyrin repeat proteins are involved in protein–protein interactions in a wide variety of organisms. They consist of tightly packed repeating modules of usually 33 amino acids. Each module forms a structural unit comprised of a β-turn followed by two antiparallel α-helices (Fig. 1a) [25]. Usually, ankyrin repeat domains consist of four to six repeats, which form a right-handed solenoid structure. The overall structure of the protein twists into a superhelix, resulting in a groove-like binding surface [18, 26].

Binz et al. exploited the modular structure of ankyrin repeat proteins for the generation of combinatorial libraries with self-compatible repeat modules [27]. Thus, as long as compatible interfaces between units are present, repeat units can be inserted, deleted or replaced without significantly affecting the overall structure [28]. In addition, these compatible interfaces will stabilize structural features inherent to a naturally occurring class of repeat proteins, thereby increasing the stability of library members [29]. From complexes of naturally occurring ankyrin repeat proteins with their targets, conserved framework residues were identified and fixed in the resulting DARPin. Surface residues potentially interacting with the target were randomized in a DARPin library format, avoiding the residues Cys to eliminate disulfide formation, and Pro and Gly as some of these residues are located in a helix [30]. The resulting libraries have an N2C (two internal repeats) or N3C (three internal repeats) format (Fig. 1b). From available 3D structures, it was shown that DARPin-target complexes closely match the library design with on average 75% of randomized positions contacting the target [31].

Hydrophilic capping repeats originating from the guanine-adenine-binding protein were included in the DARPin design at the N- and C-terminus. These proved essential for soluble protein expression: the hydrophobic core will thus be shielded from the hydrophilic environment, leading to efficient folding [32]. Indeed, most DARPins can be brought to very high protein concentrations without aggregating and show high thermodynamic stability against unfolding induced by heat or chemical denaturants [33]. Nevertheless, molecular dynamics (MD) experiments carried out on DARPins with one to five internal consensus repeat modules showed that the C-cap was the first to denature in almost all high-temperature simulations [32]. To improve the protein's stability, the C-cap was redesigned, making it more similar to the consensus. This new C-cap was proven to be much more resistant to unfolding due to better packing [34, 35].

DARPin target binding occurs through rigid-body interactions, conformational changes of the DARPin backbone do not take place. Some target epitopes however cannot accommodate

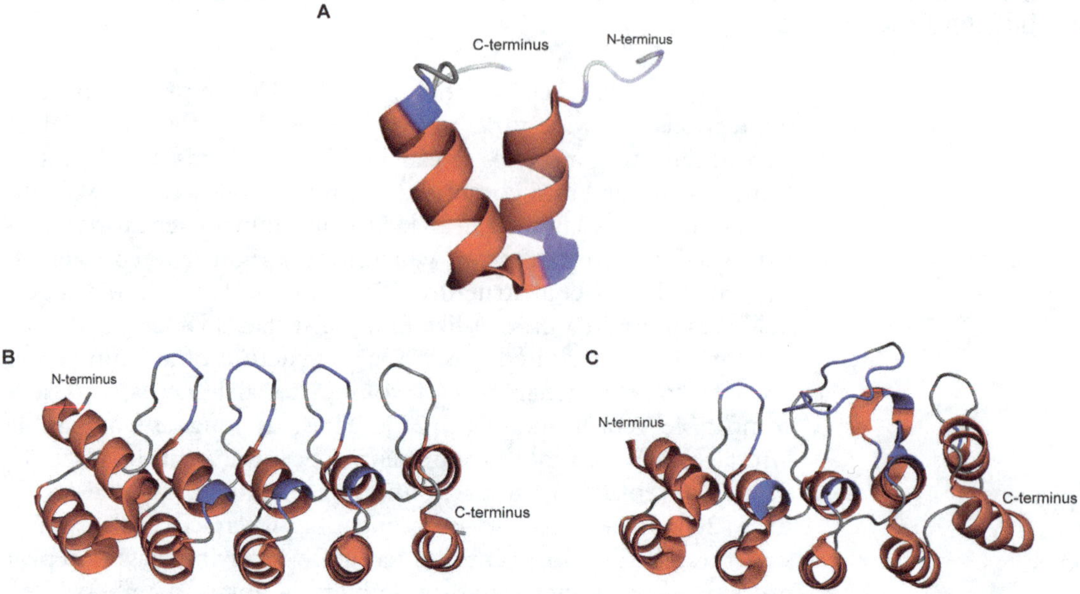

Fig. 1 Overview of 3D structures of the DARPin scaffold. (**a**) Representation of one DARPin repeat (based on PDB 1MJ0). Randomized residues per repeat are depicted in blue. (**b**) Representation of unselected N3C library member E3_5 (PDB 1MJ0). Randomized residues in the repeats are depicted in blue. (**c**) Representation of an unselected LoopDARPin (PDB 4K5C). Randomized residues are depicted in blue

this type of binding and may require more flexibility in binding. In addition, the incomplete randomized binding surface could prove a limiting factor. Inspired by an antibody's antigen-binding CDR-H3 loop, a 19-amino acid loop consensus sequence was engineered into the DARPin scaffold at the central β-turn in the second internal repeat [36]. The loop contains ten variable positions, with the top of the loop having conformational flexibility; as this loop would be surrounded by randomized β-turns of neighboring repeats, a continuous interaction surface would be created. To prevent interference with these β-turns in the surrounding repeats, the stem of the loop was kept constant. To extend the interaction surface of the LoopDARPin even further, both N- and C-capping repeats were randomized to create a LoopDARPin library. The design of LoopDARPins was validated with a 3D structure (Fig. 1c). Biophysical characterization of unselected LoopDARPin library members showed that introduction of the loop did not decrease the stability of the scaffold [36].

3 Selecting DARPins

Using synthetic DARPin libraries, binders can be selected against virtually any target with most selection methods. The majority of selections with DARPins have been carried out by ribosome display, a powerful fully in vitro method which physically links mRNA encoding the DARPin (the genotype) to the translated DARPin protein (the phenotype) via the ribosome [6, 37]. As DARPins fold well in vitro, they may be particularly suited for this selection strategy; selections typically require fewer rounds than with other protein structures, e.g., scFv fragments [38]. Specific LoopDARPins can even be selected in one round of ribosome display [36]. Selection of variants from library sizes of 10^{12} can be combined with error-prone PCR and efficient off-rate selections to increase affinity and specificity [39]. Using ribosome display, many high affinity DARPins with dissociation constants (K_d) in the low nanomolar to picomolar range have been selected against a range of different targets, e.g., kinases [40, 41], tubulin [42, 43], caspases [44–46], and HIV gp120 protein [47].

Similar to ribosome display, SNAP display is an in vitro selection strategy that makes use of the more stable DNA rather than mRNA [48]. SNAP display covalently links genotype with phenotype using a SNAP-tag in an emulsion microdroplet. The SNAP-tag itself is based on the O^6-alkylguanine-DNA alkyltransferase, an enzyme which will covalently bind DNA templates labeled with the suicide substrate benzylguanine [49]. Houlihan et al. improved the performance of DARPin-SNAP display and optimized conditions for the 10^7-fold enrichment of a HER2-binding DARPin [50]. Next, they applied SNAP display for the affinity maturation of DARPin G3-HAVD, which binds the HER2 receptor with submicromolar affinity [51]. Starting from a library with $>10^9$ variants, after four rounds of selection binders with low- to sub-nanomolar affinities were identified. The best binder, having accumulated nine mutations, showed an almost 700-fold improvement in affinity; nevertheless, a trade-off between affinity and thermal stability was observed [52].

Filamentous phage display has been extensively used for selection of peptides or antibodies with desired characteristics. The display of DARPins on the p3 phage coat protein however was shown to be problematic or even impossible, as the DARPin-p3 fusion will fold before it is secreted to a form expressed in the *E. coli* inner membrane during phage assembly. By exchanging the Sec signal sequence for a signal recognition particle (SRP)-dependent sequence, efficient display of DARPins was established, comparable to display of slow folding proteins via a Sec-dependent signal sequence [53]. Steiner et al. successfully applied SRP-phage display to select specific DARPins with subnanomolar affinities against a number of targets, including EGFR and HER2 [54].

Recently, yeast display has been added to the list of potential selection strategies for DARPins. Although yeast display is limited to smaller library sizes of 10^8 variants, the technique can be easily interfaced with fluorescence-activated cell sorting (FACS) for particular binding, selectivity or fluorescent properties. DARPins are well compatible with the yeast quality-control system of the endoplasmic reticulum, leading to high display levels [24, 55].

4 DARPins as Tools in Biochemistry

Due to their excellent biophysical and biochemical properties, DARPins have found an application as research tools in a number of experimental settings.

4.1 Crystallization Chaperones

Crystallization of complex biological macromolecules (e.g., membrane proteins, transient conformational states and intrinsically disordered proteins) may require a chaperone: this binding partner can reduce the conformational freedom by selectively stabilizing a particular conformation of the target protein. Furthermore, it can increase the surface area providing alternative crystal contacts. Chaperones are chosen on the basis of their biophysical properties, and should obviously crystallize easily [56, 57]. DARPins are ideal chaperone candidates as they have excellent biophysical properties and readily crystallize. Around 30 3D structures of specific DARPins in complex with their target proteins are available in the Protein Data Bank (for reviews, see references [14, 31, 57]).

Yet the DARPin's relatively small size and repetitive structure may limit their potential to create crystal contacts. Seeger et al. therefore designed a second-generation DARPin library based on available 3D structures of DARPin-target protein complexes to select binders against difficult targets, such as membrane proteins or soluble proteins with hydrophobic surface patches. [58]. The frequency of apolar residues at randomized positions was reduced to prevent selection of unspecific hydrophobic binders, substituting them with tyrosine residues that are specifically suited to mediate molecular contacts in protein–protein interactions [59]. The surface entropy of the DARPin backbone was reduced to favor the formation of crystal contacts; finally, an optimized C-cap with an extended randomized surface was included in the design to further improve DARPin-target interactions. Second-generation DARPins showed slightly reduced thermodynamic stability compared to the original design. To validate the library design, ribosome display selections against the membrane protein AcrB and against caspases 3 and 7 were performed. A specific enrichment of binders against these targets was already observed after two rounds of selection. Cocrystallization of the target proteins with their respective binders showed both tyrosine and tryptophan residues modulating protein–protein interactions. These findings suggest that synthetic libraries with a reduced number of apolar residues at randomized positions should be enriched in tryptophan as well. Though the test selections were only performed with three targets, the results are encouraging for a successful selection of binders against other challenging proteins.

In a different approach, Batyuk et al. modified the DARPin to generate rigid domain-domain fusions at the N- or C-terminal capping repeat, thus creating a general format for DARPins with an extended surface to improve crystal contacts [60]. Since the DARPin fold starts and ends with an α-helix, embedding either of these helices in the compatible terminal helix of a fusion partner could result in a rigid fusion of two domains. A stabilized version of TEM-1 β-lactamase (BL) [61] was identified as the most promising fusion partner: it contains both an N- and C-terminal α-helix and can be crystallized at high resolution [62]. As fusions to the N-terminal DARPin-helix were more likely to sterically hinder target binding, the N-terminal BL-helix was fused to the DARPin's C-terminal helix. Changing the length of the shared helix leads to an alteration of the relative orientation of the two domains; consequently, different crystallization chaperones can be engineered. The resulting fusion proteins were well expressed in the cytoplasm of *E. coli*. To validate the design, DARPin-BL fusion proteins were crystallized alone and in complex with their respective targets. Indeed, residues from the BL-domain provided at least 80% of the crystal contact interface area. In the complex with the target proteins, the BL-domain still provided 46.8% (in the Off7—maltose binding protein complex) and 77.1% (in the 4G61—green fluorescent protein complex) of the crystal contacts, despite making up roughly 30% of the mass of the complex. Thus, DARPin-BL fusion proteins can dominate crystal packing. Importantly, using the constant DARPin C-cap as a fusion partner, DARPins with different binding specificities can be exchanged, making this strategy a generic one [60].

4.2 Biosensors

Visualizing molecules in live cells has advanced our knowledge of molecular cell biology: single or multiple biological molecules can be tracked in vivo and their localization, trafficking, functionality and interactions can be analyzed. To this purpose, DARPins have a significant advantage over antibodies, as they do not contain any cysteine residues and can be functionally expressed in the reducing environment of the cytoplasm [27, 63].

Traditionally, live-cell imaging has been accomplished by genetically fusing the protein of interest to an autofluorescent protein, such as green fluorescent protein (GFP) [64]. Alternatively, a fluorescent probe can be chemically or enzymatically coupled to the protein of interest. Recently, fluorogen-activating proteins (FAPs) were described, based on human scFvs. These FAPs bind dyes upon which they will only then strongly fluoresce; as a consequence, background fluorescence is limited [65]. Schütz et al. expanded the FAP toolbox by applying DARPins as so-called FADAs, fluorescence-activating DARPins [55]. These FADAs were selected against the dye malachite green via ribosome display. Though the selected DARPins bound to the dye, they only exhibited weak fluorescence-activating properties. To apply selection pressure on fluorescence activation, binders were matured using yeast surface display interfaced with FACS. The best FADA, FADA-3210, showed an enhancement of fluorescence by a factor of 11,000 over the background. Elucidation of the 3D structure revealed that two FADA-3210 molecules bound malachite green as a homodimer, rigidly holding the dye in its ground-state conformation in a tight fitting hydrophobic pocket. This feature was exploited in the creation of a biosensor based on FADA-3210: a DARPin with a particular specificity can be genetically fused to FADA-3210, in such a way that this resulting bispecific DARPin will block dimerization of the FADA molecules upon binding to its target protein, thus abolishing fluorescence activation. Using an approach similar to the previously described crystallization chaperones [60], the C-terminal helix of a GFP-specific DARPin [60] was fused to the N-terminal helix of FADA-3210. Though fluorescence activation was significantly reduced, a complete loss of fluorescence activation was not observed. Hence, a more rigid design with an improved geometrical arrangement is needed to generate a sensitive biosensor. Nevertheless, as sufficient size of the target protein is the only requirement for this system, it should indeed be possible to create a generic biosensor using FADA proteins [55].

Post-translational modification (PTM) of proteins plays a key role in the regulation and propagation of intracellular signaling cascades, thus fine-tuning protein function. Binding proteins specifically targeting PTMs would therefore be useful tools to study different protein states. As a proof of principle, DARPins were selected against the nonphosphorylated (i.e., inactive) and the doubly phosphorylated, active state of the mitogen-activated protein kinase (MAPK) extracellular signal-regulated kinase (ERK2) [40]. Two conformation-specific DARPins, discriminating between the active and inactive state, were cocrystallized with their target proteins; it was shown that both DARPins bind mainly to the activation loop and an adjacent region, while the phosphorylated ERK2-specific DARPin recognizes a conformational change upon phosphorylation. The selected DARPins could be functionally expressed in the cytoplasm of mammalian COS-7 cells, and bound their targets specifically as shown by bioluminescence resonance energy transfer (BRET). In follow-up experiments, these conformation-specific DARPins were used as biosensors to study patterns of ERK2 activity in living cells [66]. Here, the DARPin specific for phosphorylated ERK2 was conjugated to a solvatochromic merocyanine dye; the fluorescence of this dye increases in a more hydrophobic environment, which is created upon binding of the DARPin to the target. Fluorescence properties and binding specificity were dependent on both the positioning and the type of dye used for labeling. The ERK-activity sensor was highly selective for phosphorylated ERK2 over ERK2, and did not bind to other MAPK family members. Phosphorylated ERK2 was predominantly found in the nucleus, consistent with previous studies. This approach could be easily adapted to study proteins with different PTMs, providing useful tools to study biological processes in situ.

5 DARPins in Diagnostic Applications: HER2 Detection

Antibodies still dominate the field of biomedical research and diagnostics, despite the fact that for these purposes usually only monovalent binding is required. Nevertheless, a diverse set of DARPins is available for diagnostic detection systems. DARPins specific for the HER2 receptor were applied in the histological analysis of breast cancer tissue; methods such as immunohistochemistry (IHC) or fluorescence in situ hybridization are used to select patients for anti-HER2 therapy [67]. In quantitative IHC, DARPin G3, which binds the HER2 receptor with picomolar affinity, could detect a positive HER2 amplification status with similar sensitivity as and significantly higher specificity than an FDA-approved monoclonal antibody against HER2 [68].

DNA-mediated affinity-based assays, such as proximity ligation assays (PLA), play an increasing role in disease diagnosis and prognosis [69]. PLA makes use of oligonucleotides conjugated to affinity reagents as reporter molecules binding to a protein target; PCR amplification and hybridization of a labeled probe to the oligos will generate a signal, enabling sensitive detection of proteins, modifications and interactions [70, 71]. Using a SNAP tag, Söderberg and coworkers were able to site-specifically label HER2-specific DARPins with one oligonucleotide per DARPin. The efficacy of the resulting PLA probes was demonstrated in in situ PLA experiments: using two DARPins with different HER2-epitope specificities on HER2-overexpressing SK-OV-3 cells, both monomeric and dimeric HER2 could be discriminated [72].

Unfortunately, IHC and PLA analyses of HER2 are single time-point samples from a single site. As information on the heterogeneity of expression or changes in expression pattern over time is not available, patients may receive a suboptimal therapy due to a misclassification of HER2 status. To overcome these limitations, Chester and colleagues labeled DARPin G3 with ^{111}In and ^{125}I for use in positron emission tomography (PET) and single-photon emission computed tomography (SPECT) imaging [73]. DARPin G3 labeled with ^{111}In achieved a high tumor-to-blood ratio within 24 h of administration to mice grafted with HER2-positive tumors. Its serum half-life was short and the radiolabeled DARPin showed low non-renal tissue uptake. Thus, this new diagnostic tool is well suited for HER2 imaging, and patients would benefit from a noninvasive diagnostic technique [73].

6 DARPins in Therapeutic Applications

Traditionally, small molecules and monoclonal antibodies have been applied as therapeutics. DARPins are finding a place in therapeutic intervention as well, yet their potential goes beyond that of small molecules and monoclonal antibodies. DARPins can be easily engineered to interact through novel cross-linking mechanisms, specifically retarget viruses without detectable off-targeting effect, or have an extended half-life. The small DARPin size ensures a short serum half-life due to renal clearance, but since engineering of DARPins is facilitated by thiol or click chemistry, a PEG molecule can easily be introduced at either the N- or C-terminus [74, 75]. Thus, the hydrodynamic radius of the resulting DARPin is increased beyond the size limit for kidney filtration.

6.1 Targeting Multiple Epitopes with Naked DARPins

IgE antibodies that bind the high-affinity IgE Fc recptor (FcεRI) found on mast cells and basophils will trigger an allergic reaction. Though an antibody is currently on the market to treat severe allergy asthma, many antibodies never reach the market due to a risk of anaphylactogenicity [76]. As DARPins do not have an Fc region, they would circumvent this risk; therefore, DARPins were selected against the IgE binding site on the FcεRIα [77]. Monovalent DARPins targeting only one epitope on the FcεRIα were not able to completely inhibit IgE binding. However, fusing two DARPins with distinct epitopes on the FcεRIα via a glycine-serine linker resulted in efficient interference with IgE-binding and inhibition of pro-inflammatory mediator release in a concentration-dependent manner [77].

In an alternative approach to decrease serum IgE levels, neutralizing DARPins were selected against a functional IgE epitope in the antibody's constant region [76]. These DARPins did not show any cross-reactivity with other human serum proteins, which is of key importance as IgE is the least abundant Ig class in humans. DARPin E2_79 was capable of inhibiting proinflammatory mediator release to the same extent as the monoclonal neutralizing antibody omalizumab. Elucidation of the 3D structure of E2_79 in complex with its target IgE-Fc showed two distinct binding sites for E2_79. Superimposition of this 3D structure with that of the IgE-Fc-FcεRIα complex showed that at site 2, three E2_79 and five FcεRIa residues make contacts <3.5 Å, causing steric hindrance and preventing IgE from binding to its receptor. In addition, quantitative kinetic modeling indicated that E2_79 facilitated IgE-FcεRIα dissociation at site 2, thus blocking an allergic reaction [78]. Recently, the disruptive efficacy of E2_79 was further improved by fusing it to nondisruptive DARPin E3_53, creating the biparatopic DARPin bi79_53 [79]. This enabled tethering of E2_79 to the IgE-FcεRIα complex, further facilitating dissociation of IgE. In addition, bi79_53 was more potent in blocking anaphylaxis in a mouse model: full inhibition of the IgE-mediated allergic response was achieved at 5 μM, whereas this effect was only achieved at 50 μM for omalizumab or E2_79 alone [79]. Thus, these disruptive IgE-inhibitors hold a promise for treatment of allergic asthma.

A combination of DARPins with different specificities in one molecular format was also applied in the targeting of the epidermal growth factor receptor (EGFR) family members; these receptors are frequently overexpressed in cancer and play a central role in tumorigenesis [80]. Boersma et al. used EGFR-specific DARPins to target the receptor as expressed on the A431 human epidermal cancer cell line [81]. They found that monovalent DARPins were able to inhibit cell proliferation, though this was a short-term effect. Fusing two EGFR-DARPins with distinct epitopes via a leucine zipper created a tetravalent, bispecific DARPin, which was capable of efficiently inhibiting proliferation. The inhibitory effect was achieved through downregulation of EGFR as a result of inhibition of receptor recycling back to the cell surface (Fig. 2) [81].

As described previously, HER2 is frequently overexpressed in breast cancer, where its overexpression is correlated with a poor prognosis [82]. HER2 drives tumorigenesis by spontaneously forming homodimers and larger receptor clusters, as well as forming ligand-independent heterodimers with other family members, particularly HER3 [83]. The effect of the monoclonal antibodies trastuzumab (Herceptin®) and pertuzumab (Omnitarg®) is unfortunately modest and resistance is rapidly acquired [84]. As this motivates research into therapeutics with new modes of actions, DARPins binding the extracellular domain of HER2 were selected via ribosome display [51, 85]. The selected monovalent DARPins covered different conformational epitopes on the HER2 target protein, though did not affect survival and proliferation of HER2-overexpressing tumor cells [86]. After determination of the epitopes of the selected DARPins [86, 87], a panel of bivalent and

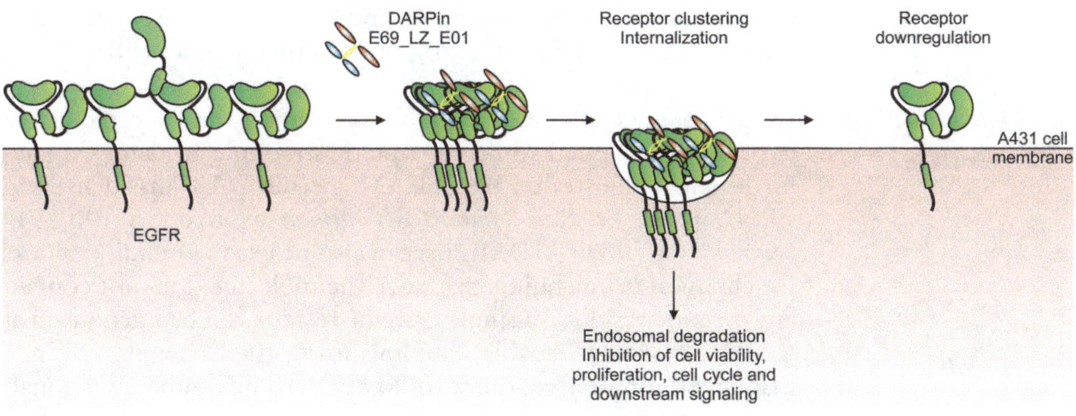

Fig. 2 Schematic overview of the activity of DARPin E69_LZ_E01. The bispecific DARPin molecule, tetravalent as it forms a homodimer via a leucine zipper, binds two distinct epitopes on EGFR, leading to receptor clustering and internalization. Instead of recycling back to the cell surface, EGFR is degraded, resulting in receptor down regulation [81]

bispecific constructs targeting epitopes on domains I and IV was generated by genetically fusing two DARPins via glycine-serine linkers of different lengths. These bispecific DARPins were not only cytostatic, but cytotoxic as well. The biological activity depended on the orientation of the monovalent DARPins in the construct as well as on the linker length used, with the shortest linker of only five amino acids showing the highest cytotoxicity. Yet this linker would be too short to connect two DARPins bound to the same HER2 molecule or engaged to two HER2 receptors in a back-to-back homodimer. Instead, the bivalent DARPin construct induced an intermolecular tilt of the ectodomains, made possible by a few disordered residues in the juxtamembrane region, acting like a pivot. Cross-linking of HER2 enforced a receptor geometry incompatible with receptor associations and trans-activation of the intracellular tyrosine kinase domains [86]. The cross-linking did not lead to an increase in internalization or degradation of the receptor in a panel of HER2-addictive cell lines. In an exhaustive analysis of the biological activity of the bispecific HER2-DARPins 6L1G and 9L1H on HER2-overexpressing cell lines, Tamaskovic et al. found that these DARPins induced dephosphorylation of both HER2 and HER3 [88]. Furthermore, DARPin 6L1G effectively induced apoptosis in vitro and in vivo via activation of the intrinsic pathway, despite being an extracellular ligand. DARPin 6L1G promoted this effect through effectively trapping HER2, consequently obstructing formation of homo- or heterodimers with EGFR and HER3. As such, 6L1G acts in fact as a pan-HER inhibitor. In contrast, trastuzumab only interferes with the ligand-independent HER2-HER3 heterodimers, leading to decreased activation of PI3K/AKT and inhibition of proliferation. However, trastuzumab-treated cells can adapt to bypass HER3 for the activation of PI3K/AKT. This study gives important insights for structure- and mechanism-based engineering of effective HER2-targeting protein therapeutics [88].

6.2 DARPins Guiding a Payload

DARPins, selected for their high target affinity and specificity, are excellently equipped to act as a guide molecule to deliver a payload. This is illustrated by EpCAM-specific DARPins, delivering siRNA [89], or *Pseudomonas aeruginosa* exotoxin A [90, 91] (reviewed in [9, 24]). DARPins can also be used as building blocks for chemical drug conjugates: since the molecule does not contain any cysteine residues, a single cysteine residue can be introduced at the C-terminus to enable coupling using thiol chemistry. When made methionine-free, the scaffold can even be doubly functionalized by introducing a single N-terminal azidohomoalanine, which is incorporated instead of the initiator methionine; this enables the use of click chemistry for site-specific conjugation [74, 92, 93]. These conjugation strategies expand the engineering space of DARPins for biomedical applications.

Specific viral delivery of a payload remains a challenge for successful gene therapy. Here, DARPins offer new approaches to generalize viral delivery. Dreier et al. developed bispecific adapters, consisting of two modules, one binding to the fiber knob of adenovirus serotype 5, the other guiding the adapter to, e.g., a tumor marker [94, 95]. As the guide DARPin can be easily exchanged for a DARPin with a different specificity, the adenovirus can be retargeted to many different cell types; therefore, the system is essentially generic [95]. Knob-specific DARPins were selected via ribosome display. These DARPins blocked interaction of the virus with the natural CAR receptor, and bound with such high affinity to their target that they remained fully bound for more than 10 days. Using the 3D structure of the trimeric knob protein in complex with three bound DARPins and computer modeling, a trimeric adapter complex was designed by making use of the trimerization module SHP from the lambdoid phage 21 [95, 96]. The specificity of the system was shown with the gene delivery of adeno-associated viral (AAV) vectors specific for HER2, EpCAM or CD4: no detectable off-targeting was observed, neither ex vivo nor in vivo [97].

6.3 DARPins in Clinical Trials

Unmodified DARPins can target their antigens efficiently as long as the affinity is in the picomolar range [24, 75, 98]. This feature was exploited in the targeting of vascular endothelial growth factor (VEGF)-A, a cytokine stimulating angiogenesis and vascular hyperpermeability [99]. VEGF-A plays a crucial role in the proliferative phase of diabetic retinopathy and age-related macular degeneration (AMD), ultimately leading to a loss of vision. Drugs targeting VEGF-A need to be frequently administered via intravitreal injection due to low efficacy and fast ocular clearance. Obviously, these treatments bring great discomfort to patients; the administration of a drug should be simplified and the frequency of the injections should ideally be reduced [100]. Therefore, DARPins targeting VEGF-A were selected in seven rounds of ribosome display. Though affinity maturation was not applied, the VEGF-A-specific DARPins showed single-digit picomolar affinity. All DARPins inhibited the interaction of VEGF-A with its receptor VEGFR-2, thereby suppressing receptor phosphorylation and endothelial cell sprout formation in vitro. Four days after intravitreal injection of 50 μg of DARPin into one eye of pigmented rabbits, an almost complete inhibition of vascular leakage was observed. The applied dose is significantly below 500 μg, the clinically used dose of the VEGF-A-specific mAb ranibizumab. Topical application of DARPins, thus circumventing an intravitreal injection, diminished corneal neovascularization by 43% in a rabbit suture model and suppressed laser-induced neovascularization in a rat model [99].

In parallel phase I/II, open-label, multicenter, dose-escalation trials the safety and preliminary efficacy of lead DARPin MP0112

(Abicipar Pegol®) was assessed on patients with diabetic macular edema (DME) [101] and AMD [102]. MP0112 has a long ocular half-life of 2 weeks, thus potentially reducing the injection frequency. Twelve weeks after a single intraocular injection of 0.4 mg, levels of MP0112 were still detectable in the patients' aqueous humor, with a median concentration of 10 nM. This concentration is several orders of magnitude above the picomolar half-maximal inhibitory concentration (IC_{50}); MP0112 effectively neutralized VEGF-A as aqueous humor levels of VEGF-A were below the limit of detection. As a result, prolonged edema reduction and improvement in vision was seen in several patients. These data support the progression of MP0112 to phase III clinical trials for the treatment of DME [103].

A group of 32 patients with exudative AMD received a single injection of MP0112 at doses ranging from 0.04 to 3.6 mg and were monitored for 16 weeks. Dose-dependent decreases in retinal thickness and leakage area were observed for patients receiving a higher dose. Visual acuity was assessed and scores were found to be stable or improved after 4 weeks of treatment compared to the baseline [102]. Nevertheless, drug-related ocular inflammation was reported in 13 of 32 patients, and attributed to impurities in the MP0112 formulation. The formulation procedure of MP0112 has since been optimized and phase III clinical trials for MP0112 in AMD are now underway [104].

7 Perspective

Since the advent of hybridoma technology, the application of mAbs has had a major impact on research, diagnostics, and medicine. Nevertheless, mAb use is limited by stability issues and large size leading to poor tissue penetration. To overcome these limitations, a great number of nonimmunoglobulin scaffolds including DARPins have been designed and developed in the past decade. DARPins have excellent biochemical and biophysical properties, exemplified by their high solubility and yield in *E. coli*, and extreme stability. Not surprisingly, they therefore offer a great alternative to antibodies as research tools, as they can easily be selected against a specific target with high affinity. An additional yet not unimportant benefit is the cheap and fast production process in *E. coli*, which provides the researcher with a seemingly endless supply of DARPin reagent. Due to the small size of DARPins (14–17 kDa) compared to human IgG molecules (150 kDa), a DARPin will penetrate and distribute more effectively into tissue. Combined with a short serum half-life, DARPins are preeminently suited as imaging reagents, and can be further engineered to improve signal-to-noise ratios.

However, the DARPin's size and kinetic properties do act like a double-edged sword as these characteristics make them less suitable as protein therapeutics. Human IgG molecules have half-lives of days to weeks; the interaction of their Fc region with the FcRn receptor can be further engineered to achieve half-lives of even several months [105]. For a DARPin to have a longer half-life, conjugation with hydrophilic molecules such as PEG is a prerequisite. As a consequence, the increase in mass eliminates one important advantage of DARPins, namely their small size and low molecular weight. In addition, the stability of the DARPin molecule could be affected. Alternatively, to overcome limitations posed by low molecular weight and fast clearance, the use of slow-release formulations or innovative drug delivery devices could be explored for the delivery of DARPins. Slow-release formulations have not been extensively used to deliver alternative scaffolds in general. Most alternative scaffolds act as antagonists, generally requiring a relatively high dose to neutralize the target molecule effectively; these doses have been considered too high to be formulated successfully in a slow-release product. However, with reports of picomolar DARPins, e.g., targeting HER2 [85] and VEGF-A [99], such considerations might no longer be applicable, and doses of up to 10 mg per administration become feasible [106, 107].

The targeting of distinct molecules with one multispecific construct has met with considerable interest in recent years. Multispecific or multimeric constructs could enhance neutralization efficacy, or increase clearance through the formation of large immune complexes. When it comes to multispecific molecules, alternative scaffolds have a clear and major advantage over traditional mAbs: novel formats are easy to produce, with properties such as solubility and stability being similar to that of a monospecific scaffold. In addition, multimerization of the scaffold can boost the effect of the monomeric binder, as seen for EGFR-DARPins [81].

Nevertheless, when the number of DARPin molecules in one construct increases, so does the number of repeating motifs and with it, the immunogenic potential. Immunogenicity, the triggering of an innate immune response, is a complex process; it is not only influenced by the structure of the therapeutic protein, but also by, among others, formulation, dosing and patient-related factors. It is difficult to predict whether a therapeutic protein would elicit an immune response: no models, neither in vivo nor in vitro, are available. Clinical studies remain the standard to assess the immunogenicity of a therapeutic protein. For DARPin MP0112, clinical data indicated that the DARPin was well tolerated in patients with DME or ADM [101, 102]. However, as the DARPin is locally administered, it remains difficult to answer questions on whether the protein would elicit a systemic immune response. Nevertheless, in this case the risk of immunogenicity would clearly be balanced by the patient benefit of receiving a single dose treatment.

Since their creation more than ten years ago, DARPins have inspired creativity and innovation through their use as research tools. Still, challenges remain, and the use of DARPins in clinical practice will require additional innovation. It will therefore be exciting to see the progress of DARPins in biomedical applications.

References

1. Köhler G, Milstein C (1975) Continuous cultures of fused cells secreting antibody of predefined specificity. Nature 256:495–497. https://doi.org/10.1038/256495a0
2. Ecker DM, Jones SD, Levine HL (2015) The therapeutic monoclonal antibody market. MAbs 7:9–14. https://doi.org/10.4161/19420862.2015.989042
3. Mullis KB, Faloona FA (1987) Specific synthesis of DNA in vitro via a polymerase-catalyzed chain reaction. Methods Enzymol 155:335–350
4. Skerra A, Plückthun A (1988) Assembly of a functional immunoglobulin Fv fragment in *Escherichia coli*. Science 240:1038–1041
5. Clackson T, Hoogenboom HR, Griffiths AD et al (1991) Making antibody fragments using phage display libraries. Nature 352:624–628. https://doi.org/10.1038/352624a0
6. Hanes J, Plückthun A (1997) *In vitro* selection and evolution of functional proteins by using ribosome display. Proc Natl Acad Sci 94:4937–4942
7. Honegger A (2008) Engineering antibodies for stability and efficient folding. Handb Exp Pharmacol:47–68. https://doi.org/10.1007/978-3-540-73259-4_3
8. Stumpp MT, Binz HK, Amstutz P (2008) DARPins: a new generation of protein therapeutics. Drug Discov Today 13:695–701. https://doi.org/10.1016/j.drudis.2008.04.013
9. Boersma YL, Plückthun A (2011) DARPins and other repeat protein scaffolds: advances in engineering and applications. Curr Opin Biotechnol 22:1–9
10. Hey T, Fiedler E, Rudolph R et al (2005) Artificial, non-antibody binding proteins for pharmaceutical and industrial applications. Trends Biotechnol 23:514–522. https://doi.org/10.1016/j.tibtech.2005.07.007
11. Thurber GM, Schmidt MM, Wittrup KD (2008) Antibody tumor penetration. Adv Drug Deliv Rev 60:1421. https://doi.org/10.1016/j.addr.2008.04.012
12. Löfblom J, Frejd FY, Ståhl S (2011) Non-immunoglobulin based protein scaffolds. Curr Opin Biotechnol 22:843–848. https://doi.org/10.1016/j.copbio.2011.06.002
13. Škrlec K, Štrukelj B, Berlec A (2015) Non-immunoglobulin scaffolds: a focus on their targets. Trends Biotechnol 33:408–418. https://doi.org/10.1016/j.tibtech.2015.03.012
14. Jost C, Plückthun A (2014) Engineered proteins with desired specificity: DARPins, other alternative scaffolds and bispecific IgGs. Curr Opin Struct Biol 27:102–112. https://doi.org/10.1016/j.sbi.2014.05.011
15. Andrade MA, Perez-Iratxeta C, Ponting CP (2001) Protein repeats: structures, functions, and evolution. J Struct Biol 134:117–131. https://doi.org/10.1006/jsbi.2001.4392
16. Pancer Z, Cooper MD (2006) The evolution of adaptive immunity. Annu Rev Immunol 24:497–518
17. Marcotte EM, Pellegrini M, Yeates TO et al (1999) A census of protein repeats. J Mol Biol 293:151–160. https://doi.org/10.1006/jmbi.1999.3136
18. Kobe B, Kajava AV (2000) When protein folding is simplified to protein coiling: the continuum of solenoid protein structures. Trends Biochem Sci 25:509–515
19. Brunette TJ, Parmeggiani F, Huang P-S et al (2015) Exploring the repeat protein universe through computational protein design. Nature 528:580–584. https://doi.org/10.1038/nature16162
20. Grove TZ, Cortajarena AL, Regan L (2008) Ligand binding by repeat proteins: natural and designed. Curr Opin Struct Biol 18:507–515. https://doi.org/10.1016/j.sbi.2008.05.008
21. Reichen C, Hansen S, Plückthun A (2014) Modular peptide binding: from a comparison of natural binders to designed armadillo repeat proteins. J Struct Biol 185:147–162. https://doi.org/10.1016/j.jsb.2013.07.012
22. Valerio-Lepiniec M, Urvoas A, Chevrel A et al (2015) The αRep artificial repeat protein

scaffold: a new tool for crystallization and live cell applications. Biochem Soc Trans 43:819–824. https://doi.org/10.1042/BST20150075

23. Speltz EB, Nathan A, Regan L (2015) Design of protein-peptide interaction modules for assembling supramolecular structures in vivo and in vitro. ACS Chem Biol 10:2108–2115. https://doi.org/10.1021/acschembio.5b00415
24. Plückthun A (2015) Designed ankyrin repeat proteins (DARPins): binding proteins for research, diagnostics, and therapy. Annu Rev Pharmacol Toxicol 55:489–511. https://doi.org/10.1146/annurev-pharmtox-010611-134654
25. Li J, Mahajan A, Tsai MD (2006) Ankyrin repeat: a unique motif mediating protein-protein interactions. Biochemistry (Mosc) 45:15168–15178
26. Parmeggiani F, Huang P-S (2017) Designing repeat proteins: a modular approach to protein design. Curr Opin Struct Biol 45:116–123. https://doi.org/10.1016/j.sbi.2017.02.001
27. Binz HK, Stumpp MT, Forrer P et al (2003) Designing repeat proteins: well-expressed, soluble and stable proteins from combinatorial libraries of consensus ankyrin repeat proteins. J Mol Biol 332:489–503
28. Tripp KW, Barrick D (2004) The tolerance of a modular protein to duplication and deletion of internal repeats. J Mol Biol 344:169–178. https://doi.org/10.1016/j.jmb.2004.09.038
29. Forrer P, Stumpp MT, Binz HK et al (2003) A novel strategy to design binding molecules harnessing the modular nature of repeat proteins. FEBS Lett 539:2–6
30. Binz HK, Amstutz P, Kohl A et al (2004) High-affinity binders selected from designed ankyrin repeat protein libraries. Nat Biotechnol 22:575–582
31. Gilbreth RN, Koide S (2012) Structural insights for engineering binding proteins based on non-antibody scaffolds. Curr Opin Struct Biol 22:413–420. https://doi.org/10.1016/j.sbi.2012.06.001
32. Interlandi G, Wetzel SK, Settanni G et al (2008) Characterization and further stabilization of designed ankyrin repeat proteins by combining molecular dynamics simulations and experiments. J Mol Biol 375:837–854
33. Kohl A, Binz HK, Forrer P et al (2003) Designed to be stable: crystal structure of a consensus ankyrin repeat protein. Proc Natl Acad Sci 100:1700–1705
34. Kramer MA, Wetzel SK, Plückthun A et al (2010) Structural determinants for improved stability of designed ankyrin repeat proteins with a redesigned C-capping module. J Mol 404:381–391. https://doi.org/10.1016/j.jmb.2010.09.023
35. Wetzel SK, Ewald C, Settanni G et al (2010) Residue-resolved stability of full-consensus ankyrin repeat proteins probed by NMR. J Mol Biol 402:241–258. https://doi.org/10.1016/j.jmb.2010.07.031
36. Schilling J, Schöppe J, Plückthun A (2014) From DARPins to LoopDARPins: novel LoopDARPin design allows the selection of low picomolar binders in a single round of ribosome display. J Mol Biol 426:691–721. https://doi.org/10.1016/j.jmb.2013.10.026
37. Dreier B, Plückthun A (2011) Ribosome display: a technology for selecting and evolving proteins from large libraries. Methods Mol Biol Clifton NJ 687:283–306. https://doi.org/10.1007/978-1-60761-944-4_21
38. Zahnd C, Amstutz P, Plückthun A (2007) Ribosome display: selecting and evolving proteins in vitro that specifically bind to a target. Nat Methods 4:269–279. https://doi.org/10.1038/nmeth1003
39. Zahnd C, Sarkar CA, Plückthun A (2010) Computational analysis of off-rate selection experiments to optimize affinity maturation by directed evolution. Protein Eng Des Sel 23:175–184. https://doi.org/10.1093/protein/gzp087
40. Kummer L, Parizek P, Rube P et al (2012) Structural and functional analysis of phosphorylation-specific binders of the kinase ERK from designed ankyrin repeat protein libraries. Proc Natl Acad Sci 109:E2248–E2257. https://doi.org/10.1073/pnas.1205399109
41. Parizek P, Kummer L, Rube P et al (2012) Designed ankyrin repeat proteins (DARPins) as novel isoform-specific intracellular inhibitors of c-Jun N-terminal kinases. ACS Chem Biol 7:1356–1366. https://doi.org/10.1021/cb3001167
42. Pecqueur L, Duellberg C, Dreier B et al (2012) A designed ankyrin repeat protein selected to bind to tubulin caps the microtubule plus end. Proc Natl Acad Sci 109:12011–12016. https://doi.org/10.1073/pnas.1204129109

43. Ahmad S, Pecqueur L, Dreier B et al (2016) Destabilizing an interacting motif strengthens the association of a designed ankyrin repeat protein with tubulin. Sci Rep 6:28922. https://doi.org/10.1038/srep28922
44. Flütsch A, Ackermann R, Schroeder T et al (2014) Combined inhibition of caspase-3 and caspase-7 by two highly selective DARPins slows down cellular demise. Biochem J 461:279–290. https://doi.org/10.1042/BJ20131456
45. Flütsch A, Schroeder T, Barandun J et al (2014) Specific targeting of human caspases using designed ankyrin repeat proteins. Biol Chem 395:1243–1252. https://doi.org/10.1515/hsz-2014-0173
46. Schroeder T, Barandun J, Flütsch A et al (2013) Specific inhibition of caspase-3 by a competitive DARPin: molecular mimicry between native and designed inhibitors. Structure 21:277–289. https://doi.org/10.1016/j.str.2012.12.011
47. Mann A, Friedrich N, Krarup A et al (2013) Conformation-dependent recognition of HIV gp120 by designed ankyrin repeat proteins provides access to novel HIV entry inhibitors. J Virol 87:5868–5881. https://doi.org/10.1128/JVI.00152-13
48. Stein V, Sielaff I, Johnsson K et al (2007) A covalent chemical genotype–phenotype linkage for *in vitro* protein evolution. Chembiochem 8:2191–2194. https://doi.org/10.1002/cbic.200700459
49. Houlihan G, Lowe D, Hollfelder F (2013) SNAP display—an *in vitro* method for the selection of protein binders. Curr Pharm Des 19:5421–5428
50. Houlihan G, Gatti-Lafranconi P, Kaltenbach M et al (2014) An experimental framework for improved selection of binding proteins using SNAP display. J Immunol Methods 405:47–56. https://doi.org/10.1016/j.jim.2014.01.006
51. Zahnd C, Pecorari F, Straumann N et al (2006) Selection and characterization of Her2 binding-designed ankyrin repeat proteins. J Biol Chem 281:35167–35175
52. Houlihan G, Gatti-Lafranconi P, Lowe D et al (2015) Directed evolution of anti-HER2 DARPins by SNAP display reveals stability/function trade-offs in the selection process. Protein Eng Des Sel 28:269–279. https://doi.org/10.1093/protein/gzv029
53. Steiner D, Forrer P, Stumpp MT et al (2006) Signal sequences directing cotranslational translocation expand the range of proteins amenable to phage display. Nat Biotechnol 24:823–831
54. Steiner D, Forrer P, Plückthun A (2008) Efficient selection of DARPins with sub-nanomolar affinities using SRP phage display. J Mol Biol 382:1211–1227. https://doi.org/10.1016/j.jmb.2008.07.085
55. Schütz M, Batyuk A, Klenk C et al (2016) Generation of fluorogen-activating designed ankyrin repeat proteins (FADAs) as versatile sensor tools. J Mol Biol 428:1272–1289. https://doi.org/10.1016/j.jmb.2016.01.017
56. Lieberman RL, Culver JA, Entzminger KC et al (2011) Crystallization chaperone strategies for membrane proteins. Methods San Diego Calif 55:293–302. https://doi.org/10.1016/j.ymeth.2011.08.004
57. Bukowska MA, Grütter MG (2013) New concepts and aids to facilitate crystallization. Curr Opin Struct Biol 23:409–416. https://doi.org/10.1016/j.sbi.2013.03.003
58. Seeger MA, Zbinden R, Flütsch A et al (2013) Design, construction and characterization of a second-generation DARPin library with reduced hydrophobicity. Protein Sci 22:1239–1257. https://doi.org/10.1002/pro.2312
59. Koide S, Sidhu SS (2009) The importance of being tyrosine: lessons in molecular recognition from minimalist synthetic binding proteins. ACS Chem Biol 4:325–334. https://doi.org/10.1021/cb800314v
60. Batyuk A, Wu Y, Honegger A et al (2016) DARPin-based crystallization chaperones exploit molecular geometry as a screening dimension in protein crystallography. J Mol Biol 428:1574–1588. https://doi.org/10.1016/j.jmb.2016.03.002
61. Kather I, Jakob RP, Dobbek H et al (2008) Increased folding stability of TEM-1 beta-lactamase by *in vitro* selection. J Mol Biol 383:238–251. https://doi.org/10.1016/j.jmb.2008.07.082
62. Minasov G, Wang X, Shoichet BK (2002) An ultrahigh resolution structure of TEM-1 beta-lactamase suggests a role for Glu166 as the general base in acylation. J Am Chem Soc 124:5333–5340
63. Brient-Litzler E, Plückthun A, Bedouelle H (2009) Knowledge-based design of reagentless fluorescent biosensors from a designed ankyrin repeat protein. Protein Eng Des Sel 23:229–241. https://doi.org/10.1093/protein/gzp074
64. Brauchle M, Hansen S, Caussinus E et al (2014) Protein interference applications in cellular and developmental biology using DARPins that recognize GFP and mCherry. Biol Open 3:1252–1261. https://doi.org/10.1242/bio.201410041

65. Szent-Gyorgyi C, Schmidt BF, Schmidt BA et al (2008) Fluorogen-activating single-chain antibodies for imaging cell surface proteins. Nat Biotechnol 26:235–240. https://doi.org/10.1038/nbt1368
66. Kummer L, Hsu C-W, Dagliyan O et al (2013) Knowledge-based design of a biosensor to quantify localized ERK activation in living cells. Chem Biol 20:847–856. https://doi.org/10.1016/j.chembiol.2013.04.016
67. Wolff AC, Hammond MEH, Hicks DG et al (2013) Recommendations for human epidermal growth factor receptor 2 testing in breast cancer: American Society of Clinical Oncology/College of American Pathologists clinical practice guideline update. J Clin Oncol Off J Am Soc Clin Oncol 31:3997–4013. https://doi.org/10.1200/JCO.2013.50.9984
68. Theurillat J-P, Dreier B, Nagy-Davidescu G et al (2010) Designed ankyrin repeat proteins: a novel tool for testing epidermal growth factor receptor 2 expression in breast cancer. Mod Pathol 23:1289–1297. https://doi.org/10.1038/modpathol.2010.103
69. Gremel G, Grannas K, Sutton LA et al (2013) *In situ* protein detection for companion diagnostics. Front Oncol 3:271. https://doi.org/10.3389/fonc.2013.00271
70. Yan J, Gu GJ, Jost C et al (2014) A universal approach to prepare reagents for DNA-assisted protein analysis. PLoS One 9:e108061. https://doi.org/10.1371/journal.pone.0108061
71. Greenwood C, Ruff D, Kirvell S et al (2015) Proximity assays for sensitive quantification of proteins. Biomol Detect Quantif 4:10–16. https://doi.org/10.1016/j.bdq.2015.04.002
72. Gu GJ, Friedman M, Jost C et al (2012) Protein tag-mediated conjugation of oligonucleotides to recombinant affinity binders for proximity ligation. N Biotechnol 30:144–152. https://doi.org/10.1016/j.nbt.2012.05.005
73. Goldstein R, Sosabowski J, Livanos M et al (2015) Development of the designed ankyrin repeat protein (DARPin) G3 for HER2 molecular imaging. Eur J Nucl Med Mol Imaging 42:288–301. https://doi.org/10.1007/s00259-014-2940-2
74. Simon M, Stefan N, Borsig L et al (2014) Increasing the antitumor effect of an EpCAM-targeting fusion toxin by facile click PEGylation. Mol Cancer Ther 13:375–385. https://doi.org/10.1158/1535-7163.MCT-13-0523
75. Tamaskovic R, Simon M, Stefan N et al (2012) Designed ankyrin repeat proteins (DARPins) from research to therapy. Methods Enzymol 503:101–134. https://doi.org/10.1016/B978-0-12-396962-0.00005-7
76. Baumann MJ, Eggel A, Amstutz P et al (2010) DARPins against a functional IgE epitope. Immunol Lett 133:78–84. https://doi.org/10.1016/j.imlet.2010.07.005
77. Eggel A, Baumann MJ, Amstutz P et al (2009) DARPins as bispecific receptor antagonists analyzed for immunoglobulin E receptor blockage. J Mol Biol 393:598–607. https://doi.org/10.1016/j.jmb.2009.08.014
78. Kim B, Eggel A, Tarchevskaya SS et al (2012) Accelerated disassembly of IgE-receptor complexes by a disruptive macromolecular inhibitor. Nature 491:613–617. https://doi.org/10.1038/nature11546
79. Eggel A, Baravalle G, Hobi G et al (2014) Accelerated dissociation of IgE-FcεRI complexes by disruptive inhibitors actively desensitizes allergic effector cells. J Allergy Clin Immunol 133:1709–1719.e8. https://doi.org/10.1016/j.jaci.2014.02.005
80. Tebbutt N, Pedersen MW, Johns TG (2013) Targeting the ERBB family in cancer: couples therapy. Nat Rev Cancer 13:663–673. https://doi.org/10.1038/nrc3559
81. Boersma YL, Chao G, Steiner D et al (2011) Bispecific designed ankyrin repeat proteins (DARPins) targeting epidermal growth factor receptor inhibit A431 cell proliferation and receptor recycling. J Biol Chem 286:41273–41285. https://doi.org/10.1074/jbc.M111.293266
82. Duffy MJ, Harbeck N, Nap M et al (2017) Clinical use of biomarkers in breast cancer: updated guidelines from the European Group on Tumor Markers (EGTM). Eur J Cancer 75:284–298. https://doi.org/10.1016/j.ejca.2017.01.017
83. Citri A, Yarden Y (2006) EGF-ERBB signalling: towards the systems level. Nat Rev Mol Cell Biol 7:505–516. https://doi.org/10.1038/nrm1962
84. Montemurro F, Scaltriti M (2014) Biomarkers of drugs targeting HER-family signalling in cancer. J Pathol 232:219–229. https://doi.org/10.1002/path.4269
85. Zahnd C, Wyler E, Schwenk JM et al (2007) A designed ankyrin repeat protein evolved to picomolar affinity to Her2. J Mol Biol 369:1015–1028

86. Jost C, Schilling J, Tamaskovic R et al (2013) Structural basis for eliciting a cytotoxic effect in HER2-overexpressing cancer cells via binding to the extracellular domain of HER2. Structure 21:1979–1991. https://doi.org/10.1016/j.str.2013.08.020
87. Epa VC, Dolezal O, Doughty L et al (2013) Structural model for the interaction of a designed ankyrin repeat protein with the human epidermal growth factor receptor 2. PLoS One 8:e59163. https://doi.org/10.1371/journal.pone.0059163
88. Tamaskovic R, Schwill M, Nagy-Davidescu G et al (2016) Intermolecular biparatopic trapping of ErbB2 prevents compensatory activation of PI3K/AKT via RAS-p110 crosstalk. Nat Commun 7:11672. https://doi.org/10.1038/ncomms11672
89. Winkler J, Martin-Killias P, Plückthun A et al (2009) EpCAM-targeted delivery of nanocomplexed siRNA to tumor cells with designed ankyrin repeat proteins. Mol Cancer Ther 8:2674–2683. https://doi.org/10.1158/1535-7163.MCT-09-0402
90. Martin-Killias P, Stefan N, Rothschild S et al (2011) A novel fusion toxin derived from an EpCAM-specific designed ankyrin repeat protein has potent anti-tumor activity. Clin Cancer Res 17:100–110. https://doi.org/10.1158/1078-0432.CCR-10-1303
91. Stefan N, Martin-Killias P, Wyss-Stoeckle S et al (2011) DARPins recognizing the tumor-associated antigen EpCAM selected by phage and ribosome display and engineered for multivalency. J Mol Biol 413:826–843. https://doi.org/10.1016/j.jmb.2011.09.016
92. Simon M, Zangemeister-Wittke U, Plückthun A (2012) Facile double-functionalization of designed ankyrin repeat proteins using click and thiol chemistries. Bioconjug Chem 23:279–286. https://doi.org/10.1021/bc200591x
93. Stefan N, Zimmermann M, Simon M et al (2014) Novel prodrug-like fusion toxin with protease-sensitive bioorthogonal PEGylation for tumor targeting. Bioconjug Chem 25:2144–2156. https://doi.org/10.1021/bc500468s
94. Dreier B, Mikheeva G, Belousova N et al (2011) Her2-specific multivalent adapters confer designed tropism to adenovirus for gene targeting. J Mol Biol 405:410–426. https://doi.org/10.1016/j.jmb.2010.10.040
95. Dreier B, Honegger A, Hess C et al (2013) Development of a generic adenovirus delivery system based on structure-guided design of bispecific trimeric DARPin adapters. Proc Natl Acad Sci 110:E869–E877. https://doi.org/10.1073/pnas.1213653110
96. Forrer P, Chang C, Ott D et al (2004) Kinetic stability and crystal structure of the viral capsid protein SHP. J Mol Biol 344:179–193. https://doi.org/10.1016/j.jmb.2004.09.030
97. Münch RC, Muth A, Muik A et al (2015) Off-target-free gene delivery by affinity-purified receptor-targeted viral vectors. Nat Commun 6:6246. https://doi.org/10.1038/ncomms7246
98. Zahnd C, Kawe M, Stumpp MT et al (2010) Efficient tumor targeting with high-affinity designed ankyrin repeat proteins: effects of affinity and molecular size. Cancer Res 70:1595–1605. https://doi.org/10.1158/0008-5472.CAN-09-2724
99. Stahl A, Stumpp MT, Schlegel A et al (2013) Highly potent VEGF-A-antagonistic DARPins as anti-angiogenic agents for topical and intravitreal applications. Angiogenesis 16:101–111. https://doi.org/10.1007/s10456-012-9302-0
100. Amadio M, Govoni S, Pascale A (2016) Targeting VEGF in eye neovascularization: what's new? A comprehensive review on current therapies and oligonucleotide-based interventions under development. Pharmacol Res 103:253–269. https://doi.org/10.1016/j.phrs.2015.11.027
101. Campochiaro PA, Channa R, Berger BB et al (2013) Treatment of diabetic macular edema with a designed ankyrin repeat protein that binds vascular endothelial growth factor: a phase I/II study. Am J Ophthalmol 155:697–704., 704.e1–2. https://doi.org/10.1016/j.ajo.2012.09.032
102. Souied EH, Devin F, Mauget-Faÿsse M et al (2014) Treatment of exudative age-related macular degeneration with a designed ankyrin repeat protein that binds vascular endothelial growth factor: a phase I/II study. Am J Ophthalmol 158:724–732.e2. https://doi.org/10.1016/j.ajo.2014.05.037
103. Smithwick E, Stewart MW (2017) Designed ankyrin repeat proteins: a look at their evolving use in medicine with a focus on the treatment of chorioretinal vascular disorders. Anti-Inflamm Anti-Allergy Agents Med Chem 16:33–45. https://doi.org/10.2174/1871523016666170502115816
104. Hussain RM, Ciulla TA (2017) Emerging vascular endothelial growth factor antagonists to treat neovascular age-related macular degeneration. Expert Opin Emerg Drugs 0:1–12. https://doi.org/10.1080/14728214.2017.1362390
105. Vaccaro C, Zhou J, Ober RJ et al (2005) Engineering the Fc region of immunoglobulin G to modulate in vivo antibody levels. Nat

Biotechnol 23:1283–1288. https://doi.org/10.1038/nbt1143

106. Stevenson CL, Rhodes CA, Prestrelski SJ (2012) Delivery of peptides and proteins via long acting injections and implants. SpringerLink:409–427. https://doi.org/10.1007/978-1-4614-0554-2_20

107. Vazquez-Lombardi R, Phan TG, Zimmermann C et al (2015) Challenges and opportunities for non-antibody scaffold drugs. Drug Discov Today 20:1271–1283. https://doi.org/10.1016/j.drudis.2015.09.004

INDEX

Andrew K. Udit (ed.), *Protein Scaffolds: Design, Synthesis, and Applications*, Methods in Molecular Biology, vol. 1798, https://doi.org/10.1007/978-1-4939-7893-9, © Springer Science+Business Media, LLC, part of Springer Nature 2018

CPSIA information can be obtained
at www.ICGtesting.com
Printed in the USA
LVHW06*1751170618
581000LV00003B/186/P

9 781493 978922